CULTURAL BOUNDARIES OF SCIENCE

CULTURAL BOUNDARIES OF SCIENCE

Credibility on the Line

THOMAS F. GIERYN

The University of Chicago Press
Chicago & London

THOMAS F. GIERYN is professor of sociology at Indiana University. He is the editor of three books, most recently of *Theories of Science in Society*.

The University of Chicago Press, Chicago 60637
The University of Chicago Press, Ltd., London

08 07 06 05 04 03 02 01 00 99 5 4 3 2 1

ISBN (cloth): 0-226-29261-4
ISBN (paper): 0-226-29262-2

Library of Congress Cataloging-in-Publication Data

Gieryn, Thomas F.
Cultural boundaries of science : credibility on the line / Thomas F. Gieryn.
p. cm.
Includes bibliographical references (p.) and index.
ISBN 0-226-29261-4 (cloth).—ISBN 0-226-29262-2 (paper)
1. Science—Social aspects. 2. Science—Philosophy. 3. Science—Cross-cultural studies. I. Title.
Q175.5.G54 1999
303.48′3—dc21 98-20394
CIP

♾ The paper used in this publication meets the minimum requirements of the American National Standard for Information Sciences—Permanence of Paper for Printed Library Materials, ANSI Z39.48-1992.

Contents

Preface

I have always been fond of maps. My mother insists that I learned how to read from a street map of Rochester, New York, where I grew up. From an early age, I spent hours drawing maps of imagined cities, crude ones at first in thick black lines, then more sophisticated efforts in colored pencil, water colors, rapidograph pens, transfer letters—and most recently, with graphics software on my Mac. I still make time for creative cartography and love to arrange lines and colors into fictional urban spaces for no other purpose than my own amusement. I also collect city street maps (more than 300 and accepting donations), less for directions than for art.

Still, it took the greatest sociologist of all time to suggest how this passion for maps might allow me to see *science* in a different and fruitful way. In the 1894 novel *Tom Sawyer Abroad,* Mark Twain narrates a conversation his hero has with Huck Finn as they float above the midwestern countryside in a fantastical balloon:

> "Tom didn't we start east?"
> "Yes."
> "How fast have we been going?"
> "Well, you heard what the Professor said when he was raging around; sometimes, he said, we was making fifty miles an hour, sometimes ninety, sometimes a hundred . . ."
> "Well, then, it's just as I reckoned. The Professor lied."
> "Why?"
> "Because if we was going so fast we ought to be past Illinois, oughtn't we?"
> "Certainly."

"Well, we ain't."

"What's the reason we ain't?"

"I know by the color. We're right over Illinois yet. And you can see for yourself that Indiana ain't in sight."

"I wonder what's the matter with you, Huck. You know by the *color?*"

"Yes—of course I do."

"What's the color got to do with it?"

"It's got everything to do with it. Illinois is green and Indiana is pink. You show me any pink down there if you can. No, sir, its green."

"Indiana *pink?* Why, what a lie!"

"It ain't no lie; I've seen it on a map and it's pink."

". . . Seen it on a map! Huck Finn, did you reckon the States was the same color out doors that they are on the map?"

"Tom Sawyer, what's a map for? Ain't it to learn you facts?"

"Of course."

"Well, then, how is it going to do that if it tells lies?"[1]

Anybody who has hung around the sociology of science lately will recognize at once what Tom and Huck are fighting over: what *is* the relationship between nature "out doors" and its representation on a map—or in a scientific fact? Perhaps nobody has ever been as naive as the old-fashioned realist Huck Finn: no gazetteer or scientific theory mimes reality literally, without mediation or translation or interpretation or contextualization. But for Huck, verisimilitude must reign, absolutely.

Sociologists of science have found lots of stuff that intervenes between reality and map, signified and signifier, object and image, referent and representation, nature and knowledge. Merton's pioneering studies of the social structure of science situated scientists in a constraining and enabling milieu of institutionalized norms, reward and evaluation systems, and communication networks—all of which were assumed to shape their representations of nature. With the emergence of the sociology of scientific knowledge in the 1970s, attention focused on social processes that were demonstrably connected to the content of scientific belief, and it soon became apparent just how substantially nature outdoors underdetermines its inscriptions. Interests, rhetorical tropes, power, identity, hands-on practices, tacit skills, instruments, experimental systems, and (as a catchall) culture are now standard ingredients in sociolog-

1. Mark Twain, *Tom Sawyer Abroad* (New York: Oxford University Press, 1996), 17–18. Bibliographic note: for primary historical documents, full references are given in footnotes when they are introduced for the first time in a chapter; for the secondary literature, short identifying references are given in the notes and full references can be found in the bibliography.

ical studies of the construction of scientific knowledge. What happens to nature in all this kitchen work depends upon the chef you ask: for some, nature is a seasoning thrown in to flavor the social meat and cultural potatoes; for others, nature is what is finally brought to the table, what gets ladled into bowls, either thick stew with chunks of social left in or thin broth after the "meat" is methodically strained out and discarded; still others never bother to pick up any nature at the market—it is social down to the bottom of the pot.[2]

After almost three decades of sober sociological and historical inquiry into how scientific knowledge gets prepared and why scientists accept some accounts of nature as provisionally and incompletely true, this much can safely be said: "Les faits ne naissent pas dans des choux."[3] Lingering debates over relativism versus realism, over the fraction of a scientific theory caused by natural versus social forces, over the existence of chairs at 30,000 feet have become stale, even a little moldy. Possibly the time is ripe for sociologists to look at science from a different vantage—not upstream at facts in their making, but downstream at their consumption. Scientists, their expertise, their claims and material artifacts eventually leave laboratories and technical journals and make their way out into the rest of the social world, where they are called upon to settle disputes, build airplanes, advise politicians, ascertain truth. And they do so with a special authority that begs for sociological explanation: Why is science so widely trusted? Why do we turn so often to scientists for help in reaching personal or policy or corporate decisions? Why do we provide copious public patronage to support more scientific research? Why is science conferred the legitimate power to define and explain nature and other realities?

The answers will not be found upstream, I suggest, but down. Nothing in the practices of scientists at their benches, nothing in their skillful mangle of gadgets or critters, nothing in the literary machinery that translates inquiry into facts on a page can alone explain why science is trusted (in so many and varied situations) to provide credible and useful accounts of nature. Or, more precisely, upstream science substantially un-

2. Merton's work is gathered up in his *Sociology of Science.* Constructivist sociologies of scientific knowledge are reviewed in Barnes, Bloor, and Henry, *Scientific Knowledge;* Lynch, *Scientific Practice,* chaps. 2–3; Shapin, "History of Science" and "Here and Everywhere." A review of both orientations is found in Zuckerman, "Sociology of Science." A sample of recent work in the field is provided by Jasanoff et al., *Handbook of Science and Technology Studies.*

3. A double translation is needed: "Facts are not born in cabbages"—or, for Americans, they don't come from storks. Bruno Latour, "Y a-t-il une science après la guerre froide?" *Le Monde,* January 18, 1997.

derdetermines the epistemic authority that marks its consumption downstream. Just as the constructivist sociology of scientific knowledge found that nature outdoors was an incomplete cause of scientific belief, so is science as practiced in labs and journals an incomplete cause of its power, prestige, and influence in society *tout court*. What is missing? If nature is socially constructed, so is science: the practices, skills, texts, achievements, and potentials of scientists are wrapped up in layers of discursive interpretations as they make their way downstream to respectful waiting publics. It is in these mediating representations of what science is or what scientists do that sociologists will find a robust explanation for the predominance of science these days in settling questions about the real.

I take one other thing from Twain. In the same way that Huck learned about Indiana from a map, people all over learn about science from maps of it. The layered interpretations that surround scientists and scientific facts with a special believability often come in a rhetorical form best described as cartographic. "Science" becomes a space on maps of culture, bounded off from other territories, labeled with landmarks showing travelers how and why it is different from regions of common sense, politics, or mysticism. These cultural maps locate (that is, give a meaning to) white lab coats, laboratories, technical journals, norms of scientific practice, linear accelerators, statistical data, and expertise. They provide the interpretative grounds for accepting scientific accounts of reality as the most truthful or reliable among the promiscuously unscientific varieties always available. Maps of science get drawn by knowledge makers hoping to have their claims accepted as valid and influential downstream, their practices esteemed and supported financially, their culture sustained as the home of objectivity, reason, truth, or utility. Maps of science get unfolded and read by those of us not so sure about reality, or about which accounts of it we should trust and act upon.

These cultural cartographies of science-in-culture are historical phenomena, with a local and episodic (rather than transcendent) existence. The same concatenation of interests, identities, discourses, and machineries that come together to make scientific knowledge have also come together to shape representations of science itself in a contextually contingent way. If I am correct that sociologists will find an explanation for the epistemic authority of science in its cartographic (re)constructions for public consumption, they will *not* find in those interpretative representations that science is any single thing (or even a small and consistent set of qualities). What science becomes, the borders and territories it assumes, the landmarks that give it meaning depend upon exigencies of the moment—who is struggling for credibility, what stakes are at risk, in front of which audiences, at what institutional arena? It is exactly this pliability and suppleness of the cultural space "science" that accounts for its long-running success as the legitimate arbiter of reality: science gets stretched and pulled, pinched and tucked, as its epistemic authority is reproduced time and again in a diverse array of settings.

Such an argument calls for detailed examinations of local and episodic constructions of science, highlighting the different cultural spaces science becomes in order to serve diverse pragmatic ends. The book ahead consists of five historical cases of the cultural cartography of science: they are written fresh but over an unbearably long period, not exactly in the same style, and arranged chronologically by when they were penned. I introduce the five cases with a chapter of theory mongering (where concepts such as "credibility contest," "epistemic authority," and "boundary-work" are defined and anchored), and the book ends with some reflexive horrors brought on by the so-called science wars. Each episode starts out with a struggle for credibility: somebody somewhere seeks to ride science into the public's trust or support or vindication. It is never easy: somebody else challenges their credentials as scientist, their skills, their procedures, their potential for making a truly better world. The maps start to fly, as contestants create distinctive cultural worlds—with discrepant locations and features for science—in order to convince those downstream that their claims about nature are credible or pertinent and that their practices are worthy of esteem or trust or patronage. There are winners and losers of course, but through it all and over historical time, the connection is reproduced between science (a fuzzy set if there ever was one) and the legitimate power to define the real.

Sociological attention is centered on how the boundaries of science are episodically established, sustained, enlarged, policed, breached, and sometimes erased in the defense, pursuit, or denial of epistemic authority. As knowledge makers seek to present their claims or practices as legitimate (credible, trustworthy, reliable) by locating them within "science," they discursively construct for it an ever changing arrangement of boundaries and territories and landmarks, always contingent upon immediate circumstances. When people outside the laboratories and technical journals dispute the authority or credibility of claims and knowledge-making practices, what does science become then and there? The selective attribution of this or that characteristic to science cannot be explained by what science "really" is at the bench or in a journal, but only by the pragmatic utility of any given borders and territories for the protection or expansion or denial of scientific authority over the facts. In other words, the question to be asked of any cultural map is not "Is it accurate?" but "Is it useful? If so, by whom, for what?"

"Science" is a cultural space: it has no essential or universal qualities. Rather, its characteristics are selectively and inconsistently attributed as boundaries between "scientific" space and other spaces are rhetorically constructed. The longstanding question, "What unique, essential, and universal features of science justify its authority in politics, law, media, advertising, and everyday reckonings of reality?" should be replaced, I suggest, by this more tractable question: "How do people sustain the epistemic authority of science as they seek to make their claims and practices credible (or useful) by distinguishing them from unworthy claims and practices of some nether region of *non*-science?" Science is a symptom of the legitimate power to decide reality—its edges and contents disputed, moved all over the place, settled here and there as decisions about truthful and reliable claims are acted upon by jurists, legislators, journalists, managers, activists, and ordinary folk. Representations of science—where it is, and where it is not—have less to do with the cultural realities they supposedly depict, and more to do with the cultural realities they sustain.

This book was written from many places, and the more I ponder the sociological importance of "place" these days, the more I believe that it matters *where* I was—when I thought, when I wrote, when I anguished. The journey has been long and twisted; no map could possibly be detailed enough to show all the people who have helped me along the way. The influence of my teachers at Columbia University is palpable still. Robert Merton and Harriet Zuckerman worried little about the substantive content of my sociology, but demanded only that I execute it as well as I could. Their lives and works offer a model of scholarly excellence—for me, and for so many others—and our friendship through years of divergent paths measures a shared commitment to the norm of organized skepticism. In New York, Peter Messeri and I argued forever the possibility of defining science, and Astrida Butners showed me how to survive the City.

My trailhead is probably best located on the banks of Cazenovia Lake, not far from Syracuse, New York, where Donald Campbell (a much missed friend) convened an extraordinary gathering of historians, sociologists, and philosophers of science. I went there a Mertonian, and left . . . something more. Steve Woolgar recognized my inchoate babblings as "boundary-work," and Karin Knorr Cetina helped me find my own way (so did Sal Restivo).

Most of the struggle happened at the Department of Sociology at Indiana University, which has provided for two decades a salutary climate of vigorous challenge and warm encouragement. I thank my local colleagues one and all (including some who are no longer local: David Brain and Ron Giere). David Zaret has heard every single idea of mine ad nauseam, though I shall never convince him why constructivists board airplanes with as much (or as little) confidence as he does. I bored my students even more with endless cultural cartographies of science, but they were kind enough not to show it: Steve Zehr, George Bevins, Anne Figert, Mitch Berbrier, Alyssa Kinker (who collected a ton of stuff for chapter 5, all to be recycled in due course), Barb Halpenny, Todd Paddock, Emanuel Gaziano, Walter Jacobs, Joe Tatarewicz, Karen Rader, Brad Hume, and everybody in Science Club. I have learned a thing or two about life from Jim Capshew—once an undergraduate in my very first class at Indiana, now a faculty colleague, always an inspiration.

Material for chapter 3 was collected while my family and I survived a blustery spring at Ard Carrach, Lydia Forbes's home in Carradale, Kintyre, Scotland. I traveled to Edinburgh's Science Studies Unit, where David Edge got me into the Combe archives, Barbara Edge told us how to stay warm, and Steve Shapin explained everything phrenological. There were beneficial side trips that year in Britain, to the reflexive fun-house at York (Mike Mulkay and Malcolm Ashmore), and to Bath (Harry Collins). The theoretical chapters were written in Ithaca, during a year at Cornell's Department of Science and Technology Studies. Sheila Jasanoff insisted that the book make an argument. Bruce Lewenstein and I (with Dougan's help) cooked up the Cornell Cold Fusion Archives that enabled chapter 4 (in which Trevor Pinch caught some sloppy errors).

For many years, *Cultural Boundaries of Science* has had a peripatetic existence—a road show, if not traveling circus. I thank my hosts and helpful audiences at the many universities where I have delivered bits and pieces in colloquia and conferences: Virginia Tech (twice!), Darmstadt, Iowa, Cornell, Minnesota, New York University, Illinois, MIT, Rensselaer Polytechnic Institute, Pennsylvania, Notre Dame, and St. Andrews Presbyterian College. Chapter 2 was helped along by a grant from the National Science Foundation, which did not mind at all a study of itself. Thanks to J. Merton England for help with the NSF archives.

The Institute for Advanced Study at Princeton provided a perfect place to write the epilogue. The Faculty of the School of Social Science and my peer Members that year were of enormous help, as we sat in the middle of the science wars. Clifford Geertz helped me dodge a few bullets, Daniel Woolf gave me a lift in his car, and Diane Vaughan was such a friend and co-conspirator. From Princeton, the manuscript made its way to the University of Chicago Press, but corporeally I never got to Hyde Park because we did the whole business by e-mail, fax, and FedEx. Chicago may only be a virtual place on this trip, but I am no less grateful to my editor Doug Mitchell and his team: Matt Howard (who may be related to Albert and Gabrielle), Carlisle Rex-Waller, Liz Demeter, and Barbara Fillon.

Back home again in Indiana, to Bloomington and my compost pile, flowers and maps (thanks, Mom, for that map of Rochester, and Dad, for adding many more to my collection, from your own travels). As if a mere lifetime of joyful surprise was not enough, Carolynne invented the book's subtitle at the eleventh hour. If only my book could provide as much intellectual comfort as her glorious sweaters provide warmth to many. When I started to write these chapters, our three sons—Nate, Patrick, and Sam—kept us up at night in need of feedings; now they do it by staying out too late at night with the car. Alas, this book about maps will not help you find your way home.

CONTESTING CREDIBILITY CARTOGRAPHICALLY

. . . as if Nature could support but one order of understandings.

Thoreau, *Walden*

Thoreau's celebration of the multiplicity of understandings—spiritual, metaphysical, political, empirical, geometrical, or commonsensical—is a curse sometimes. Just when we are most desperate to find out which among several competing understandings of nature may be trusted, we find it most difficult to sort out justifiable assertions from unworthy candidates. "Credibility contests" are a chronic feature of the social scene: bearers of discrepant truths push their wares wrapped in assertions of objectivity, efficacy, precision, reliability, authenticity, predictability, sincerity, desirability, tradition. People often take shortcuts when faced with practical decisions about how to allocate "epistemic authority," the legitimate power to define, describe, and explain bounded domains of reality.[1] Without the time or wherewithal to look at nature for ourselves (as Thoreau did), "science" often stands metonymically for credibility, for legitimate knowledge, for reliable and useful predictions, for a trustable reality: it commands assent in public debate. If "science" says so, we are more often than not inclined to believe it or act on it—and to prefer it over claims lacking this epistemic seal of approval.[2]

1. "Epistemic authority" resembles Paul Starr's "cultural authority," which he defines as "the probability that particular definitions of reality . . . will prevail as valid and true" (*Social Transformation of American Medicine,* 13). Both concepts are rooted in Weber's definition of authority (or legitimate domination) as "the probability that certain specific commands . . . will be obeyed by a given group of persons" (*Economy and Society,* 1:212).

2. "In modern societies, science is next to being *the* source of cognitive authority: anyone who would be widely believed and trusted as an interpreter of nature needs a license from the scientific community" (Barnes and Edge, *Science in Context,* 2). "The prestige of science is so great that it . . . is believed to possess such authority and be able to answer any of life's ques-

Alas, not even "science" settles things easily and unequivocally, when it comes to competing purveyors of more or less credible claims about nature.[3] Instead, new doubts and uncertainties arise: Who is a scientist? What is scientific? Precisely because so much rides on how the authority to describe and explain nature is parsed out—merely guilt or innocence, life or death, cornucopia or wasteland, utopia or nightmare—justifications and attempted persuasions via "science" rarely go unchallenged: Is your training or expertise scientific? Did you follow methodologically proper scientific procedures? Would most other scientists agree with you? Is the science in which you claim expertise pertinent to the issue at hand? Are the issues indeed ones that science can or should address? The adjudication of competing truths and rival realities is, often enough, accomplished in and through provisional settlements of the boundaries of science.

Newspapers, talk shows, and cyberspace are fat with credibility contests. Experts bearing science are enlisted everywhere to defend all sides and all opinions with putatively objective, reliable, and accurate facts. Forensic scientists question whether the DNA typing of O. J. Simpson's blood was done in a manner consistent with accepted standards of good laboratory practice. Researchers from a "Tobacco Research Institute" report that no scientific studies confirm unquestionably the causal connection between secondhand smoke from cigarettes and increased rates of lung cancer among nonsmokers. Climatologists battle with abstract models of long-range weather patterns, amid concrete policy debates over restrictions on burning soft coal or tax breaks for electric vehicles. Is premenstrual syndrome a mental illness, is RU486 safe and effective, are urine tests reliable, how serious a problem is drunk driving? Economists cannot agree on whether changes in the calibration of the gross national product would be a good thing for all of us, some of us, none of us. Medical scientists equivocate on the exact beginning and exact ending of

---

tions. . . . Thus, to invoke the symbols of science is to make policies sound, commodities desirable, and behavior legitimate" (Toumey, *Conjuring Science,* 6). On "scientism," see Cameron and Edge, *Scientific Images;* Hakfoort, "Historiography of Scientism." Michael King long ago (1971) identified the explanation of scientific authority as a key sociological problem: "[We] must discover the sources of scientific authority and examine the manner of its exercise. . . . Science is acclaimed and patronized to the extent that its intellectual authority, its capacity to give definitive answers to culturally significant questions, is acknowledged. The failure to give due priority to the problem of cognitive authority wielded by scientists has vitiated much of the sociology of science of the last three decades" ("Reason, Tradition," 16).

3. The problem of credibility has been put on the sociological agenda most forcefully by Steven Shapin, writing about the English seventeenth century in *Social History of Truth,* esp. 409–17, and "Cordelia's Love."

life (as loved ones, ethicists, religious activists, feminists, prospective organ-donor recipients press their cases). Psychologists slug it out over a woman's sexual hostility toward her lover, and whether it is caused by repressed memory of (maybe) incestuous relations with her father—next time on Oprah.[4]

It might, at first, seem that the epistemic authority of science itself is compromised or diminished by such public disagreement over scientific methods, facts, theories, and predictions—opening the way for non-scientific authorities to settle the hash (with political power, a sense of justice, religious values, inexpert personal testimony, or just random private interests). Stanley Milgram found that when two white-coated experts openly disagreed, their subjects were less inclined to obey subsequent commands to give electric shocks to each other for wrong answers—sundered epistemic authority loses force.[5] In society at large, just the opposite is more likely, at least in those cases where disputing experts all appeal to science as the tribunal of reason and truth. As each side brings science to the battle in defense of its claims, the link in principle between science and truth or reliability is sustained—even as some supposed facts and interpretations get canceled out as unscientific, false, or risky. So secure is the epistemic authority of science these days, that even those who would dispute another's scientific understanding of nature must ordinarily rely on science to muster a persuasive challenge.[6] We may be unsure about which truthsayers are really scientific or whether they enforced proper scientific procedures, but still we routinely appeal to science—whatever and whoever it is—as a first-pass source of cred-

4. The examples are not chosen at random; each has been the subject of sociological inquiries into the boundaries between science and non-. On O. J., see the forthcoming special issue of *Social Studies of Science* on DNA fingerprinting, edited by Michael Lynch; on smoke: Troyer and Markle, *Cigarettes;* on climate change: Zehr, "Accounting for the Ozone Hole" and "Centrality of Scientists"; on premenstrual syndrome: Figert, *Women and the Ownership of PMS;* on RU486: Clarke and Montini, "Many Faces of RU486"; on piss: Horstman, "Chemical Analysis of Urine"; on DWI: Gusfield, *Culture of Public Problems;* on economic models: Evans, "Soothsaying or Science?" and McCloskey, *Rhetoric of Economics;* on the beginning and ending of life: Mulkay, *Embryo Research Debate;* Casper, "Margins of Humanity"; Timmermans, "Saving Lives or Saving Multiple Identities?" and Johnson-McGrath, "Speaking for the Dead"; on repressed memories: Ofshe and Watters, *Making Monsters.* On expertise generally: Haskell, *Authority of Experts;* Martin, *Confronting the Experts.* On the "public understanding of science" and full of cartographic aperçus: Irwin and Wynne, *Misunderstanding Science?*

5. Milgram, *Obedience to Authority,* 107.

6. Writ large, this is Ulrich Beck's argument in *Risk Society:* under the weight of environmental degradation, the rational skeptical eye of science is turned back on its own Enlightenment foundations—now seen to be the foundations of our current problems, the result of industrialization and externalizing costs of pollution. Science becomes its own toughest critic.

ible understandings of nature, which brings me roundabout to this book's core argument.

When credibility is publicly contested, putatively factual explanations or predictions about nature do not move naked from lab or scientific journal into courtrooms, boardrooms, newsrooms, or living rooms. Rather, they are clothed in sometimes elaborate *representations* of science—compelling arguments for why science is uniquely best as a provider of trustworthy knowledge, and compelling narrations for why my science (but not theirs) is bona fide. Thoreau's legacy is a kind of "cultural cartography" of science, a mapping out of epistemic authority, credible methods, reliable facts—with borders and landmarks used to locate in the "culturescape" a space for science, surrounded by less believable or useful terrain.[7] These lay-of-the-culture representations of science put into a maplike discourse the grounds for choosing scientific over rival reckonings of reality, and for distinguishing genuine scientists from false prophets. On occasions when it is consequential to decide among discrepant accounts of natural reality, these maps help us feel our way. At the same time, looked at sociologically, these maplike representations become the linchpin of interpretative explanations of the quite stable and large epistemic authority of science. As individuals and organizations sift through a multiplicity of facts and theories using cultural maps drawn for them by proponents of a certain version of natural reality—choosing science while ignoring or discarding its impostors and rivals—they accomplish then and there the epistemic authority of science. Put bluntly, a sociological explanation for the cultural authority of science is itself "boundary-work": the discursive attribution of selected qualities to scientists, scientific methods, and scientific claims for the purpose of drawing a

7. Geertz, *Interpretation of Cultures,* 21. My book is just a sociological follow-up to Geertz's suggestion that "there is a great deal more to say . . . about . . . how thought provinces are demarcated." Or, as he also writes: "To analyze symbol use as social action is . . . an exceedingly difficult business," requiring one "to attend therefore to such muscular matters as the representation of authority, the marking of boundaries" (*Local Knowledge,* 154, 153). I have obviously taken to heart Shapin and Schaffer's suggestion that "the cartographic metaphor is a good one," for it "reminds us that there are, indeed, abstract cultural boundaries that exist in social space." Furthermore, "we still need to understand how such boundary-conventions developed: how, as a matter of historical record, scientific actors allocated items with respect to *their* boundaries (not ours). . . . Nor should we take any one system of boundaries as belonging self-evidently to the thing that is called 'science'" (*Leviathan and the Air-Pump,* 333, 342). Historian of science Michel Serres also mentions a "cartography of knowledge" (*Hermes,* xx). On the use of cartographic metaphors in social science generally, see Reichert, "On Boundaries"; Abbott, "Things of Boundaries"; and Silber, "Spaces, Fields, Boundaries." Wolfe says flatly that "boundaries are the stuff of sociology" ("Public and Private," 187).

rhetorical boundary between science and some less authoritative residual non-science.[8] Empirically, the contents of these maps of science become sociologically interesting precisely by their variability, changeability, inconsistency, and volatility—from episode to episode of cultural cartography, few enduring or transcendent properties of science necessarily appear on any map (or in the same place). The contours of science are shaped instead by the local contingencies of the moment: the adversaries then and there, the stakes, the geographically challenged audiences.[9]

How *does* science get represented in these credibility contests—when, where, and with what effects? This introduction addresses the question conceptually, with empirical and illustrative evidence reserved for the five (and a half) episodes that follow. I begin by examining the rhetorical form of boundary-work and its creation of a cultural space for science, then shift to consideration of the manifold consequences of these cartographic representations both for "science" and for candidate scientists. Why does boundary-work abound? I list the various occasions on which cultural cartographies of science are likely to occur, which will help to identify more specifically not only the agents and players who routinely get involved in it but why they do. All this would seem to beg the question, "So what *is* science that I should believe and trust its claim?" Perforce, I turn next to the contents of science—which, on this score, become a wildly variable set of qualities selectively chosen from "science-first-time-through" or other embodiments of "real science," then attributed to "science-the-cultural-space" in order to win a contest for credibility (though this makes the process sound more reductionistic

8. Steve Woolgar suggested the concept to me in 1981 ("Playing with Relativism"), and I first used it in 1983: Gieryn, "Boundary-Work." That paper pursued the linguistic turn in science studies then epitomized in the "discourse analysis" of Gilbert and Mulkay, *Opening Pandora's Box.* Studies that focus on the rhetoric of demarcation include Holmquest, "Rhetorical Strategy of Boundary-Work"; McOmber, "Silencing the Patient"; Taylor, *Defining Science;* Yearley, "Dictates of Method and Policy." Kerr, Cunningham-Burley, and Amos offer a Mulkayesque analysis of boundary-work among geneticists in "New Genetics."

9. The historical and contextual contingency of the boundaries of science—and their involvement in contests over legitimate knowers and knowledge—has been a perennial theme in feminist thought: "Both the boundaries and the nature of science are continuously subject to change over time and place" (Rose, "My Enemy's Enemy," 98). Donna Haraway gets "nervous" with a "social constructionism" that asserts "no insider's perspective is privileged, because all drawings of inside-outside boundaries in knowledge are theorized as power moves, not moves toward truth." She adds: "Feminists have to insist on a better account of the world; it is not enough to show radical historical contingency and modes of construction for everything" (*Simians, Cyborgs and Women,* 185, 184, 187). I quite agree: a *better* account of the social world begins with an appreciation for the contingent, constructed, and contested character of boundaries that demarcate legitimate knowers from illegitimate, fact from hope, science from politics.

than it really is). I end the chapter with a peek at the episodes ahead, suggesting how they connect theoretically and what they say about how credibility contests get resolved with myriad unsettled maps of science (and environs).

MAPPING OUT SCIENCE

It is neither by accident nor habit that representations of science in credibility contests often take the rhetorical form of maps. One can, of course, conjure science in discourses that do not evoke borders and territories, landmarks and coordinates. Science may be a game with rules and players, stakes and strategies, winners and losers; or science could be a network with nodes and channels, links and flows, dense or loose, connected or lonely; or science might be a category in a classification, with no metaphoric spatialization.[10] But when questions arise like "What is science?" or "Who is a scientist?" the answers often address the odd-sounding query, "*Where* is science?" To understand why this is so, something must be said about what maps are—and what they are good for.

Take a look at a map of science, or more precisely, a cartographic representation of Mount Science, located just above the town of Reason in the State of Knowledge, which is adjacent to the States of Fine Prospect and Improvement, across the Sea of Intemperance from the State of Plenty, all this on the other side of the Demarcation Mountains from the towns of Darkness, Crazyville, and Prejudice, and the islands of Deaf, Blind, and Folly.[11] Is the "Map of a Great Country" really a map at all? Sure: all the familiar cartographic features are here. The universe is carved up into oceans, seas, lakes, continents, islands, states, districts, and towns; coordinates are used to orient the user (the North Pole is inhabited by alcoholics and the South by teetotalers, while the journey of life moves West to East); landmarks like the Temperance Railroad are labeled to help travelers find their way; a scale is chosen to reveal Mount Science, but whatever is on it is too small to be visible. And although this map is drawn out in ink on paper, it could—like any other map—be

10. Investigations of "cultural classifications" have long been a part of anthropology (e.g., Douglas, *How Institutions Think*), the anthropologically inspired "new" cultural history (Chartier, *Cultural History;* Hunt, *New Cultural History*) as well as sociology (Ben-Yehuda, *Deviance and Moral Boundaries;* DiMaggio, "Classification in Art"; Lamont and Fournier, *Cultivating Differences;* Schwartz, *Vertical Classification;* Zerubavel, *Fine Line*).

11. "Map of a Great Country" (see below, pp. 8–9) is reprinted in Post, *An Atlas of Fantasy;* see also Hill's *Cartographical Curiosities.*

translated into spoken words (as Huck did to pink Indiana for Tom) or stored in your head as a mental image of where things are located. Nevertheless, no hiking boots, no car, no hot-air balloon will get you to Mount Science as a spot poking up out of the skin of the earth. Clearly, maps need not restrict their contents to geographic places that you can walk to, drive on, or fly over. A map is a *form* of representation, not a category of things that can be put into this form.[12]

Maps do to nongeographical referents what they do to the earth. Boundaries differentiate this thing from that; borders create spaces with occupants homogeneous and generalized in some respect (though they may vary in other ways). Arrangements of spaces define logical relations among sets of things: nested, overlapping, adjacent, separated. Coordinates place things in multidimensional space, making it possible to know the direction and distance between two things. Landmarks and labels call attention to typicalities or aberrations, reduce ambiguities about the precise location of a boundary, highlight differences between spaces of things: they are reality checks, of sorts. Most important, just as maps of earthly patches get drawn to keep travelers from getting lost, so maps of other worlds—culture, for example—are drawn or talked to help us find our way around. People navigate not just streets and highways but the culturescape: we wend our way through or around entrenched institutions, decide which rules apply where, subvert expectations by exiting, discern the signs given off by material objects, locate events in some historical narrative—and we routinely do so quickly and effortlessly. But not always: just as New Jersey roads are incomprehensible to a visiting driver without a map (and maybe even with one), so too is the cultural universe sometimes confusing, surprising, murky. And so maps of culture (like those of New Jersey) get drawn and talked and imagined plentifully.

The "Map of a Great Country" is one such map of the culturescape, on which science is a landmark, a place. What does this prototypal cultural map—different from those in the episodes to come only in its graphic visuality, and in its lessons—tell travelers about Mount Science? It shows what science is by spatially segregating this mountain range from places and spaces where it is not: cartographic contrast is the essence of cultural maps (and geographical ones too). Mount Science is not Prejudice or Superstition, towns located several borders and territories

12. On maps, see King, *Mapping Reality;* Lobeck, *Things Maps Don't Tell Us;* Monmonier, *How to Lie with Maps;* Robinson and Petchenik, *Nature of Maps;* Robinson et al., *Elements of Cartography;* Thrower, *Maps and Civilization;* Wood, *Power of Maps.* For a discussion of science as a map of nature, see Turnbull, *Maps Are Territories—Science Is an Atlas.*

"Map of a Great Country." Reprinted by permission of The Free Library of Philadelphia.

DENT SPIRITS

Gauke
Mountains of Guilt
Gout Cove
Sugar of Lead Cr.
Sly mix R.
Stench R.
Soddom
BAD WINE L.
Cheatburgh
Decoction R.
Straw coffee
Dimness
WHISKEY LAKE
Turn stomach R.
Nakedness
Barefoot
Shiver town
Bone tea
Horror
Reckless Pt.
Coast of Destruction
Ragville
Mountains of Shame
Crime
Hard scrabble
Slaughter House Cove
Barley R.
Dunoverd
Bury all
Bump pate
Craggy Harbour
Swine ville
Dare
Pin up
Self reproach Bay
Great dread
Race course
Misery
Pitch patch
Wretchedness
Whip wife
Sigh town
Bay of Remorse
Jargon burgh
Gnash teeth
Blood nose
Foul Mouth Pt.
Cave of agony
Gnaw chain
Wallow hollow
Shoals of Fruitless Effort
Sea of Accumulated Sorrows Anguish
Curse point
Gulf of Perdition
Despair
Whistford
Straits of Abstinence
Gulf of Broken Pledges
Fairmount
Victory
Content
Quarrel town
Socialvillage
Cherry valley
District of Abundance
Flourdale
Hillsdale
Tea River
Peach Orchard
High gate
STATE OF PLENTY
Corntown
No touch
Milktown
Temptation St.
Old Point Comfort
Bounty
Courage Pt.
Brooklyn
Richfield
Wheatville
Tea Bay
Butterfield
Rice field
Oxford
No waste
Shepherdstown
Comfort Bay
SEA OF TEMPERANCE
Plenty I.
Contentment Cove
Comforttown
Coffee River
Salem
Charity
Thrift
Coffee Bay
Breadville
Peace I.
New hope
Sobriety Isl.
Middletown
Quiet
Great Satisfaction
City of Canaan
Rest burg
Good Repute I.
Thankfullness
Perpetual Spring
Coast of Bliss
Longevity I.
Old age Light house
Faithfull town
Clearfield
Eaton
Spring Bay
Calm vale
Reason
Good Hope
Fairfield
Salem
Point Gladness
Clearmont
Mount Science
Pleasant Valley
STATE OF KNOWLEDGE
Patience harbour
Study River
Newhaven
Sweet hollow
Glory Harbour
FINE PROSPECT
Clearhead
Good recompence
Outlet of old age
Purity
No remorse Point
Coolchester
Milk River
Unity
Easthaven
Solid joy
Sugar Cr.
Bridgewater
Cold Spring
Spring River
Intelligence R.
Sharon
Sweet Repose
Salmon R.
Education Patent
SELF DENIAL

away. We learn about science by seeing what is far from it, or near: Blind Island is far away; much nearer are College Mountains, the town of Reason, and Intelligence River. Science is encompassed by the State of Knowledge, in turn encompassed by the Continent of Self-Denial. The State of Knowledge is illustrated and exemplified not just by Mount Science but also by landmark towns located there, big or important enough to be worthy of labeling: Education, Patent, Bridgewater, Coolchester, Clear Head, Cold Spring, Good Hope. Coordinates and compass points arrange the place of science in this universe relative to other places: North-South is drink/abstain, East-West is birth/death. As one travels through life from left to right, Always Busy and the Grove of Diligence must be traversed before one can climb Mount Science, and from there one eases downhill to the Coast of Bliss. Moving in an inebriated direction (bottom to top) takes one through Foul Mouth Point, across Whiskey Lake to the Mountains of Guilt. From this map, we learn *what* science is by finding *where* Mount Science is located—and we learn how to get there, why one might want to visit, and the costs of getting lost.

Mount Science is a prominent landmark steering life-travelers through sobriety toward eternal bliss, away from ardent spirits and destruction. In this rendering of the cultural universe, just enough is implied cartographically about "science" so that those who pass by its peaks know that they are on the correct path. The "Map of a Great Country" was drawn to describe and defend the temperate life: see the sad places one will end up through the bottle, what pleasant spots await those who resist. Science also appears on many other cultural maps which interest me more than those drawn in praise of sobriety and abstinence—such as those drawn (or more likely talked) in order to locate credible and useful accounts of nature, reliable predictions, objective methods, and trustworthy experts. These maps—as the episodes of this book will show—work in the same way as "Great Country," and with the same cartographic rhetoric. Science becomes a "cultural space": it is made locatable (and interpretable) by spatial segregations that highlight contrasts to other kinds of knowledge, fact-making methods, and expertise; boundaries define insiders and outsiders, while labeled landmarks give distinctive illustrations of each side; scale is enlarged to show internal differentiations within science or reduced to make science a single spot like Mount Science; coordinates tell us where we end up when we move away from science in various directions—toward faith to the East perhaps, politics to the West, techno-wonders to the South, error and ignorance to the North. We arrive at meaningful understandings of science (its products, people, practices, and potentials) by seeing or hearing about its place on a map, and we form images of its contents and capabilities by remembering where it has been located in spatial relation to places it is not.

Maps of all kinds proliferate, as they probably always have. No completely accurate and detailed map ever settles the lay of the land; it just begets more maps. Neither geography nor culture can be captured fully and permanently. Those who draw maps choose or make up which features of Indiana outdoors (or of science) they wish to represent, and how. Some maps are drawn as they are in order to anticipate the needs and wants of presumed users. Howard Becker tells a story about a group of tired tourists in San Francisco looking at a street map for the way back to their motel. The straightest route is charted, but then the tourists look up from the map and discover something that is not marked on it—up, up, up at one of San Francisco's daunting hills. Why wasn't that hill on the tourists' map? "The maps are made for motorists, financed by gasoline companies and automobile associations and distributed through service stations—and drivers worry less about hills than pedestrians."

Given San Francisco, there is no inherently best way to map it. It depends, for one thing, on whether the map is intended for walkers or drivers. No transcendent criteria determine that any one map is the most accurate, most user-friendly, most reliable. There are, instead, perpetually irreconcilable and revisable standards for deciding the "best" map (and so we get a lot of them, each different from the next). Becker concludes: "It seems more useful . . . to think of every way of representing social reality [maps, for example] as perfect—for something. The question is *what* it is good for."[13] Questions about accuracy—that is, a map's mimetic fidelity to the places it represents—can be answered only in terms of pragmatic utility. Seismologists need a different map of San Francisco than do tourists, drivers, pilots, market researchers—all of whose maps may be accurate as well as profoundly different.

But why are cultural maps of *science* also constantly being drawn and redrawn? In part, the process is driven by the diverse utilities of presumed users. Mount Science is mapped out one way for tempted gin-guzzlers hoping to stay away from Gloom, Distress, and Ruin, but "science" looks different on maps for school boards trying to decide whether both evolution and creation belong in biology textbooks. There is another explanation for the unending work of cultural cartographers—obvious, when one remembers that Rand-McNally and Michelin both print up

13. Becker, *Doing Things Together*, 121, 125.

their own street maps of San Francisco in order to make money. Does anybody "profit" from drawing a cultural map of science?

WHY BOUNDARY-WORK ABOUNDS

To ask of cultural cartography *cui bono?* is at the same time to ask why maps of science are endlessly remade and why they never really converge on a single supposed reality. When put abstractly, such questions quicken the pulse of theoretically inclined sociologists. What are the relations between things that comprise stratification and hierarchy (material resources, power, control, prestige, influence) and things through which people make sense (culture, meanings, interpretative frames, cognitive schema, maps)? What are the relations between the durable, distended, constraining stuff of social structure and the motivated, chosen, contingent, and pliable stuff of agency?

Such questions seem to call for declarations of allegiance, so I pledge: First, none of these ontological domains—neither material nor semiotic, neither structure nor agency—can be granted permanent and unequivocal explanatory privilege in accounting for individual choices or historical change; each couplet is connected recursively, and they are mutually constitutive. Second, inequality and interpretation are inseparably intertwined: money assumes value only within a particular cultural web of signification, and meanings become the basis for legitimating or justifying allocations and expressions of power.[14] Third, the "epistemic authority of science" exists only in its local and episodic enactment as sellers proffer truth and buyers choose to use/believe, but this all happens within structural contexts of available resources, historical precedents, and routinized expectations that enable and constrain the contents of a map and its perceived utility or accuracy in the eyes of users. Such a pledge of theoretical allegiance could get me labeled "constructivist." Let me explain.[15]

Cultural cartography is not idle play with Venn diagrams: maps of science give definitions of situations real in their consequences, both for those who rely on them and those who draw them. People and organiza-

14. Zelizer, *Social Meaning of Money*. On the general issue, "economics and social relations are not prior to or determining of cultural ones; they are themselves fields of cultural practice and cultural production" (Hunt, *New Cultural History*, 7). The point is made well in Sewell, "Theory of Structure."

15. Gieryn, "Distancing Science from Religion," further develops a constructivist orientation to science by contrasting it with a functionalist alternative.

tions use cultural maps to find workable truths about nature and (sometimes) suffer the consequences of practical choices they make based on where epistemic authority is located: Will the child get smallpox if not vaccinated? Will Miami Beach stay above water if soft coal continues to be burned or if electric cars never get off the drawing boards? Cultural maps allow us to trace out the provenance of contending natural facts and to reach decisions based on the features of where they are said to come from: objective methods, disinterested investigators, powerful instruments, Nature herself. Our reliance on cultural maps also permits us to distribute responsibility for facts of nature we accept as provisionally true and—to an extent—for the practical choices we make based on those truths. Decisions are grounded in science, or perhaps some other authority, as we become convinced that its claims are more accurate, trustworthy, and effective, but also because we can later justify those decisions (if we must) by surrounding them with the greater expertise and competence of those we trusted with the truth.

The stakes are as large for cultural cartographers themselves, for their particular version of nature hangs in the balance.[16] For example, conservation biologists convinced that declining biodiversity imperils human existence will either have their claims dismissed as wails from a politicized and misdirected Cassandra or see them translated from fact into environmental policy—depending upon where they and their claims are positioned in the culturescape by legislators, the press, corporate executives, foundation officials, social movement organizers, and citizens.[17] Evidence, models, and theories are not sufficient for these biologists to move policy and save the earth: a cartographic case for the credibility and reliability of these claims must also be put forth to persuade audiences that "our" nature (but not theirs) is the way it is really. Not just nature but one's credibility as a spokesperson for nature is contested and at risk. Losers see their claims moved out from fact to illusion, lie, ulterior motive, or faith while they (and their methods, practices, organizations, and institutions) get marginalized or excluded fully from the domain of epistemic authority reserved for science and its genuinely licensed practitioners. To the victors go the spoils of successful cultural cartography: not only do their claims become real enough for others to act on them, not only is their authority to make truth provisionally sustained, but they enjoy (for a while anyway) the soaring esteem, cascading

16. I started to use the concept of stakes after reading Cambrosio and Keating, "Disciplinary Stake."

17. Takacs, *Idea of Biodiversity*.

influence, and possibly abundant material resources (cash, equipment, bodies-and-minds) needed to make still more truthful tales. The legitimate right to have one's reality claims accepted as valid or marginally useful is no plum at all if everybody enjoys it all the time. Epistemic authority exists only to the extent that it is claimed by some people (typically in the name of science) but denied to others (which is exactly what boundary-work does).

So ordinary folks seek out cultural maps to locate credibility; fact makers produce maps to place their claims in a territory of legitimacy—but such cultural cartography also has consequences for the spaces mapped out. In credibility contests, the epistemic authority of "science" as a cultural space is chronically reproduced. Familiar reasons are given for why science is a preferred source of knowledge about nature even as the allocation of that legitimacy among practitioners, methods, and claims remains to be disputed, negotiated, and maybe eventually settled by mappers and audiences. Starting out as an authoritative but otherwise largely featureless terra incognita, science is then given particular (but nonaligned) borders and territories, landmarks and labels, in order to enhance the credibility of one contestant's claims over those of other authorities or scientists manqué. The epistemic authority of science is in this way, through repeated and endless edging and filling of its boundaries, sustained over lots of local situations and episodic moments, but "science" never takes on exactly the same shape or contents from contest to contest.[18]

18. This constructivist explanation for the epistemic authority of science stands in contrast to (1) Durkheimian efforts to explain the credibility of scientists in terms of functional necessity: "That [science] was born indicates that society needed it. . . . For the harmonious development of a complicated social order the co-operation of reflective intelligence soon became indispensable. . . . It was, then, a result of certain collective ends that science was formed and developed" (Durkheim, *Moral Education,* 70); (2) political economic efforts to explain the authority of science by its benefits for capital or another ruling class: "Science cannot escape capital and has been subsumed under the dialectic of the production of needs and capital (use value and exchange value)" (Aronowitz, *Science as Power,* 40); (3) state-centered theories to explain the authority of science by its utilities for governance (the legitimation of policies, for example): "The power of science has become part of the power of the modern state because it gains power in the hands of policymakers" (Mukerji, *Fragile Power,* 202; Mukerji's argument is more sympathetic to constructivism than this brief extract might imply); (4) attempts to explain the the cultural authority of science in terms of its alignment with or embodiment of other widely endorsed values, as when Yaron Ezrahi suggests that science and technology are "harnessed" "to ideologically defend and legitimate uniquely liberal-democratic modes of public action" (*Descent of Icarus,* 1); (5) explanations that identify some single watershed event ("the" scientific revolution, the bomb, Sputnik) that ratchets in a degree of epistemic authority that science subsequently never loses. These other possibilities have their merits, but share two flaws: first,

In this sense, it is a little misleading to speak of the "epistemic authority of science" as if it were an always-already-there feature of social life, like Mount Everest. Epistemic authority does not exist as an omnipresent ether, but rather is enacted as people debate (and ultimately decide) where to locate the legitimate jurisdiction over natural facts. Such spatialized allocations—to science provisionally and workably bounded, or to some other putative worthy source of knowledge and guidance—are local and episodic, extant then and there for interpretative finding-one's-way or for practical fact-based decisions but also for seizing/denying the spoils of credibility contests. The cultural space of science is a vessel of authority, but what it holds inside can only be known after the contest ends, when trust and credibility have been located here but not there.[19]

The stakes—authority, jobs, fame, influence, nature—create big incentives for some cultural cartographers to (re)draw the boundaries of science one way, just as others then have good reason to counter with maps of their own. Little wonder that the shelf life of any particular representation of the boundaries of science is short to vanishing, even if the epistemic authority of science-in-the-blank endures. The spaces in and around the edges of science are perpetually contested terrain: cultural maps are the interpretative means through which struggles for powerful ends are fought out—the right to declare a certain rendition of nature as "true" and "reliable." Prima facie evidence for the permanently unsettled cultural space of science is the ubiquitous boundary-work that can only be hinted at in the five episodes to come. The universe of such credibility contests divides into three genres, each an occasion for a different sort of boundary-work.

Expulsion. The first genre defines a contest between rival authorities,

---

they presume that "science" has a stable and distributed meaning for all institutions, classes, or states seeking to use it, thus obviating analysis of its contingent interpretative constructions; second, they attach too little theoretical significance to the idea that scientists (and many others) have a stake in the seizure or defense of epistemic authority, and that they adjust meanings of science to pursue that stake. These nonconstructivist interpretations assume that something immanent about scientific practice or scientific knowledge causes powerful forces in society to grant science epistemic authority: agency (i.e., boundary-work) gets dwarfed by impersonal structural forces.

19. Roger Chartier has a fine sense of the back-and-forth relationship between institutionalized cultural categories and those momentary, contingent, discursive, and pragmatic enactments of spatial classifications that make up boundary-work: "We [must] accept the schemata found in each social group or milieu (which generate classifications and perceptions) as true social institutions," but at the same time it is important to realize that "behind the[ir] misleading permanence . . . we must recognize not objects but objectifications that construct an original configuration each time" (*Cultural History*, 6, 46).

each of whom claims to be scientific. All sides seek to legitimate their claims about natural reality as scientifically made and vetted inside the authoritative cultural space, while drawing a map to put discrepant claims and claimants outside (or, at least, on the margins). Real science is demarcated from several categories of posers: pseudoscience, amateur science, deviant or fraudulent science, bad science, junk science, popular science.[20] Boundary-work becomes a means of social control: as the borders get placed and policed, "scientists" learn where they may not roam without transgressing the boundaries of legitimacy, and "science" displays its ability to maintain monopoly over preferred norms of conduct. Expulsion often pits orthodox science against heterodox, mainstream against fringe, established against revolutionary—but of course the issue in dispute is who and what belongs on which side. Neither side wishes to challenge or attenuate the epistemic authority of science itself, but rather to deny privileges of the space to others who—in their pragmatic and contingent judgment—do not belong there.

Expansion. Boundary-work also takes place when two or more rival epistemic authorities square off for jurisdictional control over a contested ontological domain.[21] Those speaking for science may seek to extend its frontiers, or alternatively, spokespersons for religion, politics, ethics, common sense, or folk knowledge may challenge the exclusive right of

20. As a sampler of studies on various kinds of non-science, see, on the possible fraud of Sir Cyril Burt, Gieryn and Figert, "Scientists Protect Their Cognitive Authority"; on the paranormal, Hess, *Science in the New Age;* on popular science, Hilgartner, "Dominant View of Popularization"; on parapsychology and fraud, Pinch, "Normal Explanations"; on borderline science, Wallis, "Science and Pseudo-science" and *On the Margins of Science;* on the boundary between science and common sense, Derksen, "Are We Not Experimenting Then?"; on politics and science, Moore, "Organizing Integrity."

21. The idea that professions—or institutions, like science—compete for jurisdictional control over the provision of vital tasks is developed in Abbott, *System of Professions,* a work that has been of enormous utility for my concoction of a cultural cartography of science, and which I consider further in Gieryn, "Boundaries of Science." However, Abbott flimflams a bit on matters at the core of constructivist challenges to all varieties of structuralist sociology. He admits rightly that "human realities are both subjective and objective," but then calls "objective" the actual practices of doctors with their patients, and calls "subjective" representations elsewhere of those first-time-through practices. Even though these representations of professional jurisdictions are real and consequential—"to investigate the subjective qualities of jurisdictions is thus to analyze the mechanisms of professional work itself"—Abbott's parsing out of "subjective" and "objective" belies his wish to see in the actual practices of doctors or lawyers or accountants the weightier "cause" of their professional dominance: "Once a group enters the competition, what matters for us is not what it claims to be, but what it actually is. . . . Autonomous change in the objective character of the task transformed the profession" (*System of Professions,* 82, 39). I am inclined to put greater weight on how downstream interpretations construct the actual practices of scientists in the course of settling issues of epistemic authority.

science to judge truths. On these occasions, the interpretative task is not to distinguish real science from ersatz, but rather to distinguish science from (or identify it as) one of the less reliable, less truthful, less relevant sources of knowledge about natural reality.

Protection of autonomy. A slightly different kind of boundary-work results from efforts of outside powers, not to dislodge science from its place of epistemic authority, but to exploit that authority in ways that compromise the material and symbolic resources of scientists inside. When legislators or corporate managers seek to make science a handmaiden to political or market ambitions, scientists put up interpretative walls to protect their professional autonomy over the selection of problems for research or standards used to judge candidate claims to knowledge. In the same way, when the mass media take upon themselves the task of distinguishing genuine scientific knowledge from putatively less responsible claims, scientists whose claims were made suspect will redraw the cultural map to restore a monopoly over such cartographic efforts to those inside science—autonomy of a different kind. Finally, scientists will draw boundaries between what they do and consequences far downstream—the possible undesired or disastrous effects of scientific knowledge—in order to escape responsibility and blame (which often come coupled with intrusive demands for accountability or restriction).[22]

The boundaries of science have not, historically, been set in amber because—in the first instance—nature does not allow but one order of understandings, and therefore those serving up discrepant realities can draw discrepant cultural maps to legitimate their claims as uniquely credible and useful. As contestants for credibility pursue, deny, expand, constrict, protect, invade, usurp, enforce, or merely justify the epistemic authority of science (however bounded and landmarked), cultural maps get drawn and drawn again. But what determines victory in any specific contest? There are no fixed or general determinants of persuasive cultural cartography ("science" becomes the winners' map only after the dust has settled)—in part because of a "cartographic regress." When people face multiple and discrepant claims about nature that are located in different epistemic spaces, they need still other maps to assign authority over the task of mapping. "Second-order" boundary-work uses a map drawn primarily to locate authoritative accounts of nature also to locate credible cultural cartographers—and often to dismiss a rival's map as unskilled or

22. Anne Figert and I develop this point for the Challenger explosion, in "Ingredients for a Theory of Science in Society." The locus classicus is Vaughan, *Challenger Launch Decision.*

misleading or deceptive.[23] This interpretative layering of boundary-work is not a rare event but probably the norm in credibility contests, as the episodes ahead will suggest. The authority to represent both nature and culture, both facts and maps, is simultaneously contested, negotiated, and (in the end) distributed.

WHAT SCIENCE IS

Perhaps it is clearer now why those involved in credibility contests have strategic reasons to draw lots of culturescapes with variously located places for authentic and authoritative understandings of nature. But is it really the case that cultural spaces for science are drawn incommensurably each time out, or that competing maps during one credibility contest have no likenesses whatsoever? Has there been no convergence at all, over the three or more centuries of science and of its representation, toward a small and consistent set of features able or necessary to distinguish scientific knowledge, methods, and practitioners from everything else in the cultural world? Are there no limits on what characteristics may be attributed to science in order to endow it with distinctive epistemic authority or to restrict, exploit, or reallocate its supposed legitimacy over questions of natural fact?

Doesn't the referent science "behind" all those representations of it—the science happening first time through in laboratories and field stations, in journals, at professional meetings—constitute a reality to which contestants for credibility can appeal in order to justify their own distinctive mappings of epistemic authority? Unfortunately not. If all relationships between referent and representation were merely mimetic, so that representations reflected (better or worse) the thing itself, credibility

23. No particular map can carry the day, on all occasions, for all audiences: such a definitive map of science, with fixed and unambiguous boundaries and absolutely stable landmarks, would make all subsequent efforts to redraw science-in-culture into tilts at windmills. Some maps no doubt appear less "real" as depictions of the cultural terrain of objectivity, investigative skill, authenticity, expertise, and practical utility. For example, maps that appear weirdly idiosyncratic—like those upside-down geographical maps where South America is on top (weird, at least to North American eyes)—are less likely, ceteris paribus, to convince others that this is where epistemic authority is really located (if only because map readers suspect that some hidden motive or plain ineptitude is causing the map to look nothing at all like their baseline expectations). Similarly, a map that arrays cultural practices and institutions in ways unflattering or downright insulting to the epistemic travelers' own line of work—for example, locating "lawyers" in the "Valley of the Ambulance Chasers" in a court presentation—will probably diminish its perceived accuracy and utility (at least for the attorneys and judges).

contests would not be the chronic and ubiquitous phenomena they have always been: just look straight at nature to know who is telling the truth about it. Even Thoreau spotted the impossibility of that! The problem is not that there is no "real science" behind the cartographic representations, but that there are too many "real sciences." And even when all those sciences are added up, they still together do not allow either the sociologist or the players themselves to know a priori what science will look like on the next occasion for its mapping. But neither does it make much sense to think of science-on-the-map as just made up any old way. Boundary-work is constrained by the several "real sciences," but not determined by any or all of them.

Sociological constructivism is not nihilism: an absence of hills from that map of San Francisco does not mean that they do not exist (try telling that to the tired tourists looking up). Nor can everything on a map be found outdoors: the border between Illinois and Indiana, so clear on Huck's mental map, is invisible to hikers in the woods crossing over from one state to the other. Whatever science might be first time through, it is both too much and too little to determine its place on a cultural map. Maps must not only simplify, distill, and reduce their referents, but then reconfigure, distort, and embellish them. It is not immediately obvious where to look for science-the-real-thing in order to check out the accuracy of a certain cultural map—or at what, at whom, and when. Science—as practiced, as written up in technical papers, as regulated by norms of research conduct—has a robustness, a plenitude, a scale that defies complete mapping. Selections from this real science must be made by cultural cartographers, and they are—strategically. And even if an exhaustively authoritative rendering of science as practiced could be drawn, it would not tell travelers lost in the wash of contending truths to find out what they want to know: there is nothing on a scientific instrument, a fact, a statistic, or white lab coat that says "true" or "trustworthy" or "credible." Such labels are added to the reality of first-time-through science in the course of making the map useful for getting around—like inventing different colors to distinguish the states visually.

Moreover, science-the-referent is not embodied only in these first-time-through practices, instruments, research materials, facts, and journals; it has several other realities too. Science gets sedimented as an ordinarily tacit space on most everybody's mental map of culture, a bit of the cognitive schema we use everyday to navigate material and symbolic lands. Science also exists in codified bureaucratic procedures, as when university catalogs divvy up the universe of learning into natural science, social science, and humanities.[24] And science exists as a "cartographic legacy," accumulated residues of previous instances of boundary-work, as when a creased and dog-eared map gets unfolded rather than drawn fresh.[25] Any of these real sciences may be used to legitimate the next new map as accurate, but always tenuously so. One's tacit images of science suddenly become both explicit and doubtable when others draw for us different boundaries and ask for assent; codified pigeonholes for bureaucrats are always surrounded by interpretative flexibility (where does "history" belong?);[26] it is always possible in principle to challenge the pertinence or applicability of some previously authoritative cultural map to the immediate credibility contest at hand. In none of these embodiments—schema, procedure, precedent—is science black boxed.[27]

24. Geertz writes that "grand rubrics like 'Natural Science,' 'Biological Science,' 'Social Science,' and 'Humanities' have their uses in organizing curricula, in sorting scholars into cliques and professional communities, and in distinguishing between broad traditions of intellectual style." But I think he underestimates the utility of cultural maps—however "distorting"—for allocations of epistemic authority, as he continues: "When these rubrics are taken to be a borders-and-territories map of modern intellectual life . . . they merely block from view what is really going on out there where men and women are thinking about things and writing down what they think" (*Local Knowledge,* 7).

25. D'Alembert's Enlightenment "Mappemonde"—his *Preliminary Discourse* to Diderot's *Encyclopedia*—drew legitimacy by selective comparisons to an earlier "topography of knowledge" by Francis Bacon in *The Advancement of Learning.* See Darnton's "Philosophers Trim the Tree of Knowledge," in his *Great Cat Massacre,* 195. Darnton's chapter is a superb illustration of the real consequences of cultural cartography. He shows that if "setting up categories and policing them is . . . a serious business," the subsequent "pigeon-holing is therefore an exercise of power" (193, 192). The connection between spatialized representations of knowledge and the exercise of power is an enduring theme for Michel Foucault: "The formation of discourses and the genealogy of knowledge need to be analyzed, not in terms of types of consciousness, modes of perception and forms of ideology, but in terms of tactics and strategies of power . . . deployed through . . . demarcations, control of territories and organizations of domains" (Foucault, *Power/Knowledge,* 70–71, 79). For a more recent collection of studies on power and boundaries, see Nader, *Naked Science.*

26. The idea that any object—natural, made-material, social, cultural—is surrounded by interpretative flexibility (and therefore has the potential for a diversity of meanings and evaluations) is a staple of constructivist thought. It is developed paradigmatically (for the case of bicycles) by Pinch and Bijker, "Social Construction of Facts and Artifacts."

27. The idea that objects move along a gradient of stabilization belongs to Latour (*Science in Action* and *Aramis*). My suggestion that "science" assumes multiple embodiments of longer than momentary stability and wider than local dispersion—all of them "real"—owes much to Bourdieu on cultural reproduction (*Outline of a Theory of Practice,* 78–87) and Giddens on structuration (*Constitution of Society,* 25–66). In calling these embodiments "real," I am not suggesting that representations of science are unreal (mere phantasms without social consequences) or that they are not an integral part of "doing" science. Zerubavel considers this issue as well: "Boundaries are mere artifacts that have little basis in reality," though "our entire social order rests on the fact that we regard these fine lines as if they were real." While it may be true

Rather, these real sciences, along with all the first-time-through stuff, remain available for interminable cultural cartographers to pick and choose whatever they find advantageous to attribute to their version of science-space.

"Real science" thus comprises a reservoir of meanings selectively used to draw culture and locate epistemic authority: the repertoire of qualities attributable to science during any credibility contest is neither infinite nor small, neither fixed by these various embodiments nor completely disconnected from them. Cultural cartography sits at the juncture of given realities and strategic representations: science is already present but constantly made up. Even the scant few episodes of cultural cartography examined in the chapters ahead turn up a surprisingly large number of qualities and characteristics used to bound and locate science in distinctive ways: science is practically useful but useless; quantitative and qualitative; experimental and observation based; holistically homogeneous throughout and texturally variegated; finite and infinite (in terms of what can be known scientifically); politically or ethically engaged and detached; driven by theory and by data. These are just several of many coordinates used in the cultural cartography of science, but never consistently so. Nor is there a discernable direction in the long-run history of boundary-work toward one pole or another. Real science and its boundaries on cultural maps are supple and pliable things, like warm putty, but not so elastic that they may stretch endlessly in every direction.

One must look into the contingencies of each local and episodic contest for credibility in order to find out what science becomes then and there, to discover the many places it can be located on the cultural maps that become interpretative grounds for extending or denying epistemic authority. Instead of being a more or less accurate copy of one or another real science, these cultural maps are contextually tailored selections from a long menu, where "context" is defined by the players and stakeholders, their goals and interests, and the arena in which they operate.

Players and Stakeholders

The skin and innards of science will vary depending on who draws the map—and against whom, and for whom. Credibility contests create at

---

that "boundaries are normally taken for granted and, as such, usually manage to escape our attention," those occasions where the location of borders and territories becomes the focus of explicit discursive negotiation—as in my five episodes—are sociologically strategic for examining the play of agency and structure in the domain of culture (Zerubavel, *Fine Line,* 2).

least three social roles: (1) the contestants who draw maps seeking to attach epistemic authority to their claims about nature; (2) those who rely on those maps—and form their own cartographic images—as they allocate credibility in the course of making practical decisions based on supposed reliable knowledge; and (3) the many people affected by allocations of epistemic authority—and especially by the institutional decisions grounded in such allocations—though they may not themselves be in a position to parse out epistemic authority. Properties attributed to science on any occasion depend largely on the specifics of its "other," on who or what is being excluded from the cultural space of "science." If the other (who may also be a foil) is a rival source of epistemic authority, boundary-work will highlight those unique properties or accomplishments of science that make it a distinctively superior way of knowing or form of knowledge. For example, science may appear as the linear fount of panacea technologies when the excluded space is religion; its detachment from market pressures may be used to distinguish science from engineering or applied research and development. Boundary-work to exclude an impostor "scientist" will focus attention on the poser's failure to conform to expected methodological or ethical standards variously mapped out as necessary for genuine scientific practice: proper instrumentation, credentials, peer review, objectivity, and skepticism. Cultural maps drawn from "outside" the scientific space take on still a different character: rival epistemic authorities seeking to curtail the reach of scientific methods will point to their inadequacy or liability for certain kinds of issues—subjective matters, political, ethical, or artistic. Scientists whose legitimacy qua scientists is challenged sometimes stretch the bounds of legitimate research practices or stretch the boundaries around those granted legitimacy to *decide* the legitimacy of claims or claims making.

The makeup of immediate audiences for cultural cartography, seeking the route to credible knowers and authentic claims, will also shape episodic maps of science. Scientific knowledge and practice can be made to appear thoroughly accessible and just like common sense, as untutored lay people (an "other" cultural space) are invited to see for themselves the putative validity of a theory. Or, alternatively, scientific knowledge and practice may become impenetrably esoteric when mapped out before those seeking to impose political or ethical constraints on the unbridled search for truth. In every instance, the specific selection of qualities attributed to science then and there are those perceived to be most convincing rhetorically for establishing the desired boundaries and allocations of epistemic authority.

Goals and Interests

Boundary-work is strategic practical action.[28] As such, the borders and territories of science will be drawn to pursue immediate goals and interests of cultural cartographers, and to appeal to the goals and interests of audiences and stakeholders. Insider scientists use boundary-work to pursue or protect several different "professional" goals. Among the most common cartographic tropes is this: if the stakes are autonomy over scientists' ability to define problems and select procedures for investigating them, then science gets "purified," carefully demarcated from all political and market concerns, which are said to pollute truth; but if the stakes are material resources for scientific instruments, research materials, or personnel, science gets "impurified," erasing the borders or spaces between truth and policy relevance or technological panaceas. The sociological question is not whether science is really pure or impure or both, but rather how its borders and territories are flexibly and discursively mapped out in pursuit of some observed or inferred ambition—and with what consequences, and for whom?

Boundary-work is surely instrumentalist. But is it merely so? Is it fair to accuse scientists seeking to protect autonomy or secure research funding of duplicitous behavior, as they concoct a representation of themselves tailored to maximize whatever returns are on the line? Can boundary-work—and the specific contours of science then and there—be *reduced* to interests? Too crude by half: interests are not preformed and fixed forces (fully knowable and articulatable by cartographers or their audiences) that lie behind cultural maps, any more than the several embodiments of "real science" determine (in an unmediated way) the contents of its occasional representations. Boundary-work brings social

28. Scientists use several strategies "to secure academic respectability," one of which is to "impose a demarcation of the scientific and the non-scientific which is designed to forbid any inquiry liable to question the bases of its respectability." In other words, "these representations [of science and non-science] are *ideological strategies* and *epistemological positions* whereby agents . . . aim to justify their own position and the strategies they use to maintain or improve it, while at the same time discrediting the holders of the opposing position and strategies" (Bourdieu, "Specificity of the Scientific Field," 19, 22, 40). The pattern is illustrated in controversies over vitamin C and cancer (Richards, "Politics of Therapeutic Evaluation") and over fluoridation (Martin, *Scientific Knowledge in Controversy*).

interests and real science together in the mapping, and on these cultural maps both get articulated, altered, appreciated, denied, deployed, reconstructed, and translated in and through the cartographic process. Those features chosen for attribution to science (or to its others) are chosen strategically—it makes little sense to argue that cultural cartographers are indifferent about how epistemic authority is allocated or that they would deliberately prefer tactics designed to lose it. Neither are they omniscient or deceitful: the best-drawn maps of culture sometimes fail to secure credibility for one's claims, and more often than not, I suspect, scientists really believe that their representations of science tell it like it is. Interests are like destinations: cultural maps provide interpretative tools for trying to get there, but just as destinations sometimes change or get more precise in the middle of a trip, so may interests.

Arena

Cultural cartography takes place in a variety of institutional or organizational settings, and the declared contents of science may also be affected by the peculiarities of each of these arenas. Boundary-work would be expected in settings where tacit assumptions about the contents of science are forced to become explicit: where credibility is contested; where regnant assumptions about boundaries suddenly appear murky or inapplicable; and—most important—where allocations of epistemic authority are decided and consequentially deployed. Legislative and judicial forums, along with the media and corporate boardrooms,[29] are ripe spots for picking juicy episodes of cultural cartography: the practices and claims of scientists move to these arenas from bench or journal, mediated by maps attaching credibility to them or denying it. Here is where practical decisions are negotiated and settled—public policies, investments of funds for making new knowledge, guilt or innocence, technological futures, sick children and healthy planets—on the basis of provisional allocations of epistemic authority contested cartographically. Insofar as science "has" epistemic authority, it exists at that cartographic moment. In subsequent practical decisions, which offer no guarantees about where science will be placed or how epistemic authority will be allocated next time, tacit understandings or procedural routines or tattered maps fail. Even at the workplaces of science-first-time-through, the boundaries oc-

29. On the history of media representations of science, see LaFollette, *Making Science Our Own*.

casionally move from tacit convention to explicit contention: the division of laboratory labor among professor, lab technician, postdoc, and grad student may be one such moment.[30]

God and good sociology is in the details. Only inside the episodes is it possible to examine how the available embodiments of "real science" (lab, instrument, journal, statistic, memory trace, bureaucratic niche, old maps) and the local and episodic contingencies of immediate occasions (players, stakes, interests, arena) come together then and there to make "science" into so many cultural spaces.

TOTTING UP THE EPISODES

Why should any of this be surprising to sociologists, or a matter of theoretical debate? Ours is, after all, the discipline in which the social construction of anything and everything comes with mother's milk. Is science different from (say) race, class, or gender in its contingently negotiable character; as a site for struggles over power, resources, and control; as constrainingly real yet incessantly remade in representations just as real in their consequences? Science remains a tough nut for constructivists: it is precisely because of its apparently secure grounding in reason, logic, empirical evidence, and dispassionate debate that sociological efforts to make these virtues into rhetorical weapons or momentary discursive accomplishments are perceived as threats to the firmament of civilization. Many believe that the epistemic authority of science is *justified* (not just episodically won) by the unique, necessary, and universal elements of its practice—behaviors, dispositions, methods, rules, tools, and languages that simply work best to make truth.

Sociologists, historians, and philosophers have long pursued the essence of science, and some have found what they believe to be the sine qua non of its accuracy, utility, credibility, and authority. This hunt for "principles of demarcation" is a mapping out of science for analytical purposes—just to find out philosophically, sociologically, or historically its unique, invariant, essential, and universal features, but not necessarily to settle any credibility contest immediately at hand. What *really* makes science *science* (not just endless maps of it)? For philosopher Karl Popper, science is unique among knowledge systems in requiring its assertions to be falsifiable: there must, in principle, be some empirical obser-

30. A historical case of boundary-work to distinguish scientist from technician is discussed in Shapin, "Invisible Technician."

vation with the potential to make assertions false. For the sociologist Robert Merton, science is distinguished by its institutionalized norms—communism (the injunction to share findings promptly), universalism, disinterestedness, and organized skepticism—binding on scientists' behaviors and judgments, internalized by them through socialization, reinforced by a system of rewards and sanctions, and functionally necessary for the efficient pursuit of certified knowledge. For historian Thomas Kuhn, science distinctively moves in an oscillation between paradigmatic calm—periods of widespread consensus over what can be known, what isn't already known, and how methodologically to fill in the puzzle—and occasional revolutions that turn reality upside down until a new but incommensurate paradigm emerges.[31] Scientists themselves have also sought to give their trade an essential and definitive shape, in books with telltale titles like *Science—Good, Bad, and Bogus* and *Fads and Fallacies in the Name of Science* (Gardner), *Beyond Velikovsky* (Bauer), *At the Fringes of Science* (Friedlander), or *Science, Nonscience, and Nonsense* (Zimmerman). But unlike the analytical Popper, Merton, or Kuhn, their mission is bluntly ideological: to justify and protect the authority of science by offering principled demarcations from poachers and impostors (New Age, UFOlogy, parapsychology, and, lately, postmodernism)—object lessons for us to tuck away (they hope) on our mental maps of culture, recalled next time the precise location of credibility is unclear.

Plainly, these analytic or ideological attempts to define the essential ingredients of science as such is an enterprise wholly unlike my own. One might challenge such efforts on empirical grounds, mustering evidence to show that scientists are neither falsifiers, skeptics, nor puzzle solvers. My tack is different, in effect, a shift of the dependent variable: those who seek essentialist demarcation criteria should not assume that these explain the epistemic authority of science. Essentialism as a theory of scientific authority would argue that the conditions *necessary* for the production of valid and reliable knowledge are *sufficient* to explain why science has emerged historically as the so often preferred chronicler of na-

31. In Gieryn, "Boundaries of Science," I discuss these essentialist attempts at demarcation in greater detail. A distinction between essentialist and nominalist (constructivist) demarcations of science is discussed in Woolgar, *Science—the Very Idea,* 12. For the record: Popper on falsifiability, *Logic of Scientific Discovery,* 40ff.; Merton on social norms of science, *Sociology of Science,* chap. 13; Kuhn on paradigms, *Structure of Scientific Revolutions.* Kuhn, by the way, seems to have overlooked the sociological significance of how science gets represented, presuming "no" answers to two rhetorical questions that I would answer with a hearty "yes": "Can very much depend on a *definition* of science? Can a definition tell a man whether he is a scientist or not?"(*Structure of Scientific Revolutions,* 160).

ture. I disagree: credibility in the culturescape is not decided in tinkerings at the lab bench or in the refereeing of a manuscript or in the machinations of instruments, statistics, or logic. It is analytically presumptuous to think that people and organizations in society at large possess an understanding of science fixed clearly by actual practices in the lab,[32] or by scholarly tries at demarcation, or by idealized reminiscences. Epistemic authority is decided downstream from all that, as claims float through layers of cartographic interpretations where credibility is attached or removed. To miss the interpretative work that creates contexts for decisions about who to trust with reality is to lose the sociological handle on what is happening. When considered as a cultural space constructed in boundary-work, science becomes local and episodic rather than universal; pragmatic and strategic rather than analytic or legislative; contingent rather than principled; constructed rather than essential.

One could also dismiss Popperian or Mertonian or Kuhnian demarcationism because it privileges analysts' representations of science when it might be more interesting (sociologically) to watch how people in society negotiate and provisionally settle for themselves the borders and territories of science.[33] Social constructivism stands in the long tradition of interpretative social science from Weber to Mead to Schutz to Geertz—all of it focused on actors' understandings of things in their worlds.[34] Still, there

32. Christopher Toumey identifies a "paradox of respect without comprehension," which to me says that a sociological explanation of the cultural authority of science cannot be based on ordinary folks' direct apprehension and understanding of scientific practices or texts first time through (*Conjuring Science,* 7).

33. In 1974, Barnes anticipated the value of this constructivist orientation to the authority of science: "We should not seek to define science ourselves; we must first seek to discover it as a segment of culture already defined by actors themselves. . . . It may be of real sociological interest to know how actors conceive the boundary between science and the rest of culture, since they may treat inside and outside very differently" (*Scientific Knowledge and Sociological Theory,* 100). Three years later, Mendelsohn suggests that the problem has not yet been addressed: "The institutional and social processes which established the boundaries of that which would ultimately be included within science remain unstudied" ("Social Construction of Scientific Knowledge," 6). A review of early research taking up these suggestions is in Shapin, "History of Science": in contrast to philosophical or essentialist "prejudgment[s] of what counts as science and what does not," a constructivist "approach involves how historical actors *themselves* defined what belonged to science (or 'natural philosophy,' or whatever term or cultural domain was indicated) and what did not" (177).

34. "Interpretive explanation . . . trains its attention on what institutions, actions, images, utterances, events, customs, all the usual objects of social-scientific interest, mean to those whose institutions, actions, customs and so on they are" (Geertz, *Local Knowledge,* 23). Also, analysts face a social reality that has been "preselected and preinterpreted . . . by a series of common sense constructs," so that our sociological "constructs [are] of the second degree, namely constructs made by actors on the social scene" (Schutz, *Collected Papers,* 1:6).

is little reason to suppose that actors' everyday meanings and analysts' scientific accounts are hermetically sealed in separate rooms, and this was probably obvious well before Giddens gave name to the idea of the "double hermeneutic."[35] Interpretations and representations leak out everywhere and make themselves available for scholarly and ideological projects willy-nilly. So it is with academic accounts of the essence of science: Popper's falsifiability later turns up in court as a coordinate used to locate creationism not in science but religion (no observation could challenge the inerrancy of the Bible, it was suggested by those inclined toward evolution).

Actors' and analysts' accounts may intersect in other ways as well. Some sociological interpretations of science (perhaps the constructivist one offered here) might be read as working against the professional interests of scientists, thus triggering a second-order cartographic squabble: Who really has the epistemic authority to map science? To display (as I shall) the strategically flexible cartographic work that accounts for always provisional assignments of epistemic authority to science is deeply troubling for scientists who would rather have their credibility grounded more securely in immanent and enduring features of scientific knowledge-making than in its momentary rhetorical constructions. Some of these scientists, or their apologists, could respond to such constructivism by drawing their own cultural maps that put sociologists of science like me outside, denying us the insiders' authority to speak for and about science. Predictably, this leads to countermaps from sociologists, respatializations of credibility . . . I am not making this up! The "science wars"—supposedly pitting scientists against those who study them for a living—are really just another credibility contest to decide where science is and who has the legitimate power to map its contours. Indeed, my epilogue brings cultural cartography home to roost, by taking up implications of this reflexive predicament: boundary-work becomes an inescapable practice, it would seem, for those who study boundary-work.

The five chapters preceding the epilogue will make good, I hope, on the promissory notes of this conceptual introduction. The episodes taken together display, at an empirical level, several sociological patterns out of which one might build a cartographic understanding of the epistemic authority of science.

First, each case centers on a credibility contest, in which disputes over

35. Giddens, *Constitution of Society*, 284.

nature are settled in and through disputes over culture. Cartographic depictions of the distribution of epistemic authority provide interpretative grounds for people to decide which accounts of nature may be reliably trusted and deployed, and—in turn—which sources of knowledge are esteemed and materially supported (and, alternatively, whose knowledges scorned).

Second, the texts and talk that constitute boundary-work lend themselves to a sociological reading as cultural maps. Their rhetorical style is cartographic, or at least the familiar features of geographic maps—borders, territories, labeled landmarks, scale, and coordinates—make it easier (when transposed to cultural terrain) to grasp how the epistemic authority of science is haggled over and decided. Nobody involved in these episodes actually gets out a pen and draws a map of the culturescape, but rather the discourse through which they talk or write about science and its environs exploits and evokes cartographic imagery.

Third, the episodes accumulate a longish list of qualities attributed to science (and to its others) for the purpose of distinguishing reliable knowers or knowledge from suspect brands. Such provisionally defining features of science are neither consistent among themselves (for example, scientific inquiry becomes at once the pinnacle of objectivity and inescapably subjective) nor historically convergent (representations of science as subjective are not progressively replaced over time by objective ones—episodic oscillation is more like it).

Fourth, the characteristics then and there attributed to science and to its surrounds are shaped by local and occasional circumstances. Only by exploring empirically the following contingencies can a sociologist hope to understand why science becomes what it does amid struggles for credibility: Who are the players (the contestants for epistemic authority, the audiences who face practical choices necessarily grounded in how cultural maps are read)? What are the stakes in negotiations of credibility and utility, for diverse (and sometimes unseen and unheard) players? In what organizational or institutional arenas does the boundary-work occur?

Fifth, the episodes suggest how all this cultural cartography is connected to outcomes—both for the players situationally involved and for the trans-situational cultural space of "science." Cultural maps are shown to be variously useful for enrolling audiences—offering more or less appealing and sensible justifications for who should have authority over natural knowledge. Maps matter for those whose claims are accorded or denied the authority of science, and maps matter too for the culturally reproduced link between "science" and knowledge that is authoritative, credible, reliable, and trustworthy.

These five threads are woven into the five episodes, but each case draws them through in a distinctive way—highlighting something different (theoretically, historically) about the cultural cartography of science. John Tyndall (chapter 1) provides a vanilla case of boundary-work as expansion. From his scientific bully pulpit at London's Royal Institution in the mid-nineteenth century, Tyndall makes the case for science over and against the competing cultural authorities of religion and mechanics (for us, engineering or applied science). He battles for the legitimate power to define and explain nature, and for government patronage, public respect, a larger presence in educational institutions. Is scientific research or collective prayer more effective in halting cholera? In building better lighthouses for Britain's long coast, does basic research have anything to add to the hands-on wisdom of mechanics? It is precisely because Tyndall must work two fronts in the battle to expand science that we see just how flexibly it can be staked out. He uses one set of coordinates to locate the border between science and religion, but quite another to bound science from mechanics. In the first instance, practical utility, empiricism, and skepticism define science. In the second, science becomes pure knowledge grounded in experiment and theory; its principal utility is as a source of discipline and clear thinking; and it is far removed from market concerns. The point is not that one or both representations are mere fictions: science is arguably all of these things (and much more). The point is rather to watch how features of scientific practice and knowledge are selectively deployed in a contingent contest for epistemic authority and resources among multiple makers of belief.

"Social" science has forever stood in a cartographically ambiguous place with respect to the sciences of nature: Does the boundary go between or around them both? The question is not merely academic, nor has it ever been answered once and for all.[36] Twice this century, the mat-

36. A paradigmatic illustration is found in James Q. Wilson's recent plea that social scientists not be allowed to serve as expert witnesses in the courts. His boundary-work is taut: social science is not "real science," but "often merely 'pretend' science." "Real science involves testing theories by repeated independent experiments . . . and is most likely to be produced (although not always) when scholars study inanimate objects, as do biologists and chemists." But social science at its best is "much less empirically exact," offering statistical generalizations rather than determinist predictions; at its worst, it is pursued by "advocates" who "care deeply about mobilizing public opinion." The connection between cultural cartography (exclude the social

ter reached the floor of the United States Congress: In the late 1940s, legislators contemplated the terrain of what would become the National Science Foundation, and found themselves in hot dispute over whether the social sciences belonged inside; in the mid-1960s, the possibility of a stand-alone National Social Science Foundation was debated with only slightly less heat. Chapter 2 is not about a credibility contest narrowly drawn, but rather this: Should the credibility of natural science expand to embrace an altogether different ontological domain—society and culture? The stakes for social scientists are plain: official embodiment of these disciplines as scientific, to say nothing of the federal patronage that would tag along. Boundary-work on this occasion took a "same-or-different" format. Some argued that natural and social science are the "same": both are practically useful, rely on common objective methods, produce the same quality of reliable knowledge. Such maps were drawn in the 1940s by those who favored inclusion of social science in the incipient National Science Foundation, and again in the 1960s by those who argued that social science—by then having worked its way inside the foundation—should stay right there. Others argued that the two are "different," although landmarks in the two territories varied from the 1940s to the 1960s as contingent circumstances and stakes changed. After World War II, those who wanted to keep social science outside the proposed foundation offered invidious comparisons to natural science: next to physics or chemistry, sociology and economics were politically charged, methodologically flawed, and no better than common sense. But fifteen or so years later, proponents of a National Social Science Foundation celebrated the demarcation of the social from the natural sciences. The former disciplines had developed their own distinctive methodologies useful for attacking controversial social problems with 1960s gusto, and they no longer needed to legitimate themselves as immature siblings of physicists. The outer bounds of science stretch to fit the social sciences—but also shrink to exclude them—in testimonies of congressional witnesses and in impassioned pleas by politicians. Social science becomes a fungible thing, constructed diversely to serve various policy and professional ambitions.

George Combe's failed bid in 1836 to secure the chair of logic and metaphysics at the University of Edinburgh (chapter 3) may simply be read as the exclusion of his "obviously bogus" phrenology from science

---

sciences) and the denial of credibility (they are not experts, and have no place as such in court) could not be more plain (Wilson, "Keep Social-Science 'Experts' Out of the Courtroom").

and its relocation to pseudoscience. The banishment to non-science of expert readings of one's character from skull bumps was indeed the result of the Edinburgh Town Council's decision to give the chair instead to William Hamilton, a proponent of Scottish "common sense" philosophy. But it was not the cause of their decision, for the scientific status of phrenology was exactly the issue taken up in the many testimonials delivered to the town council by supporters of Combe—and in those for Hamilton as well. Each candidate made science into a distinctive cultural space that would not only locate their métier at its epicenter, but also enlist rafts of allies who would then sing their praises to the town council. This is where Combe stumbled: in order to make phrenology attractive to social reformers, penologists, educators, and physicians, he radically redrew the entire Scottish culturescape (and the place of science therein) in ways that might have seemed threatening (or at least confusing) to those asked to vote on his candidacy for the university chair. Combe blurred then secure lines between philosophy of mind and anatomy of brain, between pure science and its applications, between science and religion, and between experts and Everyman. In effect, the credibility contest at hand (is phrenology true? is it practically and reliably useful?) was settled by the town council's preference for Hamilton's cultural map—at once familiar to them as embodied in the bricks and mortar of early-nineteenth-century Edinburgh and congruent with their political interests.

Almost a century and a half later, the boundaries of good science were still unsettled as scientists—and almost the whole world in 1989—pondered the veracity of cold fusion and the credibility of its discoverers Stanley Pons and Martin Fleischmann (chapter 4). The eventual falsity of cold fusion was decided (provisionally and never universally), not by chemical data or physical theories alone, but through interpretative maps that depicted Pons and Fleischmann's science as pathological. Their claim became incredible as their science went bad. But the two struggled mightily for a time to save cold fusion—in part, by moving the adjudication of facts outward from science tightly drawn to a more encompassing playing field populated also by journalists and politicians. This episode brings to the fore questions about who has the authority to draw legitimate cultural maps of science—only scientists themselves, or the media too, or even Congress? Pons and Fleischmann tried to keep cold fusion alive first by inviting the press into their labs to decide for themselves whether their procedures exemplified good laboratory practice: early reports in the media pictured the codiscoverers as very normal scientists (not pathological). Then Pons and Fleischmann took their case to Capitol Hill, where congressional support for a $25 million cold fusion research institute in Utah hinged on how the border between science and technology was drawn: Does an energy panacea move sequentially from science to technology or does it grow all at once in both territories simultaneously? In other words, journalists were asked to bound good science from bad, and Congress was asked to locate the line between science and technology. One week later, however, order was restored both to nature and culture at the Baltimore meeting of the American Physical Society—where journalists and politicians were excluded from deliberations over the boundaries of good science; where cold fusion was transported from credible scientific knowledge to magic; where Pons and Fleischmann were moved out to the nether-land of pathological science because of their secretiveness, their lust for fame and interest in fortune, their unwillingness (or inability) to scrutinize their pet fact with the skepticism putatively displayed by all good scientists. The authority of scientists to set their own boundaries was protected as it was instantiated, for the time being.

The final case moves beyond the episodic moment to consider a lifetime of boundary-work in the person of Sir Albert Howard—along with his wife and collaborator, Gabrielle (chapter 5). Working in England, the West Indies, and India during the first half of the twentieth century, the Howards—trained as botanists—were hybridizers of plants but also hybridizers of culture. In their technical papers and political tracts, science is bred with peasant skills, market economics, imperialist politics, and grassroots social movements—all in response to changing practical problems: how to feed Indian villagers? how to make profits for British grain merchants? how to design a new experimental research station? how to ensure human health and save the earth through organic farming? If Albert Howard is remembered for anything these days, it is as the father of modern composting. But *how* his compost pile and its humus become, by turn, indigenous wisdom of Asian cultivators, then the stuff of science, later the mark of a neo-vitalist pre-Liebigian fount of human salvation (that is, not really scientific at all) is the crux of this cultural cartographic episode. The Howards' science was as imperialistic as their native country—holistic and expansive, rather than fragmented and specialized—but their story ends ironically. As Albert Howard (alone, back in Britain, following Gabrielle's death) seeks to use his scientific authority to convince the whole world of the evils of chemical fertilizers, pesticides, and herbicides, he finds his own legitimacy as a credible interpreter of nature challenged by other scientists committed to the benefits of such modern agricultural practices.

The five episodes taken together leave yawning gaps in a sociological understanding of the shifting cultural spaces for science. The cases are all historical rather than contemporary, more or less resolved rather than ongoing contests for credibility. My tiny sample is not designed to be representative of any universe of social phenomena, but just suggestive of how the sensitizing concepts of cultural cartography may be useful for interpreting the rising epistemic authority of scientists as the legitimacy to define reality is contested, claimed, and denied on occasions widely scattered in time and place. None of these cases takes place in an arena where boundary-work involving science has become acute just now: the courts.[37] I have, moreover, given greater attention to boundaries between science and something else altogether than to those dividing the various sciences and scholarly disciplines.[38] Obviously, the kind of cultural cartography developed here would be useful for studying contested authority in other institutional and professional domains, but I have resisted that temptation.[39] And I have—by design—centered on occasions when the borders and territories of science become problematic and manifestly in need of discursive mapping out, so that too little is said about when and how some cartographies get stabilized as unquestioned tacit assump-

37. On boundary-work and science in the courts, see Gieryn, Bevins, and Zehr, "Professionalization of American Scientists," on the creation versus evolution controversy. (The definitive history of this cartographic squabble is Numbers, *Creationists.*) For other studies of science and the law, see Jasanoff, *Science at the Bar;* and also Cambrosio, Keating, and Mackenzie, "Scientific Practice in the Courtroom"; Smith and Wynne, *Expert Evidence;* Solomon and Hackett, "Setting Boundaries between Science and Law"; and for a historical perspective Hamlin, "Scientific Method and Expert Witnessing." On cultural cartography in the promulgation of science-connected public policy, see Jasanoff, "Contested Boundaries in Policy-Relevant Science" and *Fifth Branch;* Shackley and Wynne, "Representing Uncertainty"; Zehr, "Accounting for the Ozone Hole."

38. Much research sympathetic to the cultural cartographic approach has been inspired by calls for interdisciplinarity or multidisciplinarity or just nondisciplinarity: see Messer-Davidow, Shumway, and Sylvan, *Knowledges;* Klein, *Crossing Boundaries.* Some historians of science are alert to power games behind the disciplinization of scholarly and scientific inquiry: "Disciplines are political institutions that demarcate areas of academic territory, allocate the privileges and responsibilities of expertise, and structure claims on resources" (Kohler, *From Medical Chemistry to Biochemistry,* 1). Disciplinary demarcations within anthropology are explored by Watson, "Social Construction of Boundaries."

39. For cultural cartographic studies having nothing to do with science, see Becker, *Art Worlds;* Beisel, "Constructing a Shifting Moral Boundary," on obscenity; Brain, "Discipline and Style," 854–55, on architecture; Burns, "Politics of Ideology," on Catholic responses to liberalism; Lamont, *Money, Morals, and Manners;* Nippert-Eng, *Home and Work;* Winch, *Mapping the Cultural Space of Journalism.*

tions or as uncontested old maps or as organizational rules and regulations. Some readers will regret the inattention here to issues of identity politics and identity epistemics; struggles for credibility and effects of scientific authority are deeply gendered, for example, and play themselves out increasingly on a multicultural terrain of uneven advantage.[40]

Lots of work ahead.

40. For a fine example, focused on "credibility struggles" between experts and lay populations at risk in AIDS research, see Epstein, *Impure Science;* also Indyk and Rier, "Grassroots AIDS Knowledge."

Good as Epstein's analysis is, I find no evidence in it at all for his bald assertion at the outset that "the very reliance on experts to adjudicate disputes tends to undercut the authority of expertise generally" (6). It is profoundly important, I believe, to distinguish the attachment of credibility or epistemic authority to a cultural space (a cartographic representation of practices, skills, accumulated knowledge of science) from its contingent allocations and denials to actual actors in particular settings. To be sure, the interpretative grounds for expertise are constantly changing along with how authority is distributed, but after only several centuries or more of such credibility contests (like the controversies over AIDS research), I see little warrant for concluding that just now "the authority of science passes its apogee"—as Cooter and Pumfrey do ("Separate Spheres and Public Places," 238).

Chapter One

JOHN TYNDALL'S DOUBLE BOUNDARY-WORK: SCIENCE, RELIGION, AND MECHANICS IN VICTORIAN ENGLAND

British science in the nineteenth century saw one triumph follow upon another: Charles Lyell's uniformitarian theory in geology, John Dalton's atomic theory, Michael Faraday's research on electromagnetism, James Clerk Maxwell's theory of light, William Thomson (Lord Kelvin)'s work on thermodynamics, Alfred Russell Wallace and Charles Darwin's discovery of a theory of natural selection, and Herbert Spencer's application of that theory to human societies. And from the start of the century, scientists enjoyed an increasingly bountiful infrastructure for pursuing science. The number of scientific chairs at British universities increased from 44 in 1810 to 110 in 1850.[1] In 1870, the House of Commons appropriated £6,000 for a Natural History Museum at South Kensington, and in the early 1870s Oxford built its Clarendon Labs and Cambridge its Cavendish.[2] By the end of the century, historian Frank M. Turner reports that scientists "had established themselves firmly throughout the educational system and could pursue research and teaching free from ecclesiastical interference."[3] Government expenditure on nonmilitary scientific research went from £70,115 in 1859–60, to £346,528 in 1879–80, and by 1914 public money for civil scientific research reached £2 million, or an

1. Morrell and Thackray, *Gentlemen of Science*, 547, table A2.

2. MacLeod, "The Ayrton Incident," 46. On the gradual transformation of Oxbridge from a place where elites study classics to an institution for (among other modern things) the study of empirical scientific research, see Ben-David, *Scientific Growth*, 120. For a historical review of the shifting fortunes of science education, see Cardwell, *Organization of Science in England*.

3. Turner, "The Victorian Conflict between Science and Religion," 376. This essay has been gathered up with others pertinent to John Tyndall and Victorian science in Turner's *Contesting Cultural Authority*.

unprecedented 3.6 percent of the total civil expenditure.[4] Some of that funding took the form of "Exhibition fellowships," research grants drawing on the "profits" of the Crystal Palace Exhibition of 1851; the first recipient was a young Ernest Rutherford. Even the cultural authority of scientists seemed to be sufficiently secure that its debate passed from the list of pressing issues: "After about twenty years [ending, roughly, in 1880] of exciting themselves in trying to keep the Truth alive, the late Victorians decided that Socialism and Imperialism, the income tax and pensions, Teddy Roosevelt and Rudyard Kipling were more interesting subjects. . . . It was agreed that the truth in religion is no relation to truth in science, and vice versa. . . . The preacher, like the physicist, is a specialist."[5]

A different story emerges from the wails and moans of public scientists of the day. Beginning in 1830 with Charles Babbage's *Reflections on the Decline of Science in England and on Some of Its Causes,* advocates for science complained of public neglect of their efforts and accomplishments, of the leisurely and sporadic introduction of science into curricula at schools and universities, and of the unwillingness of government to provide adequate patronage for research and facilities. Much of Babbage's diatribe is pointed at the increasingly anachronistic Royal Society for its failure to provide professional leadership and to lobby for resources needed to sustain careers in science. But some of the blame is laid on the government, its commissions, and its control of education, as Babbage enlists Sir Humphrey Davy and John Herschel as allies who agree "that science has long been neglected and declining in England."[6]

Babbage finds neglect and decline in three parts of scientists' professional stake: opportunities for gainful employment, curricula for proper education in science, and authority in governmental deliberations requiring exact knowledge of natural phenomena. On remuneration for a life in science, he laments: "The pursuit of science does not, in England, constitute a distinct profession [and] it is therefore on that ground alone, deprived of many of the advantages which attach to professions. . . . It appears that scarcely any man can be expected to pursue abstract science

4. MacLeod, "Science and the Treasury," 116 and 161; cf. MacLeod, "'Bankruptcy of Science' Debate." Government funding for science at this time is discussed by Alter and Davies, *Reluctant Patron.*

5. Susan Faye Cannon, *Science in Culture,* 280.

6. Babbage, *Reflections on the Decline of Science,* v; supportive testimony from Davy and Herschel, vii–ix.

unless he possess a private fortune, and unless he can resolve to give up all intention of improving it."[7] Educational opportunities for would-be scientists are inadequate: "Some portion of the neglect of science in England may be attributed to the system of education we pursue. A young man passes from our public schools to the universities, ignorant almost of the elements of every branch of useful knowledge." Finally, Babbage worries that scientists are not available (or they are overlooked) for advising government on technical matters: "As there exists with us no peculiar class professedly devoted to science, it frequently happens that when a situation, requiring for the proper fulfillment of its duties considerable scientific attainments, is vacant, it becomes necessary to select from among amateurs, or rather from among persons whose chief attention has been bestowed on other subjects, and to whom science has been only an occasional pursuit."[8]

Working conditions for scientists evidently did not improve much after Babbage's complaint, if one listens to these publicists. J. Norman Lockyer, solar spectroscopist and founding editor of *Nature,* remarked in 1873 that "there is absolutely no career for the student of science in this country. True scientific research is absolutely unencouraged and unpaid."[9] A year later, Darwin's bulldog Thomas Henry Huxley complained that "no amount of proficiency in the biological sciences will 'surely be convertible into bread and cheese.'"[10] Keeping to the same culinary metaphor, Huxley later said: "Science in England does everything—but pay. You may earn praise but not pudding."[11] Even at the end of the century, Lockyer looked back not with pride but dismay: "From time to time since this journal was started in 1869, it has been our duty to insist upon our relative deficiencies in regard to the advancement of science and the higher scientific instruction. . . . The present British intellectual equipment is not only not superior but inferior to that of other countries. . . . We must have a profound change of front on the part of the ministry and

7. Ibid., 10–11, 38. An anonymous critic (Gerard Moll) chides Babbage for his greediness: "But although Mr. Babbage objects to the term 'wealthy' applied to a nobleman, he is far from disliking wealth in the hands of men of science" (reprinted in Babbage et al., *Debates on the Decline of Science,* doc. 2, 27).

8. Babbage, *Reflections on the Decline of Science,* 3, 11–12.

9. Quotation in Cardwell, *Organization of Science in England,* 151

10. Quotation in Mendelsohn, "Emergence of Science as a Profession," 32. On Huxley's hortatory for science, see Desmond, *Huxley;* Knight, "Getting Science Across"; White, "Making a 'Man of Science.'"

11. Quotation in Basalla, Coleman, and Kargon, *Victorian Science,* 9.

the personnel of Government departments, only very few of whom have had any scientific education and who at present regard all scientific questions with apathy."[12]

John Tyndall joined this chorus from his post of nineteen years as superintendent at the Royal Institution in London.[13] This chapter examines Tyndall's arguments to justify scientists' requests for greater public attention and support. He faced two impediments: the enduring cultural authority of Victorian religion and the practical achievements of British engineering, mechanics, and manufacturing. Tyndall's campaign for science took the rhetorical form of "double" boundary-work: he attributed selected characteristics to science that effectively demarcated it from religion and—with different attributions—from mechanics, providing a rationale for the superiority of scientists in designated intellectual and technical domains. The simultaneous juxtaposition of science/religion and science/mechanics illustrates the flexible and not always consistent constructions of science that have nevertheless served the profession as an effective ideology justifying increased support of scientific research and education.

TYNDALL'S LIFE AND PUBLIC SCIENCE

Tyndall was born in 1820 to Irish Protestant parents of modest means.[14] His first, short career was as draftsman and civil engineer, but after a year teaching mathematics and drawing at one of the few schools of the day to have a laboratory, Tyndall left at age twenty-eight to pursue science at the University of Marburg. After earning a doctorate in mathematics, Tyndall joined the physics laboratory of Karl Hermann Knoblauch, who was extending Faraday's work on diamagnetism. Collaboration with Knoblauch led to Tyndall's first scientific paper, on the behavior of crystalline bodies between the poles of a magnet, in an 1851 issue of *Philosophical Magazine.* Upon his return to England, Tyndall found scarce employment for scientists, and he supported himself by translating, writing, examining, and by lecturing to the public—a task for which he proved gifted (much later, Crowther observed that Tyndall would have

12. From a 1901 editorial in *Nature*, reprinted in ibid., 490, 495.

13. On the Royal Institution, see Berman, *Social Change and Scientific Organization.*

14. This biographical sketch draws freely on Burchfield, "John Tyndall"; Eve and Creasey, *Life and Work of John Tyndall;* MacLeod, "John Tyndall." On Tyndall's rhetorical efforts, see Bartlett, "Preaching Science."

made a "splendid type for television science").[15] His fortunes picked up after election to the Royal Society in 1852, and especially after he benefited from the patronage of Michael Faraday, who became his role model in everything scientific. At age thirty-three, Tyndall was appointed by Faraday to the position of professor of natural philosophy at the Royal Institution, with an annual salary of £100 supplemented by another £80 from lectures and translations. With Faraday's death in 1867, Tyndall succeeded him as superintendent of the Royal Institution, a position he held until retirement in 1886 at age sixty-six.

Tyndall contributed much to scientific knowledge, but perhaps more to its popular appeal. His research from 1853 to 1874 focused on questions of physics: How does compression affect crystalline structures, what are the effects of solar and heat radiation on atmospheric gases, and why is the sky blue? An interest in airborne dust particles led Tyndall, after the mid-seventies, to examine organic matter in the atmosphere and to defend Pasteur in the spontaneous generation controversy. Tyndall's contributions have been doubly honored with eponyms: the Tyndall effect describes scattering of light particles in the atmosphere, and Tyndallization describes the use of discontinuous heating for sterilization. Roy MacLeod writes in his entry on Tyndall in the *Dictionary of Scientific Biography:* "This formidable capacity to move from electromagnetism through thermodynamics and into bacteriology was the hallmark of Tyndall's genius."[16] Still, these contributions cannot overshadow Tyndall's role as statesman for science. He delivered hundreds of popular lectures to large audiences of aristocrats and workingmen, while regularly contributing articles on science to the popular press. Tyndall's copious writings and lectures for popular audiences make up the materials examined here as boundary-work.

15. J. G. Crowther, *Scientific Types,* 157. The concept of "public science" was introduced in 1980 by historian Frank Turner, "Public Science in Britain," but it has been largely neglected since, according to a recent review of works on "popularization" (Cooter and Pumfrey, "Separate Spheres and Public Places"). For an exemplary study of how science was defined for its publics in the early Victorian period (just before Tyndall but pertinent to my discussion of George Combe's phrenology in chapter 3), see Yeo, *Defining Science.* For a study of popular science journals in the 1860s, see Barton, "Just before *Nature.*"

To judge from a letter by Huxley to Tyndall dated February 25, 1853, it was less important for Tyndall (in his public lectures) to teach scientific knowledge than to teach about science and its place in Victorian culture: "What they want, and what you have, are clear powers of exposition—so clear that people may think they understand even if they don't. That is the secret of Faraday's success, for not a tithe of the people who go to hear him really understand him" (quotation in Thompson, "John Tyndall and the Royal Institution," 10).

16. MacLeod, "John Tyndall," 522

Late in life, Tyndall married Louisa Charlotte Hamilton, a thoroughly devoted wife who had the misfortune, one morning in 1893, of confusing magnesia with chloral and inadvertently administering a fatal dose of the latter to her husband. She survived him for forty-seven years.

Tyndall's biography sheds light on his activities as public scientist. He was part of a generation of British scientists—one of the earliest—to make a life and career out of science without benefit of the wealth and privilege of the "gentleman amateur" typical earlier. As had many of his cohort, Tyndall had tasted the relatively well-supported German institutions for scientific education and found such support wanting upon his return to Britain. Clearly, if he wanted to do science he had to promote science, and temperamentally he was up to the fight. Early on, during the 1840s railway-building boom in England, the contentious Tyndall was dismissed as civil engineer because he protested unfair treatment of Irish workers on the project. He later made a name for himself by attacking Faraday's theories of electromagnetism, though not with the aggressiveness that would have prevented Faraday from becoming Tyndall's British mentor. Tyndall also battled with the Edinburgh physicist P. G. Tait over a priority dispute, and in the midst of the controversy, Tait tried to score a point by writing to *Nature* in 1877 that very few can do public science "without thereby losing their claim to scientific authority. Dr. Tyndall has, in fact, martyred his scientific authority by deservedly winning distinction in the popular field. One learns too late that he cannot make the best of both worlds."[17] There is much irony here: for all his efforts to place boundaries around science, Tyndall is excluded by Tait from real science for his admitted success on the popular front.

The fighting spirit emerges especially in Tyndall's dealings with religion. As president of the British Association for the Advancement of Science, Tyndall's address at the Belfast meeting in 1874 made "so bold a claim for the intellectual imperialism of the modern scientific inquiry" that theological foes and scientific friends were outraged.[18] Surely this vehement dismissal of the cultural importance of religion is related in complicated ways to Tyndall's roots as minority Protestant in the Catholic County Carlow of Ireland. Tyndall's equally powerful disdain for the pecuniary motives of mechanicians and manufacturers, and for their disregard of the importance of "pure" scientific knowledge, is likewise related to his professional journey from engineering as a young man to science in maturity. Tyndall was an Irishman who left for England, a Protestant

17. Quotation from Crowther, *Scientific Types*, 184

18. Turner, "John Tyndall," 172. On the metaphysics of the Belfast Address, see Barton, "John Tyndall, Pantheist."

who left for secular life, and an engineer who left for science: such denials of roots are sometimes accompanied by passionate commitments to the adopted statuses.

There is little need to stoop to psychologistic interpretations to account for Tyndall's rhetoric as public scientist. The message is explained less by Tyndall's character than by his hope to create for science a cultural space unfettered by competing authorities of religion and mechanics. But it is precisely Tyndall's obstinate and contentious character that makes him so appealing as a subject for these sociological inquiries into scientists' cultural cartography: his case for science is unambiguously constructed and vociferously argued; at least in public, Tyndall had few of the ambivalences and hesitancies that nagged other statesmen for science in the later nineteenth century.

Tyndall's rhetoric exemplifies a pattern prevalent in public science. His goals are *economic:* to secure for scientists various professional resources (increased employment opportunities, increased attention to science in curricula of schools and universities, increased government or private patronage for scientific research and facilities). His tactics are *cultural:* to construct and to broadcast an interpretative context that defined—in contextually distinctive ways—the meaning of science and its relationships to religion and technological progress. To secure professional resources for scientists, it was necessary to establish a "public appreciation for science and its contributions to the welfare of the nation."[19] To fortify a public appreciation of science, it was necessary to construct a widely diffused image of science that would make claims for its enlarged support self-evident and self-justifying. To judge from Tyndall's boundary-work, the authority of religion and the spectacular accomplishments of British manufacturing were impediments to improving the working conditions of scientists, for he often chose them as foils when he gave meanings to science in front of public audiences.

SCIENCE AS NOT-RELIGION

The endless controversy surrounding religion and science reached a crescendo in the decade following publication of Darwin's *The Origin of Species* in 1859. The debate was not simply a contest between two theo-

19. Turner, "Victorian Conflict," 363. Much of Tyndall's evangelism for science was channeled through an invisible college of publicists known as the X Club, with members including Huxley, Spencer, Joseph D. Hooker, George Busk, Edward Frankland, John Lubbock, Thomas Archer Hirst, and William Spottiswoode. On the "X Club" and related issues, cf. Jensen, "The X Club"; MacLeod, "The X Club"; Thompson, "John Tyndall and the Royal Institution."

ries of natural history; it was a battle for cultural and epistemic authority between two occupations, with substantial symbolic and material resources at stake.[20] The cultural authority of long-standing religious beliefs, reinforced every Sunday from the pulpit, created resistance toward scientific explanations of natural phenomena. Tyndall found his company among the "scientific naturalists," whose extreme position on science/religion was not shared by all scientists.[21] For Huxley, Joseph Dalton Hooker, Henry Maudsley, Herbert Spencer, and Tyndall, science could no longer be pursued for religious ends, nor justified by them. The "accommodating" posture adopted by Bacon in the seventeenth century—in essence, the glorification of God through scientific exploration of his handiwork in nature—was no longer tenable for scientific naturalists. Their posture was more aggressive: science was to replace religion as a

20. "It was this shift of authority and prestige . . . from one part of the intellectual nation to another that caused the Victorian conflict between religious and scientific spokesmen" (Turner, "Victorian Conflict," 359). This is scarcely the place to review the gigantic literature on the boundary science/religion; two recent works, full of references, are Brooke, *Science and Religion,* and Drees, *Religion, Science.* On Tyndall's place in this conflict, see Kim, *John Tyndall's Transcendentalism.*

21. Tyndall's boundary-work is not analyzed here as typical of the Victorian scientific community. Scientists such as Wallace and Sedgwick continued well beyond midcentury to reconcile their naturalistic inquiries with religious faith. On others who straddled this fence, cf. Turner, *Between Science and Religion;* Brock and MacLeod, "The Scientists' Declaration"; and Paul White's spot-on study of Huxley, which argues that "religious symbols and discourse were central to the meaning of scientific practice" ("Making a 'Man of Science,'" 156). White is attuned to boundary-work: "I intend to examine how such boundaries were constructed and maintained—in other words, how science came to be interpretable. . . . Through such public and private acts, boundaries of inclusion and exclusion were constructed and reinforced that were at once both intellectual and social" (6–7, 8).

On one small point would I quibble with Turner's interpretation of nineteenth-century public science, where he suggests that "naturalistic publicists sought to expand the influence of scientific ideas for the purpose of secularizing society rather than for the goal of advancing science internally. Secularization was their goal; science, their weapon" (*Between Science and Religion,* 16). These goals are not easily disentangled: a secularized society is, ipso facto, one in which the cultural authority of scientists is enlarged (and thus, in which the perceived legitimacy of their requests for material support and encouragement is better established) while that of religion is diminished. And Turner is quite right that at the level of rhetoric, scientists justified their case in terms of the benefits of secularization for society as a whole and not for the pockets of scientists. Tyndall's tone was thoroughly disinterested, but it would be naive to argue that he (like fellow publicists) was not aware that such a disinterested image would help to sell science and increase their resources (personal and collective) for research and education. Turner is also probably correct if he limits his observations to individual motivations of the publicists: it would be unacceptably reductionistic to see their boundary-work as merely fictions conjured up to make scientists powerful and wealthy. Tyndall did not structure his public talks and writings just to boost the resources of scientists (I have no grounds for doubting the sincerity of his beliefs about science), but neither was he unappreciative of the fact that this happy circumstance could result from his efforts.

superior source of natural knowledge. Insofar as religion claimed authority over "knowledge," its beliefs and dogma could only prevent the rational and empirical understanding of nature.

Two moments in Tyndall's selling of science clearly illustrate his conversion of the philosophy of scientific naturalism to a professional ideology. In 1872, Tyndall found himself embroiled in the "prayer gauge" debate, which was sparked by an article challenging Christians of the nation to conduct an experiment to determine the physical efficacy of prayer. It was then the custom for the British prime minister or Privy Council to ask a high official of the Anglican Church to call for a national day of prayer as a response to national crises. Public prayers were called as hoped-for solutions to cattle plagues in 1865, a cholera epidemic in 1866, and a case of typhoid suffered by the young Prince of Wales in 1871. To Tyndall, public prayers were an affront to science and reason. They represented a "concrete form of superstition whereby clergy with the approval of the state could hinder the dispersion of scientific explanations of natural phenomena or claim credit for the eradication of natural problems that were solved by the methods of science."[22] (When Prince Edward recovered from typhoid, clergymen pointed to the effectiveness of the nation's prayers.) Tyndall encouraged an experiment in which a selected hospital would be made the focus of national prayer, with a comparison of mortality rates before and after the day of supplication. The experiment was never conducted, but the furor provoked by its mere proposal is evidence of the extent to which the enduring prestige and authority of religion in that society impeded the acceptance of rational and empirical explanations of physical and biological events.

The church too held power over some educational institutions and on occasion used it to stall introduction of science into the curriculum. During Tyndall's presidency of the British Association for the Advancement of Science in 1874, the Catholic Church in his native Ireland rejected a request from laymen to include physical science in the curriculum of the Catholic university. Perhaps it was this move that led Tyndall, in the Belfast Address, to his unequivocal denial of the authority of religious belief over natural phenomena: "And grotesque in relation to scientific culture as many of the religions have been and are—dangerous, nay, destructive to the dearest privileges of freemen . . . we claim, and we shall wrest,

22. Turner, "Rainfall, Plagues, and the Prince of Wales," 48. At stake was the "cultural leadership in a modernized English society," for "these prayers raised the question of whether scientists, physicians, and technical experts, or the clergy should set the tone for public consideration of the health and physical well-being of the nation."

from theology the entire domain of cosmological theory. All schemes and systems which thus infringe upon the domain of science must, in so far as they do this, submit to its control and relinquish all thought of controlling it."[23]

Tyndall's scientific naturalism was a strategic rhetoric for accomplishing the professional ambitions of Victorian scientists, as they tried to remove lingering threats to the cultural authority of science. In drawing a boundary between science and religion, Tyndall emphasized for his audiences three distinguishing features: practical utility, empiricism, skepticism.[24]

Practical Utility

Science is depicted as practically useful for inspiring technological progress to improve the material conditions of the nation; by contrast, the "utility" of religion is limited to aid and comfort in emotional matters. Tyndall here attempts a functional demarcation of science and religion. He links science to the industrial progress and mechanical inventions of the day, as in a lecture to schoolmasters on elementary magnetism: "I might, of course, ring changes on the steam-engine and the telegraph, the electrotype and the photograph, the medical applications of physics, and the various other inlets by which scientific thought filters into practical life." The "of course" signals Tyndall's hope that his audience will take for granted that these technological achievements were dependent on prior discoveries of scientific knowledge. The metaphors employed by Tyndall often heighten the contrast between the practically useful scien-

23. In addition to its availability in part 2 of John Tyndall's *Fragments of Science* (New York: P. F. Collier and Son, 1905), 145–214, the Belfast Address has been reprinted in Basalla, Coleman, and Kargon, *Victorian Science,* where the above extracts will be found, 474–75. On the aftermath of the Belfast Address, see Brown, *Metaphysical Society,* 238.

24. Turner rightly notes that the mission of public scientists was, in part, to establish their authority over the definition of science: "There are usually disputes between . . . professionals and outsiders who wish to impose their own definition on the group or who presently carry out the social function that the professionalizing group wishes to share or to claim as its own exclusive domain" ("Victorian Conflict," 360). In addition, Tyndall's boundary-work at science/religion tried the narrower task of excluding from "professional" science the many amateur practitioners (such as a botanist/parson) whose primary vocation was theology. It was only in 1860 that "genuinely" scientific members outnumbered others in the ranks of the Royal Society. See Mendelsohn, "Emergence of Science as a Profession," 28. This trend is reflected in statements by scientific publicists: Galton observed in 1870 that "the pursuit of science is uncongenial to the priestly character" and, at about the same time, Huxley remarked that no one "is or can be both a true son of the church or a loyal soldier of science" (quotations are from Turner, "Victorian Conflict," 365 and 370).

tist and the practically irrelevant minister: "That the knowledge brought to us by these prophets, priests and kings of science is what the world calls 'useful knowledge,' the triumphant application of their discoveries proves."[25] Religion and theology are located in a distant cultural place, nearer to poetry, art, and "dead languages" than to modern practical science. In the Belfast Address, Tyndall suggests that "it will be wise to recognize [theologians, preachers] as the forms of a force, mischievous if permitted to intrude on the region of objective knowledge, over which it holds no command, but capable of adding, in the region of poetry and emotions, inward completeness and dignity to man."[26] The audience is asked to see that science yields the knowledge upon which Britain's material prosperity is built, and that religion yields emotions to feed the soul. If science and religion have different utilities, neither can succeed when the other interferes: "[Poetry and science] minister to different but to equally permanent needs of human nature; and the incompleteness of which I complain consists in the endeavor on the part of either to exclude the other. There is no fear that the man of science can ever destroy the lilies of the field; there is no hope that the poet can ever successfully contend against our right to examine, in accordance with scientific method, the agent to which the lily owes its glory. There is no necessary encroachment of the one field upon the other."[27] The "fields" in the final sentence are (of course) not those in which the lilies grow, but fields in which different meanings of lilies are grown, scientific or poetic. "Testimony as to natural facts is worthless when wrapped in this atmosphere of affections; the most earnest subjective truth being thus rendered perfectly compatible with the most astounding objective error."[28] "It is against this objective rendering of emotions—this thrusting into the re-

25. Tyndall, *Fragments of Science,* part 1, 102.

26. Ibid., part 2, 209. In an 1869 address, Tyndall pursues the same theme: "Here the dead languages, which are sure to be beaten by science in the purely intellectual fight, have an irresistible claim. They supplement the work of science by exalting and refining the aesthetic faculty" (106). The potential for a blurring of these distinctive utilities—practical and emotional—is ever present: "Facts like those discussed by Rumford naturally and rightly excite the emotions. . . . But in dealing with natural phenomena, the feelings must be carefully watched. They often lead us unconsciously to overstep the bounds of real knowledge, and to run into generalizations which are in perpetual danger of being overthrown" (John Tyndall, *Heat: A Mode of Motion,* 6th ed. [New York: D. Appleton, 1883], 110).

27. John Tyndall, *New Fragments* (New York: D. Appleton, 1898), 77.

28. Tyndall, *Fragments,* part 2, 19–20. Speaking of Genesis, Tyndall writes: "It is a poem, not a scientific treatise. In the former aspect it is forever beautiful; in the latter aspect, it has been, and it will continue to be, purely obstructive and harmful. To *knowledge* its value has been negative" (224).

gion of fact and positive knowledge of conceptions essentially ideal and poetic—that science, consciously or unconsciously, wages war."[29]

Empiricism

Science is empirical in that its road to truth is experimentation with observable facts of nature; religion, on the other hand, is metaphysical because its truths depend on unseen, spiritual forces assumed without verification.[30] In the midst of the "prayer gauge" controversy, Tyndall observes that in science, "to check the theory we have simply to compare the deductions from it with the facts of observation. . . . But while science cheerfully submits to this ordeal, it seems impossible to devise a mode of verification of their theories which does not rouse resentment in theological minds. Is it that, while the pleasure of the scientific man culminates in the demonstrated harmony between theory and facts, the highest pleasure of the religious man has been already tasted in the very act of praying, prior to verification, any further effort in this direction being a mere disturbance of his peace?"[31] Eight years later, Tyndall returns in metaphor to the same contrast: "Science, which is the logic of nature, demands proportion between the house and its foundation. Theology sometimes builds weighty structures on a doubtful base." And again: "To the sublime conceptions of the theologian succeeded the desire for exact knowledge characteristic of the man of science."[32] Elsewhere, Tyndall adopts an even more extreme empiricism, using the metaphor of a "photographic plate" to refer to the blank readiness of the scientific mind to "receive impressions from the light of truth."[33] His language is classic Baconian inductivism: "But in no case in framing theories does the imagination create its materials. It expands, diminishes, moulds and refines . . . materials derived from the world of fact and observation. . . . But even

29. Ibid., 393.

30. Turner sees this empiricism as part of a "positivist epistemology" that was both a cause and a weapon in the struggles between science and religion ("Victorian Conflict," 364). However, he overlooks the cartographic flexibility that is the centerpiece of the sociological analysis presented here, as he suggests that the scientific naturalists chose "empiricism over idealism, . . . the logic of Mill [induction from observation] over that of Whewhell [deduction from abstract theory]" (*Between Science and Religion,* 20). These choices were useful only in the construction of a boundary at science/religion, but idealism and Whewhell's deductive logic were used with greater effectiveness at science/mechanics, as I will suggest below.

31. Tyndall, *Fragments,* part 2, 47–48.

32. Tyndall, *New Fragments,* 13, 198.

33. Tyndall, *Fragments,* part 2, 29.

the theory as it stands did not enter the mind as a revelation dissevered from the world of experience." Scientific knowledge becomes, when Tyndall pursues this line, a literal replication of nature: "You will now observe that such logic connects our experiments in simply a transcript of the Logic of Nature."[34] Clearly, when Tyndall wishes to bound science from religion, it is the world of observable fact that is the essence of scientific knowledge: "Facts thus dealt with exercise an expansive force upon the intellect—they widen the mind to generalization. . . . Facts looked at directly are vital; when they are passed into words half the sap is taken out of them."[35] Facts are the source of theories, and science does not work nearly so well the other way around: "If the arithmetical processes of science be too exclusively pursued, they may impair the imagination and thus the study of physics is open to the same objection as philological, theological or political studies, when carried to excess."[36]

When Tyndall uses empiricism to build a boundary between science and metaphysical religion, the domain of science is restricted to observables or to the potentially observable: "As regards knowledge, physical knowledge is polar. In one sense it knows, or is destined to know everything. In another sense, it knows nothing. . . . Science knows nothing of the origin or destiny of nature." The question of First Causes—among the hottest of topics in the science/religion debates of the nineteenth century—is not properly scientific because it cannot be answered with empirical observation. "If you ask [the scientist] whence is this 'Matter' of which we have been discoursing—who or what divided it into molecules . . . he has no answer. Science is mute in reply to these questions." In other words, "the natural philosopher, as such, has nothing to do with purposes and designs. His vocation [is] to inquire what nature is, not why

34. John Tyndall, *Six Lectures on Light Delivered in America in 1872–1873,* 3d ed. (New York: D. Appleton, 1901), 95, 119.

35. Tyndall, *Fragments,* part 1, 384, 365. Tyndall praised the physicist Thomas Young, "who did not think abstractly. A philosophical fact, a difficult calculation, an ingenious instrument or a new invention would engage his attention; but he never spoke of morals, or metaphysics, or religion" (*New Fragments,* 283). Young (1773–1829) is most remembered for his revival of a wave theory of light.

36. Tyndall, *Fragments,* part 1, 310–11. This is a curious juxtaposition of cultural realms: philology, theology, and politics are (or, at least, were) not arithmetical to any degree, so it is difficult to understand how physics could become more like them if its arithmetical side is excessively pursued. One interpretation is that Tyndall sees mathematical formalisms as taking a step away from "vital" empirical facts. The implication, then, is that the dangers to physics of excessive arithmetic is the risk of losing touch with empirical reality, a problem imputed to the cultural pursuits of philology, theology, and politics.

she is."[37] This restriction of science to observables is lifted when Tyndall turns to the boundary between science and mechanics, as we shall see.

Skepticism

Science is skeptical because it respects no authority other than the facts of nature; religion is dogmatic because it continues to respect the authority of worn-out ideas and their creators. Tyndall ties science to the progressive and the modern, placing religion with the traditional and the unchanging. He first sets up the skeptical scientist: "The first condition of success is patient industry, and honest receptivity, and a willingness to abandon all preconceived notions, however cherished, if they be found to contradict the truth. . . . If a man be not capable of this self-renunciation—this loyal surrender of himself to Nature and to fact—he lacks . . . the first mark of a true [natural] philosopher."[38] The capability for self-renunciation was not widespread in medieval Europe under religious dominance: "It was a time when thought had become abject, and when the acceptance of mere authority led, as it always does in science, to intellectual death." The same dogmatism continues in the religion of Tyndall's own time, as he notes in a diatribe against observation of the Sabbath: "But the most fatal error that could be committed by the leaders of religious thought is the attempt to force into their own age conceptions which have lived their life, and come to their natural end in preceding ages. Foolishness is far too weak a word to apply to any attempt to force upon a scientific age the edicts of a Jewish lawgiver."[39] The dogmatism inherent in religion has only deleterious effects on the pursuit of science: "A favorite theory—the desire to establish or avoid a certain result—can so warp the mind as to destroy its powers of estimating facts." Apparently, Tyndall avoided such temptations, for just three years before his death, he looked back confident that he had practiced what he had preached: "In matters of science, I was always able, in the long run, to make prejudice yield to reason."[40]

The skepticism of scientists extends to their willingness to admit error,

37. Tyndall, *Fragments*, part 2, 57, 96; *Heat,* 111. This restriction of science to observables conflicts with the "scientific imperialism" offered by Tyndall at Belfast. An interpretation of this apparent waffling is offered below.

38. Tyndall, *Fragments,* part 1, 307. It is perhaps with religion in mind that Tyndall points out that "the discipline of suspending judgement is a common one in science, but not so common as it ought to be elsewhere. . . . We ought to learn to wait" (ibid., part 2, 107).

39. Ibid., 157; Tyndall, *New Fragments,* 33, 36.

40. Tyndall, *Fragments,* part 2, 19; *New Fragments,* 375.

and to correct it. This is so even for the iconic Darwin, who "has shown himself to be fallible; but here, as elsewhere, he has shown himself equal to that discipline of surrender to evidence which girds his intellect with such unassailable moral strength."[41] The admission of scientists' fallibility further distinguishes science from religion: Tyndall invites his Victorian audiences to recall the last time a religious figure of equal stature admitted that a theological pronouncement was once, but no longer, true. Tyndall's rhetorical goal is not to deny the existence of universals, but to locate them in religion, and to show that science is skeptical even about its own truths, changing them as new facts emerge. The link of science to the changing sector of Victorian culture and society is achieved, as religion is located in another cultural sector that remains forever the same.

SCIENCE AS NOT-MECHANICS

Victorian mechanicians and engineers presented a different obstacle to the expansion of scientific authority and resources.[42] Practical inventions of Victorian craftsmen and manufacturers—steam engines, telegraphs—did almost as much to stall the entry of science into universities and government as the competing cultural authority of the church. Many Britons believed that technical progress in the Industrial Revolution did not depend on scientific research, and some, like William Sewell, believed that science impeded the flowering of practical technology: "Deep thinking [is] quite out of place in a world of railroads and steamboats, printing presses and spinning jennies."[43] Many would have accepted the implicit anti-intellectualism of Samuel Smiles, who wrote in 1874: "One of the most remarkable things about engineering in England is that its principle achievements have been accomplished, not by natural philosophers nor by mathematicians, but by men of humble station, for the most part self-educated. . . . The great mechanics . . . gathered their practical knowledge in the workshop, or acquired it in manual labor."[44] If such arguments were widely shared, then the need for greater financial support of scientists and enlarged scientific education would go unappreciated by the British public and its politicians. The arguments were countered by public scientists such as Huxley, who remarked in 1850 that "practical

41. Tyndall, *Fragments,* part 1, 226.

42. For a more recent episode in the border war between science and engineering, see Kline, "Construing 'Technology' as 'Applied Science.'"

43. Quotation in Houghton, *Victorian Frame of Mind,* 114.

44. Quotation in Robinson and Musson, *James Watt and the Steam Revolution,* 1.

men [still] believed that the idol whom they worship—rule of thumb—has been the source of past prosperity, and will suffice for the future welfare of the arts and manufactures. They were of the opinion that science is speculative rubbish; that theory and practice have nothing to do with one another; and that scientific habit is an impediment, rather than an aid, in the conduct of ordinary affairs."[45] Tyndall's boundary-work sought to establish an interpretative context in which such opinions were impossible.

Moreover, as engineers began to "professionalize" by claiming expertise over certain technical issues, they sometimes confronted scientists who tried to assert their own authority in overlapping territory.[46] From 1866 until his 1882 resignation in protest, Tyndall served as "scientific" adviser to the Board of Trade on the question of how best to illuminate Britain's lighthouses. Although the lighthouses had traditionally been an engineering matter, Tyndall argued that the engineers who advised the Board "had closed their minds to external innovation" and expressed "diffidence toward the encouragement of new scientific ideas."[47] Tyndall believed that informed policy required more fundamental research, while engineers were apparently content to reach decisions with extant knowledge. In the end, Tyndall's recommendations were ignored in favor of the engineers', who "were already in positions of high civil authority. . . .

45. Quotation in Houghton, *Victorian Frame of Mind,* 113.

46. Bernal looks at this boundary-work from the other side: "The general tendency in engineering industry was to separate itself as an independent profession from the general trends of academic science" (*Science and Industry in the Nineteenth Century,* 31).

47. MacLeod, "Science and Government in Victorian England," 31, 15. Macleod quotes Tyndall on his break with the Board of Trade: "I warned them loyally and repeatedly as to the inevitable result of that course; but they deemed me a partisan, and finding at the Board of Trade a President whose views for the moment chimed in with their own [the engineers], they united with him in proceedings which ended in the wreck of a relation of seventeen years standing" (31). The charge of partisanship against Tyndall may have resulted from the claim by his adversaries that he had taken the side of a fellow Irishman. MacLeod concludes that Tyndall did not deserve the label "partisan," and that he in fact emerged as a disinterested advocate of the need for more scientific research on lighthouse illumination (37). Tyndall surely realized the consequences of his resignation in terms of arguments for the participation of scientists on such government boards. Scientists were not always participants in lighthouse boards. David Brewster, in the midst of the "declinist" controversy of 1830, noted that the three British lighthouse boards "have engineers, secretaries, and treasurers who receive good salaries . . . , yet, by a fatality which impends over every British institution, not one of all the numerous members and officers of these three scientific boards is a man of science, or is even acquainted with those branches of optics which regulate the condensation and distribution of that element which it is their sole business to diffuse over the deep" (reprinted in Babbage et al., *Debates on the Decline of Science,* 322). Clearly, in 1830, scientists had not yet convinced the state that their expertise was not only relevant but necessary for deliberations of lighthouse policy.

Practical men who had braved the brute force of nature to fashion pillars of stone and mortar had a strong emotional case against speculative men of ideas."[48]

Technological progress—even with the high-tech achievements of today—is never prima facie justification for support of basic scientific research. The dependence of technological advancement upon scientific research is not established by the steam engine or semiconductor; such dependency is a particular *interpretation* of events that must be constructed and disseminated if it is to be believed.[49] That task, in the latter nineteenth century, fell to Tyndall and his fellow statesmen for science. His work at the boundary science/mechanics would be complicated by the need to demonstrate not only the dependence of technology on science, but also the "cultural" utility of science beyond practical invention. An inconsistent set of characteristics are attributed to science when Tyndall uses these two arguments to distinguish science from mechanics, and such inconsistencies are magnified when attributes of science at this boundary are compared to characteristics of science attributed at the boundary science/religion.

Five characteristics of science are said to distinguish it from mechanics, engineering, or manufacturing: it is a fount of technology; it is experimental; it is theoretical; it is disinterested; and it is a "means of culture."

Fount of Technology

Scientific inquiry and knowledge is the fount of knowledge on which the technological progress of inventors and engineers depends. The boundary between science pure and applied is obscured by Tyndall as he tries to establish the dependency of technology on science: "There exists no category of . . . science, to which the name of applied science could be rightly given. We have science, and the applications of science, which are

48. MacLeod, "Science and Government in Victorian England," 15.

49. The dependence of technical invention on scientific research has been a matter for noisy argument among historians of science; see, for a review, Mathias, "Who Unbound Prometheus?" These historiographical debates are tangential to the sociological discussion pursued here, for each historian typically attempts a definitive account of the science/technology link. The goal is to establish (within the interpretative frame of history of science) that dependency did or did not exist at a designated time and place. My different goal is to watch contemporary ideologues argue the issue before much wider audiences with more at stake professionally than merely winning an argument with fellow scholars. In chapter 4, I describe how the boundary between science and technology was manipulated by Pons and Fleischmann in their bid to convince the U.S. Congress that it should invest big money in a research institute for cold fusion. For another instance of dispute at this cultural border, see Gieryn, "Science Pure and Applied."

united together as [a] tree and its fruit." To use boundary-work to create a distinctive region of "applied science" would by any logic simultaneously create a region of "pure science" whose practical utility would be difficult to defend. Such an image of pure science could confirm, in the minds of laymen, the anti-intellectualism implied by Sewell and Smiles. The idea that essential knowledge flows from science to mechanics is clear during Tyndall's lecture tour to America in 1872–73: "Before your practical men appeared on the scene, the force had been discovered, its laws investigated and made sure, the most complete mastery of its phenomena had been attained—nay, its applicability to telegraphic purposes demonstrated—by men whose sole reward for their labors was the noble excitement of research, and the joy attendant on the discovery of natural truth." Mechanicians (and, it seems, Sewell and Smiles) typically do not appreciate the extent to which the capacity for invention and production rests upon scientific knowledge: "The professed utilitarian . . . admires the flower, but is ignorant of the conditions of its growth. . . . Let the self-styled practical man look to those from the fecundity of whose thought he, and thousands like him, have sprung into existence. Were they inspired in their first inquiries by the calculation of utility? Not one of them."[50] A familiar cartographic metaphor emerges again in the address to American audiences: "Behind all our practical applications, there is a region of intellectual action to which practical men have rarely contributed, but from which they draw all their supplies." Science shall continue to be a fount of technology, as Tyndall asks his readers to look back on the steam engine and forward to anticipated achievements: "We cannot for an instant regard them as the final achievements of Science, but rather as an earnest of what she is yet to do." To make good on this promissory note, the vision of science must be extended to those in the factories: "Who can say what intellectual Samsons are at the present moment toiling with closed eyes in the mills and forges of Manchester and Birmingham? Grant these Samsons sight, and you multiply the chances of discovery and with them the prospects of national advancement."[51]

50. Tyndall, *Six Lectures on Light*, 223, 221–22; *Fragments*, part 1, 312.

51. Tyndall, *Six Lectures on Light*, 222; *Fragments*, part 1, 313, 314. This last quotation has several points of interest. First, Tyndall exploits a religious metaphor—the blinded Samson—to describe one who is bereft of the vision of science; audiences are invited to ponder again the contrast between the religiously blinded and the scientifically sighted. Second, the promise of science is attached to the wider national interest in technological advancement. Tyndall is careful throughout his boundary-work to justify his claims for greater support of scientists in terms of broad interests rather than in terms of the sectarian private interests of scientists as a profes-

This kind of boundary-work established the dependency of technological invention on scientific discovery, but it avoided the possible interpretation that science and technology could be collapsed into a single cultural space. Tyndall's careful articulation of the simultaneous dependence and independence of science and technology was a crucial part of his case for eliciting support for the autonomous pursuit of science.

Experimentation

Scientists acquire knowledge through systematic experimentation with nature; because mechanicians and engineers rely on mere observation, unsystematic trial and error, and common sense, they cannot understand or explain their practical successes or failures. In 1884, Tyndall contrasts the experimental method of science to mere observation: "The physical investigator desires not only to observe natural phenomena but to recreate them—to bring them, that is, under the dominion of experiment. From observations we learn what Nature is willing to reveal. In experimenting, we place her in the witness-box, cross-examine her, and extract from her knowledge in excess of that which would, or could, be spontaneously given."[52] Having defined experimentation as "a question put to nature," Tyndall suggests that its pursuit requires talents and training different from those required for mere observation: "I can see, the evidence of men, who, however keen and clever as observers, are not rightly trained as experimenters. These alone are aware of the precautions necessary in investigations of this delicate kind." And six years later, to a different audience, Tyndall reiterates: "Good observers frequently prove

sion. Finally, Tyndall urges an enlarged scientific education not just for aspiring professionals but for those entering any walk of life, including workingmen in factories. He refers to the need to increase highly specialized scientific education beyond the workingmen's institutes; on this, he disputed his good friend Huxley, who pushed a liberal arts form of science education. Tyndall here is not implying that any workingman could be a scientist, in the sort of "democracy" of authority that was present in Bacon's ideology of science, and Combe's as well (chapter 3). By the later nineteenth century, such views had disappeared in England in favor of scientific "elitism." Workingmen could—with science's light—assist the professionals' efforts by carrying out their toils informed by a familiarity with scientific method.

52. Tyndall, *New Fragments,* 213. Use of the "witness box" metaphor echoes a similar use in 1867 (cf. Tyndall, *Fragments,* part 2, 37). Huxley amplifies Tyndall's argument by suggesting the importance of laboratories for the experimental road to knowledge: "In truth, the laboratory is the fore-court of the temple of philosophy; and who so has not offered sacrifices and undergone purification there, has little chance of admission into the sanctuary" (quotation in Turner, *Between Science and Religion,* 64). Although Huxley is here working the science/mechanics boundary, his metaphorical use of sacred terms also evokes the idea that science-in-the-laboratory is an alternative route to the truth also pursued (on different roads) by religion.

indifferent experimenters." There is a profound difference between commonsense observation and informed scientific analysis: "Science has been defined as 'organized common sense,' by whom I have forgotten; but unless we stretch unduly the definition of common sense, I think it is hardly applicable to the world of molecules."[53]

The failure to experiment can lead the mere observer to error, as Tyndall notes in an extract immediately following a discussion of the religiously devout, who share that capacity for mistaken beliefs about nature: "A similar state of mind was long prevalent among mechanicians. Many of those, among whom were reckoned men of consummate skill, were occupied a century ago with the question of perpetual motion." It was only with science, and with discovery of the law of conservation of energy, that the mechanicians' error became the scientists' truth. The same point is made in a lovely passage from an 1876 discourse in Glasgow on the science of fermentation and the mechanical art of brewing beer: "It might be said that until the present year no thorough and scientific account was ever given of the agencies which come into play in the manufacture of beer, of the conditions necessary to its health, and of the maladies and vicissitudes to which it is subject. Hitherto, the art and practice of the brewer have resembled those of the physician, both being founded on empirical observation. By this is meant the observation of facts, apart from the principles which explain them, and which give the mind an intelligent mastery over them. The brewer learned from long experience the conditions, not the reasons, of success. . . . Over and over again his care has been rendered nugatory; his beer has fallen into acidity or rottenness, and disastrous losses have been sustained, of which he has been unable to assign the cause."[54]

Theory

Science is said to be theoretical. Mechanicians are not scientists because they do not go beyond observed facts to discover the causal principles that govern underlying unseen processes. It is their theoretical yearnings that set scientists apart from mechanicians: "I call a theory a principle or conception of the mind which accounts for observed facts, and which

53. Tyndall, *Fragments,* part 1, 41; part 2, 272; *New Fragments,* 81.

54. Tyndall, *Fragments,* part 2, 71, 267. Elsewhere, Tyndall again puts nineteenth-century medicine in the category of applied arts: "Indeed, previous to the discoveries of recent times, medicine was not a science, but a collection of empirical rules dependent for their interpretation and application upon the sagacity of the physician" (*New Fragments,* 197).

helps us to look for and to predict facts not yet observed."[55] Throughout his public science, Tyndall emphasizes to his audiences that scientific theory pushes knowledge beyond the sensible, empirical world: "One of the most important functions of physical science, considered as a discipline of the mind, is to enable us by means of the sensible processes of Nature to apprehend the insensible." The implication is that the practical mechanician halts at the facts: "The outward and visible phenomena are the counters of the intellect; and our science would not be worthy of its name and fame if it halted at facts, however practically useful, and neglected the laws which accompany and rule the phenomena."[56]

Tyndall's choice of words and metaphors sometimes seems odd for one who elsewhere speaks the language of naive empiricism: "Hypotheses in science, though apparently transcending experience are in reality experience modified by scientific thought and pushed into an ultra-experiential region."[57] "The theory is the backward guess from fact to principle; the conjecture, or divination regarding something, which lies behind the facts, and from which they flow in necessary sequence."[58] "We can have no explanation of the objects of experience, without invoking the aid and ministry of objects which lie beyond the pale of experience."[59] Tyndall is perhaps aware that terms such as "divination," "ultra-experiential," and "ministry" have the potential to blur the boundary between science and religion that he elsewhere worked so hard to construct. The rhetorical problem is confronted in this extract from an 1882 address: "[Man] cannot accept as final the phenomena of the sensible world, but looks beyond that world into another which rules the sensible one. From this tendency of the human mind systems of mythology and scientific theories have equally sprung. . . . [But] the principle force of science is expended in endeavoring to rend the veil which separates the sensible world from an ultra-sensible one."[60]

55. Tyndall, *Fragments,* part 2, 437.

56. Ibid., part 1, 80; 95–96.

57. Ibid., 226.

58. John Tyndall, *Faraday as a Discoverer,* 5th ed. (London: Longmans Green, 1894), 141. For other comments along the same line, from different periods and to different audiences, see Tyndall, *Fragments,* part 1, 298 (from 1854) and 77 (from 1865); *Six Lectures on Light,* 211 (from 1872–73); *Heat,* 32 (from 1883).

59. Tyndall, *Heat,* 33.

60. Tyndall, *New Fragments,* 78. An emphasis on the movement of scientific inquiry into the realm of the unseen compromised the boundary between empirical science and the nether-sciences of psychical research. The romantic Frederic W. H. Myers, head of the Society of Psychical Research, blurs the boundary between his inquiries and those of "science" when he observes that "science, while perpetually denying an unseen world, is perpetually revealing it"

One inconsistency in Tyndall's construction of science is now apparent. When Tyndall distinguished science from religion, he emphasized its empirical, inductive nature; when he here distinguishes science from mechanics, Tyndall emphasizes the abstract and theoretical components of science. Neither is the "correct" description, but they do evoke very different images of "science," especially when one has never set foot in a scientific laboratory or read a scientific journal. Tyndall's flexibility in assigning characteristics to science is the result of his effort to build boundaries on different fronts: attributions of empiricism will not separate science from the commonsense observations of mechanicians, but experiments and theories will; attributions of abstract, metaphysical theories will not separate science from religion, but empirical observation will. Science becomes—on different rhetorical occasions—whatever best distinguishes it from one or another cultural realm and professional competitor.

Disinterestedness

Scientists seek discovery of facts and theories as ends in themselves; mechanicians seek inventions to further personal profit. In an 1879 discourse on the electric light, Tyndall observes: "Two orders of mind have been implicated in the development of this subject: first, the investigator and discoverer, whose object is purely scientific, and who cares little for practical ends; secondly, the practical mechanician, whose object is mainly industrial. . . . The one wants to gain knowledge, while the other wishes to make money." Science is now equated by Tyndall with pure science: "This, then, is the core of the whole matter as regards science. It must be cultivated for its own sake, for the pure love of truth, rather than for the applause or profit that it brings."[61] Neither workmen nor manufacturers are likely to change their habits of production to conform to changing knowledge, if such changes impede profits: "The slowness

(quotation in Turner, *Between Science and Religion,* 121). The theoretical point should not be overlooked: characteristics attributed to science in the erection of one boundary (here, the abstractedness of science as a way to distinguish it from mechanics) always have the potential to be used to relocate boundaries between science and other cultural domains (here, the abstractedness of science is used to blur or possibly obliterate the boundary between psychic research and "real" science). Tyndall did some boundary-work at science/psychics: he even attended a seance and, in true Tyndallesque fashion, looked under the table. For a sociological analysis of the relationship between science and spiritualism, see Garroutte, *Language and Cultural Authority.*

61. Tyndall, *Fragments,* part 2, 472–73; *Six Lectures on Light,* 214.

with which improvements make their way among workmen is ascribed to the influence of habit, prejudice, suspicion, jealousy, dislike of change, and the narrowing effect of the subdivision of work into many petty occupations. But slowness is also due to the greed for wealth, the desire for monopoly, the spirit of secret intrigue exhibited among manufacturers. Between these two the philosopher steps in, whose business is to examine every operation of Nature and art, and to establish general theories for the direction and conducting of future processes."[62]

Scientists, it is said, do not succumb to the temptations of the industrial marketplace: "The edifice of science has been raised by men who unswervingly followed the truth as it is in nature; and in doing so had often sacrificed interests which are usually potent in this world."[63] That fortitude in resisting the easy money was essential for Faraday's accomplishments as a scientist: "If [Faraday] had allowed his vision to be disturbed by considerations regarding the practical use of his discoveries, these discoveries would never have been made by him." "He knew that his discoveries had this practical side [for example, in the development of electricity], but he steadfastly resisted the seductiveness of this side, applying himself to the development of principles."[64] So as not to undo his earlier description of science as the fount of technological progress, Tyndall carefully discriminates between the motives of individual scientists and the diverse consequences of their work: "That scientific discovery may put not only dollars into the pockets of individuals, but millions into the exchequers of nations, the history of science amply proves; but the hope of its doing so never was, and it never can be, the motive power of investigation."[65]

This line of boundary-work accomplishes two aims, each important to Tyndall's public construction of an image for science that would make its enlarged support self-evident. First, he demarcates science from mechanics and manufacturing in terms of their distinct motives: knowledge

62. Tyndall, *New Fragments,* 136. Tyndall's description of natural philosophy as a "business" heightens the contrast between pure science and its commercial applications.

63. Tyndall, *Fragments,* part 2, 403. In 1890, Tyndall pursued the same idea of scientists' denial of pecuniary gain as a motive for their work: "History showed that they thought less of worldly profit and applause, and practiced more of self-denial than any other class of intellectual workers" (*New Fragments,* 355).

64. Tyndall, *Faraday as a Discoverer,* 41 and v. Faraday faced a choice of either discovery or wealth: "He said [that] he was forced . . . to finally decide whether he should make wealth or science the pursuit of his life. He could not serve both masters, and he was therefore compelled to choose between them" (180).

65. Tyndall, *Six Lectures on Light,* 216.

for scientists, money for mechanicians. Second, Tyndall defuses a possible implication of his public science, namely, that statesmen of science are simply justifying greater wealth for science as an end it itself. Tyndall establishes the tone of disinterestedness as a way of reassuring his audiences that the goal of public science is not the improvement of the professional fortunes of scientists, for if that was the goal, Faraday would have indeed become rich by developing himself the electric light and other by-products of his research on electromagnetism. Tyndall puts his request for public attention and support in a different context: It is because scientists pursue pure knowledge that they need the patronage not directly available to them from the industrial marketplace.

"A Means of Culture"

Science need not justify its work by pointing to its technological applications, for science has nobler uses as a means of intellectual discipline and as the epitome of human culture. "But is it necessary that the student of science should have his labors tested by their possible practical applications? What is the practical value of Homer's *Iliad*? You smile, and possibly think that Homer's *Iliad* is good as a means of culture. There's the rub. The people who demand of science forget, or do not know, that it also is great as a means of culture—that the knowledge of this wonderful universe is a thing profitable in itself, and requiring no practical application to justify its pursuit."[66] To his American audiences, Tyndall tries to sell science as something *more* than the fount of technological invention: "It is mainly because I believe it to be wholesome, not only as a source of knowledge but as a means of discipline, that I urge the claims of science upon your attention. . . . Not as a servant of Mammon do I ask you to take science to your hearts, but as the strengthener and enlightener of the mind of man."[67]

These attributions seem inconsistent with Tyndall's attempts elsewhere to depict science as the fount of technology. If utilitarian consequences are often mentioned to justify increased resources for scientific

66. Tyndall, *Fragments,* part 2, 101. Earlier in the same speech, Tyndall suggests that "throughout the processes of Nature we have interdependence and harmony; and the main value of physics, considered as a mental discipline, consists in the tracing out of this interdependence and the demonstration of this harmony" (95).

67. Tyndall, *Six Lectures on Light,* 217 and 245. But Tyndall sees "discipline" itself as a practical goal: "If you demand practical ends, you must, I think, expand your definition of the term practical, and make it include all that elevates and enlightens intellect, as well as all that ministers to the bodily health and comfort of men" (*Faraday as a Discoverer,* 40).

research, why does Tyndall also present an image of pure science as a means of high culture and intellectual discipline? For three reasons, Tyndall demarcated the merely practical mechanician from the more than practical scientist. First, if science was justified only in terms of potential industrial applications, government officials could argue that profits from scientifically inspired innovations would repay private industrialists who invested in scientific research. Gladstone, prime minister for much of this period, remarked in 1872: "A fair field and no favor is the maxim of English administration. A field so fair, so extensive and so promising that all industry may find its place, and such absence of favor that one as well as another may hope for success. If, under these conditions the State does nothing for science, it cannot be helped, nor need it be lamented, considering how little science stands in need of the aid."[68] By emphasizing that science has cultural values beyond practical utility—virtues not likely to be appreciated and financially supported by profit-seeking industrialists—Tyndall presented an alternative case for government grants to scientists.

Second, descriptions of science as merely practical in an industrial sense would not have persuaded Oxford and Cambridge Universities to enlarge their science curricula and facilities. As part of the education of Britain's cultural and political elite, science was less appealing as a means to make money and more appealing as the exploration of truth and as a source of intellectual discipline. With this audience in mind, perhaps, Tyndall lectured in 1854: "And if the tendency should be to lower the estimate of science, by regarding it exclusively as the instrument of material prosperity, let it be the high mission of our universities to furnish the proper counterpoise by pointing out its nobler uses—lifting the national mind to the contemplation of the 'increasing purpose' which runs through the ages and widens the thought of men."[69]

Third, in distinguishing science from engineering, Tyndall made a case for the autonomy of science from direct control by industry or business. If science is indeed noble as a means of culture, it follows that scientists should be free to pursue hot theoretical leads not necessarily likely to yield technological innovations or commercial success. Historian Everett Mendelsohn points to the need for Tyndall to separate (on some occasions) science from its applications: "That science assumed a utilitarian garb in the nineteenth century should surprise no one; that it did not

68. Quotation in MacLeod, "Science and the Treasury," 135.
69. Tyndall, *Fragments,* part 1, 320.

become the 'handmaiden' of industry is the real wonder."[70] Science did not become the handmaiden of industry in part because science shed its utilitarian garb (in public science at least) in a way that made it difficult for audiences and consumers to believe that science had merely technological applications.

Another inconsistency in Tyndall's construction of science becomes apparent, and it too results from the need to build boundaries on two fronts. To emphasize science as a means of culture, and even to link it to the epic poems of Homer, would have risked confusion or ambiguity at the science/religion boundary (religion was located in a cultural space occupied by poetry and the emotions). But if Tyndall wished to make a case for science that did not rest on its technological applications, such attributions were most useful. Science is thus both pure and applied. What is interesting sociologically is that each description—giving rise to very different images— was used at different boundaries to create meanings of science that justified greater public attention and support.

TWO BOUNDARIES, MANY SCIENCES

How did John Tyndall's boundary-work aid the cause of Victorian scientists in the pursuit of increased public attention and enlarged occupational opportunities? Tyndall and other public scientists offered an interpretative context, a cartographic framework of meaning, that on its face justified scientists' appeals for greater support. The battle fought was cultural; the stakes were economic. Tyndall did not approach his publics and their government hat in hand; the anticipated returns to scientists were screened behind apparently disinterested definitions and descriptions of science and its complicated relationships to religion and mechanics. He redrew boundaries of the Victorian cultural landscape that would reconfigure the place of science in the cultural marketplace. An increased demand for scientists and their products required public articulation of the relationship between science and two cultural regions that functioned, as they still have the potential to do, as competitive sources of "supply" in the marketplace of beliefs. Ideally, Tyndall hoped for an autonomous cultural space for science, so that beliefs or assumptions about religion or mechanics could not be used to prevent increased attention to the professional needs of scientists.

Achievement of such a cultural niche required a representation in the

70. Mendelsohn, "The Emergence of Science as a Profession," 42.

form of boundary-work, in which audiences (real or potential consumers) learned not simply what science is, but why and how science is not-religion and not-mechanics. If the rhetorical form of Tyndall's public science was boundary-work, its content was a flexible, sometimes inconsistent set of characteristics attributed to science. The primary reason for Tyndall's interpretative flexibility in his construction of science was the need to build boundaries on two fronts. In distinctive ways, both religion and mechanics competed with Victorian science for cultural authority and for occupational resources. Yet the set of attributions effective for articulating the boundary between science and religion would not be effective for articulating the boundary between science and mechanics, and (of course) vice versa.

Tyndall selected from different characteristics of "science" to build each boundary: scientific knowledge is empirical when contrasted with the metaphysics of religion, but it is theoretically abstract when contrasted with the commonsense, hands-on observations of mechanicians; science is justified by its practical utility when compared to the merely poetic functions of religion, but science is justified by its nobler uses as a source of pure culture and discipline when compared to engineering. Alternative repertoires were available for Tyndall's descriptions of science: selection of one or another repertoire was guided by its effectiveness in constructing a boundary that rationalized scientists' requests for enlarged authority and public support. His public science exploited ambivalences inherent in the meaning(s) of science—the unyielding tension between basic and applied research, and between the theoretical and empirical aspects of inquiry—by selecting for attribution to science one or another set of characteristics effective for demarcating science from two of its cultural competitors: from religion on some occasions, mechanics on other occasions.

To look at Tyndall's boundary-work against the "spectacular" theoretical and empirical accomplishments of British scientists in the nineteenth century, and against the equally "spectacular" industrial achievements to which scientists have been linked, raises one of the many Victorian ironies. Did not science *sell itself*, either through its "obvious" discoveries or through its practical achievements? If the answer is yes, the endless exhortations of public scientists such as John Tyndall are inexplicable. Why did they bother? They "bothered" because their discoveries and related inventions needed to be interpreted—reconstructed within a cultural landscape that encompassed not just science but religion and mechanics—before they could become self-evident justifications for increased attention and public support of science. The public needed to be told that these discoveries were profoundly important for Western civilization, and they needed to be told that the industrial future of Britain depended on discoveries waiting to be made. As unreconstructed events, such discoveries and inventions were not sociological explanations of the rising cultural authority and professional fortunes of Victorian scientists. They were, instead, rhetorical resources to be exploited by scientists trying to construct for their audiences and consumers a meaningful and pragmatic map of the cultural terrain.

Chapter Two

THE U.S. CONGRESS DEMARCATES NATURAL SCIENCE AND SOCIAL SCIENCE (TWICE)

Natural science will in time subsume under itself the science of man, just as the science of man will subsume under itself natural science: there will be one science.

Karl Marx, *Economic and Philosophical Manuscripts of 1844*

Suppose we take this chapter's epigraph as just a hypothesis. Was Marx right to argue that social science and natural science would converge? One can approach the question from at least three directions. From the philosopher's armchair, demarcation criteria can be conjured, then laid upon the practices and claims of scientists in a way that creates or dissolves distinctions between the "social" and "natural" varieties (much as Popper does to exclude Marxism from science for its putative lack of falsifiability). Or the inquiring ethnographer can seek out what it is that natural and social scientists do when they make knowledge first time through, and resolve Marx's hypothesis empirically. And then there is the route that I prefer: examine settings where historical actors negotiate representations of science—social, natural—amid struggles over the allocation of resources and credibility. While philosophers dream about rules to impose order on irreducible cultural messiness, while ethnographers expose polysemic practices always amenable to diverse *re*presentations as pragmatic circumstances warrant, I choose to examine the boundaries around or between natural and social science as rhetorical accomplishments advancing somebody's interests. Analytically, these cartographic exercises are as autonomous from the practices in scientific laboratories or professional journals as from the demarcationist musings of philosophers.[1]

1. Contrast this position to that of Karin Knorr in 1981, when she sought to provide "a reconsideration of the distinction as made customarily in discussions of social science methodology . . . based not only on the above-mentioned philosophical investigations, but also on empirical findings which emerge from observations of the actual research process in the natural and

Boundary-work gets especially interesting when it happens in places of power, for the demarcation games played out there often have large consequences for the symbolic and material conditions of scientific work. Such is the case with the fifty-year exercise in cultural cartography known otherwise as the National Science Foundation (NSF): does social science belong in here—or not? Twice that question has occupied the sustained attention of members of the United States Congress, who—in the late 1940s and again in the mid-1960s—were asked to draw borders around and between the various sciences in order to decide which fields would become bona fide programs in the NSF. The stakes for scientists of all stripes could not be more obvious or consequential. To be included within the NSF would stamp a discipline as "science" for purposes of receiving federal funding for basic research—just as some measure of legitimacy as "science" would be denied those disciplines and fields left out. Would Congress make Marx right for once by joining the sciences of man to the sciences of nature?

In the first instance, in the waning days of World War II, legislators were asked to chart a course for science that would enable it to lead the rest of society toward peace and prosperity.[2] The importance of these deliberations for the professional health of scientists was not lost on Vannevar Bush—once engineering dean at MIT, then in Washington to head the Carnegie Institution, and appointed by President Roosevelt in 1941 to direct the Office of Scientific Research and Development (OSRD).

---

technological sciences." A layer is missing: neither philosophical, epistemological, and metatheoretical exercises in rule enforcement nor empirical observations of actual practices (and texts) can substitute for a sociological study of how "all that" is represented in contexts consequential for allocations of authority, credibility, and material resources. See Knorr, "Social and Scientific Method," 28; Kuklick, "Boundary Maintenance," offers a constructivist perspective on the social sciences' borders and territories.

2. This brief history of a complicated period draws on much good scholarship: Boyer, "Social Scientists and the Bomb"; Bronk, "National Science Foundation"; Chalkey, *Prologue;* Dickson, *New Politics of Science,* 25–30; Dupree, *Science in the Federal Government,* 369–75; England, "Dr. Bush Writes a Report"; Gerstein, "Social Science as a National Resource"; Graham, *Federal Utilization of Social Science;* Kevles, "National Science Foundation" and *Physicists,* esp. 334–66; Klausner, "Bid to Nationalize American Social Science"; Kleinman, *Politics on the Endless Frontier;* Kleinman and Solovey, "Hot Science/Cold War"; Koontz, "Social Sciences"; Lomask, *Minor Miracle;* Lyons, *Uneasy Partnership;* Miller, "Social Sciences"; Turner and Turner, *Impossible Science;* Wang, "Liberals, the Progressive Left, and the Political Economy"; Wolfle, "Making a Case for the Social Sciences"; Young, "Organization for Research in the Social Sciences." For a discussion of the relationship between sociology and biology prior to World War II, see Gaziano, "Ecological Metaphors as Scientific Boundary Work"; for a general discussion of the social sciences prior to World War II, see Ross, *Origins of American Social Science.*

The organization of OSRD reflected Bush's ability to convince Roosevelt that the war might well be won if the federal government were to provide research funding to scientists based largely in universities with only the minimum constraint imposed by wartime secrecy. With armistice imminent, Bush and the scientific establishment hoped to use the physicists' triumph at Hiroshima to extend this professionally salutary arrangement for basic research (government patronage for basic research with the barest political control or immediate public accountability) into the peacetime years ahead—a view that took shape in the influential report prepared by Bush and others entitled *Science—the Endless Frontier.*[3]

This image of science-government relations became the focus of intense debate in the mid- to late 1940s. As early as 1942, Senator Harley Kilgore, a New Deal liberal from West Virginia, drew on the Depression era discourse of public planning and centralized control to propose alternative legislation that would make science into a public resource, guided democratically toward societal betterment. Such New Deal rhetoric and ideals did not fade out completely during the war: in 1947, President Truman vetoed legislation that would have created a science bureaucracy much closer to Bush's *Endless Frontier* than Kilgore's old New Deal. But Kilgore's force was dissipating fast, in part because of sustained and better-organized efforts of the scientific establishment to promote the virtues of autonomous basic research directed by scientists themselves, which also attracted allies from the private corporate sector (the National Association of Manufacturers) and from members of Congress now worried more about Korea and the Cold War than about central planning and public accountability. Enabling legislation for a National Science Foundation—funded meagerly, but organized as Bush had envisioned—passed Congress and was signed into law by Truman in 1950.

Where did the social sciences fit into debates over "the science legislation"? During the Depression, social scientists had enjoyed their most intimate and influential relations with Washington politicians, called on

3. Vannevar Bush, *Science—the Endless Frontier.* (Washington: Office of Scientific Research and Development, 1945; reprinted by the National Science Foundation, 1980). Dupree suggests that the war permanently transformed the relationships between American science and government. The budget for scientific research was ratcheted in at $1 billion minimum, and the bulk of it went to weapons-related initiatives even as peace broke out. Bush was less worried about what would happen to classified military research or the primarily applied aspects of work on atomic energy than about establishing a government agency for basic research that would funnel patronage to scientists at universities without the security restrictions under which scientists had chafed during the War (Dupree, *Science in the Federal Government,* 373–74).

as they were to techno-manage society out of unemployment, poverty, and stagnation.[4] Kilgore's New Deal roots no doubt contributed to his presumption that science of course included social science: so many of the pressing problems to be addressed during the recovery would be economic and sociological, and evidence suggests that most rank-and-file natural scientists agreed that social science should be included in the science legislation. In contrast, Bush saw little warrant to include social science in any federal agency designed to support basic research, and in President Roosevelt's 1944 letter requesting studies leading up to *The Endless Frontier,* no mention is made of them. The reasons for Bush's reluctance to embrace social science were various: perhaps he genuinely believed that there could not be a bona fide science of social phenomena; perhaps he doubted that such research could contribute materially to postwar problems of military security and technological development; perhaps Bush harbored lingering resentment of social scientists stemming from their relatively large political influence in the 1930s (when physical scientists felt decidedly unloved);[5] or perhaps he was afraid that in a political climate increasingly disenchanted with "central planning," social science might be so easily confused with socialism that any science legislation that included such disciplines was doomed to failure. Whatever Bush's reasons, the 1940s stage is now set for some heavy cultural

4. On science policy during the Depression, see ibid., chap. 18.

5. "In the 1930s, influential social scientists had helped defeat scientists' request for government research funds; and for many years in the 1950s, influential scientists responded in kind to social scientists' requests for funds from the National Science Foundation" (Orlans, *Contracting for Knowledge,* 235; cf. Karl, "Presidential Planning and Social Research," and Greenberg, *Politics of Pure Science,* 106).

The absence of social science from Roosevelt's letter requesting *The Endless Frontier* may have been due to Bush's predrafting efforts: "At the time I made sure that the letter confined its attention to the natural sciences, and I suggested that if it was intended to make a similar inquiry into the social sciences this should be done through other channels" (Vannevar Bush to D.C. Josephs, September 19, 1946, Bush Papers, Library of Congress, box 60, file 1416). His fears about social science torpedoing any science legislation were expressed in a letter to James E. Webb of the Bureau of the Budget: "There was great discussion as to whether the social sciences should or should not be included. . . . I believe that there is great danger in introducing this matter in a bill for federal support of science until it has been thoroughly considered." He goes on: "There are pitfalls because the name of social scientist is often seized upon by those who are not scientists at all but special pleaders" (December 27, 1946, Library of Congress, box 85, file 1912). In a subsequent letter to Bush from John Teeter we learn the reason for excluding social science: "No direct mention of social science is made since it is recognized [that] any attempt to include it would kill the bill" (January 15, 1947, Bush Papers, Library of Congress, box 85, file 1912). Zachary reports that Bush "only sparingly" helped historians who sought him out for interviews "partly because he disliked the inexact methods of history" (*Endless Frontier,* 392).

cartography: did sociology, economics, psychology, and anthropology "really" belong in the proposed new science foundation, and on what grounds could each side justify their discrepant cultural maps?

When that same question again reached Capitol Hill almost twenty years later, the circumstances of debate had changed dramatically. The NSF organic legislation of 1950 had no specific mandate to provide immediate funding for basic research in the social sciences. Instead, the foundation began life with a "permitted, but not mandatory" posture toward what were described as "and other sciences": the governing body of the NSF—the National Science Board—was permitted to expand the domain of the foundation into the social sciences when and if later study of the matter warranted such a move. This idea had emerged in July 1946 as a compromise amendment to the NSF bill (S.R. 1850) proposed by Senator Thomas Hart to placate those who disagreed bitterly on whether the social sciences should be explicitly mandated inside what would become the NSF. Although Hart had formally removed a previously proposed Division of Social Science from subsequent discussion, his amendment became a toe in the door of the NSF that would open ever wider throughout the 1950s.

NSF historian J. Merton England has traced out the earliest foundation moves toward providing support to the social sciences, which accelerated dramatically after the appointment of sociologist Harry Alpert as study director for the social sciences in 1953.[6] Later that year, the Anthropological and Related Sciences Program was added to the Division of Biological and Medical Sciences, and in 1954, a Demography and Human Ecology Program was added to the Mathematical and Physical Sciences Division. Modest funding was provided to social scientists for projects that could—in sometimes precariously stretched ways—be made to "converge" with the problems and methodologies of the natural sciences. By 1958, this back door to the NSF was moved out front with the creation of the Social Science Research Program, which became the Office of Social Science in 1959 and the Division of Social Science in 1961. After that, budgets for social science research increased precipitously during the heady post-Sputnik days of federal largesse.[7]

6. England, *Patron for Pure Science.* Additional studies of social science—inside the NSF and beyond—between the late 1940s and the mid-1960s include Heims, *Constructing a Social Science;* Larsen, *Milestones and Millstones;* Lyons, "Many Faces of Social Science"; Riecken, "Underdogging"; Sherwood, "Federal Policy for Basic Research."

7. My sources for this second historical period include Brezina, *Congressional Debate on the Social Sciences;* Harris, *Social Science and National Policy;* Larsen, *Milestones and Mill-*

All fields of research benefited handsomely from worries that advances in Soviet science and technology could put the United States to shame or, worse, at military risk. Vannevar Bush's Endless Frontier became John F. Kennedy's New Frontier: federal funding to universities for basic scientific research increased from $433 million in 1960 to $1.6 billion in 1968, much of it procured with only lip service paid to the immediate practical problems it would address. But an important policy shift took place near the middle of this period of rising scientific fortune and political influence, coincident with the second occasion on which Congress deliberated at length the scientific status of social science. After 1965, Lyndon Johnson's Great Society redefined science as a tool for working directly on practical social problems. "Relevance" was used to justify efforts to steer money from basic research into programs of applied research that directly addressed growing concerns about transportation, crime, urban dissent, health, and poverty. Johnson sought to make science more accountable to the agenda of public problems of the day, as he and his successor, Richard Nixon, believed that "the principle task now facing government was not to discover more new knowledge, but to find ways of putting what already existed to good use."[8]

Although some of these practical issues (health, military defense, innovative technologies capable of boosting the domestic economy) were linked to biological and physical sciences, many of the rest fell into the bailiwick of sociology, economics, and psychology. Some asked whether a social scientific assault on practical problems of poverty, unemployment, substandard housing, decaying cities, and burgeoning crime rates could really best be carried out by the National Science Foundation, which provided funds exclusively for basic research, and which from its inception had been created for and administrated by natural scientists. In 1965, Project Camelot—an ill-fated research investigation defined by some angry Chileans as a covert attempt by the United States to destabi-

stones; Lyons, *Uneasy Partnership;* Stine, *History of Science Policy;* Zuiches, "Organization and Funding of Social Science."

Kleinman and Solovey divide this period in two. They suggest that from 1953 through Sputnik, NSF officials were extremely cautious in their support of social science research, avoiding like the plague any project even remotely controversial (sex, religion, politics). After Sputnik in 1957, the foundation justified increased support for social science by emphasizing how such research could combat a variety of social ills at home and communist expansion abroad ("Hot Science/Cold War," 118–24).

Two NSF reports capture much of the unsettled thinking about social science in the late 1960s: *Knowledge into Action* and *Behavioral and Social Sciences.*

8. Dickson, *New Politics of Science,* 29.

lize a left-leaning political regime under the guise of value-free science—created enough political embarrassment to bring the matter of government-financed social science to a head.[9]

How should government organize its funding programs so that social science research can be explicitly directed to practical problems at home and abroad—and in a way that its results will be sufficiently "scientific" (objective, reliable) to steer and to legitimate future public policies and programs? In 1966–67, Senator Fred Harris (D-Oklahoma) put this question in the form of legislation calling for the creation of a National Foundation for Social Science (NFSS, but also known as NSSF), an agency for funding "relevant" basic research in the social sciences but administratively autonomous from the NSF. Time for more boundary-work: the possibility of an NFSS depended on a cultural map that demarcated the natural and social sciences in ways that made it seem sensible to move the latter disciplines out from under the aegis of the NSF. Harris and his allies used a "rhetoric of difference" that reiterated in interesting ways the arguments put forth twenty years earlier by those who believed that social science had no place at all in the incipient science foundation. His opponents evoked a "rhetoric of similarity": Science is "one" whether it studies nature or society, and it is best to fund it all through a common government agency—this, an echo of 1940s attempts by Kilgore and friends to get social science inside.

My task in this chapter is to exhibit and make sense of the cultural cartography that comprised these two moments: both the decision to create a National Science Foundation and the debates surrounding the possibility of hiving off a National Foundation for Social Science elicited map upon map, with diverse renderings of spaces and territories, boundaries and landmarks. "Science" was variously constructed in the representational practices of people seeking one or another public policy, with huge outcomes for scientists' professional wherewithal. Cultural spaces such as "natural science," "social science," or simply "science," were mapped out in discrepant and changing ways by players with different interests, facing different historical contingencies. Karl Marx's epigraphic hypothesis could not, cannot, be adjudicated in any definitive way, because (as the evidence will show) social and natural sciences are shaped simultaneously to converge *and* diverge—as circumstances make one or the other representation pragmatic for the pursuit of some identifiable

9. On Project Camelot, see Rieff, *On Intellectuals;* Herman, *Romance of American Psychology;* Horowitz, *Rise and Fall of Project Camelot* and "Life and Death of Project Camelot"; Sjoberg, *Ethics, Politics, and Social Research.*

goal. The struggle for material resources and for scientific legitimacy occasioned these two episodes of cultural cartography, as it does all the others in this book.

What empirical evidence would evince such flexibility in constructions of boundaries between natural and social science? At each historical moment, congressional committees and subcommittees held hearings that brought to Washington scientists of diverse persuasions armed with cultural maps of all kinds and purposes. Their published testimony—along with an occasional staff report or committee print or floor debate—are the data analyzed in this chapter. My organization of the material is not chronological but analytical: I begin with the 1940s argument for the inclusion of social science in the NSF, and then show cartographically how the same cultural maps—one space for science—were used twenty years later to justify the continued funding of social science research through the NSF. After that, I return to the postwar years and pick up the alternative boundary-work: maps putting borders and distance between social and natural science were used in the 1940s to keep the former fields from inclusion in the proposed NSF, and similar cartographic renderings turned up again in the 1960s as a rationale for why an NFSS was desirable.

Before the empirical analysis, a bit more should be said about the political contexts in which these policy discussions were played out. There are several noteworthy similarities between postwar debates on the "science legislation" and Great Society dreams about the relevance of social science to urgent national needs. In both instances, for example, although greatest enthusiasm for social science funding came from politicians easily classified as liberal—Kilgore in the 1940s; Harris, Kennedy, and others in the 1960s—almost none of the legislators who spoke up flatly rejected the desirability of some federal support for such disciplines.[10] The issue hinged on how—administratively, organizationally—such funding should be routed from taxpayers to social researchers: NSF, NFSS, or something different altogether. Moreover, the funding itself was, in a sense, a side issue on both occasions. In the 1940s debates over postwar science policy, the hottest-button issues did not concern social science directly but focused on how the NSF was to be governed (in a

10. For example, a 1966 House subcommittee headed up by Congressman Daddario prepared a report on the NSF that noted historically inadequate levels of funding for the social sciences (House Committee on Science and Astronautics, *The National Science Foundation: Its Present and Future,* report of the Subcommittee on Science, Research, and Development, 89th Cong., 1st sess., Committee Print, serial M).

nutshell: would the National Science Board and the NSF director be accountable primarily to Congress or to the scientific profession?) and patents (another nutshell: would profits on products developed through research supported by the NSF return to the foundation for public use?). In the mid-1960s, as I have noted, the fallout from Project Camelot meant that politicians were as much concerned about the perceived scientific legitimacy of government-supported social science as they were about the levels of its funding.

Finally, social scientists themselves seemed to be divided and somewhat in disarray during both sets of deliberations. After the war, most prominent social scientists argued that social science belonged in whatever would become the NSF, but many of my predecessors were plainly ambivalent about how vociferously they should make that case. After the Hart Amendment in 1946 ("permissive, not mandatory"), leaders of social scientific professional associations evidently felt that the science bill itself would be legislatively jeopardized if they continued to insist on the inclusion of their disciplines. They chose to back off, perhaps yielding to the greater power of natural scientists in the Inter-Society Committee for a National Science Foundation, perhaps because of their own inability to mount an effective lobbying campaign, or perhaps because they were shrewd enough to anticipate that, once established, the NSF would be more likely to see social science as worthy of federal funding than a Congress moving swiftly in a McCarthyite direction. Twenty years later, political enthusiasm for an NFSS waned in part because social scientists themselves did not speak for its creation with one voice. Among those who testified at congressional hearings, more social scientists spoke against the new foundation as for it.[11] Those opposing the NFSS believed that long-term funding and perceived legitimacy of social science research would be better protected against shifting political winds if the disciplines were tied to the less easily challenged fields of physics, chemistry, and biology. Social scientists who saw merit in an NFSS portrayed

11. The actual distribution of the fifty-three witnesses who testified at a 1967 hearing was tallied up this way in a congressional staff study:

No attitude expressed about an NFSS	12
In favor of its creation	14
Opposed to its creation	20
Unclear, undecided, equivocal	5
Unclassifiable position	2

(Senate Committee on Government Operations, Subcommittee on Research and Technical Programs, *Staff Study: The Use of Social Research in Federal Domestic Programs*, part 4, "Current Issues in the Administration of Federal Social Research," April 1967, 6).

natural scientists not as allies but impediments, to both the amount of and the control over funding for their research—at a time when the size of the NSF budget was still growing but at a noticeably diminished rate.

Ironically, despite dissent and disorganization among social scientists at each moment, all's well that ends well. Even without specific inclusion in the enabling NSF legislation of 1950, funding for basic research in social science started up just three years later and accelerated rapidly after that. And despite the eventual failure to establish an NFSS, Congress emended the NSF's organic legislation to include "official" mention of social science just one year after Senator Harris gave up his fight, in 1968. When political winds turned again with the Reagan administration of the 1980s, NSF officials successfully defended their social science programs against an executive assault that could have left them penniless. It remains to be seen whether the foundation can continue to provide shelter for social science sufficient to withstand the conservative budget-gutting mania of the mid-nineties. I'll return to these political prospects for social science funding at the end of the chapter.

BRINGING SOCIAL SCIENCE WITHIN: THE 1940s

Those who sought to have the social sciences included in what would become the National Science Foundation offered three arguments, each erasing the boundary between natural and social science. First, it was argued, social science is practically useful in the same way that natural science is practically useful: basic research in all the sciences yields knowledge that can be applied effectively toward technical solutions of social problems. Second, production of knowledge in social science is accomplished with the same methodology as one would find in natural science: both use observation and mathematical analysis to reach objective facts. Third, those who came down on the side of inclusion maintained that science is one discipline regardless of whether it studies atoms or humans—because that is the way reality is packaged (to divide up and insulate various inquiries impedes comprehensive understanding).

Complementary Utility

From the seventeenth century to yesterday, scientists have used the language of instrumental utility to legitimate their practices and to justify requests for material support from other sectors of society. Science *works,* as John Tyndall told audiences who believed that prayer was an efficacious a solution to cholera (chapter 1), and as George Combe (chapter 3) told audiences skeptical of the facticity of his phrenology. Still, the rhetorical effort needed to connect science to the solution of critical human problems is far from trivial. Whether something is a problem, whether it is a serious problem, whether a solution takes the form of a certain technology, and whether science is the fount of such technological panaceas—all this is a matter for social construction, denial, negotiation, and eventual settlement in the history books. Scientists work hard to define as paramount those problems ripe for solution with technologies that their knowledge now (or soon) enables.

After 1945, the utilitarian case for the creation of a new federal agency to support basic scientific research was mainly built on the wartime accomplishments of physicists, and mainly for their successful construction and deployment of an atomic bomb.[12] However, the instrumental language of "science works" was not in itself sufficient to mold the nascent NSF into the kind of agency envisaged by the leaders of establishment science. Vannevar Bush, Isaiah Bowman (president of Johns Hopkins), and others were strongly committed to the pursuit of an NSF that would combine government patronage with professional autonomy, in the hope that pure scientific research could return to university campuses and get out from under the oppressive constraints of wartime secrecy and military direction. Without sacrificing the idea that science had given us the ability to win wars, cure dread disease, and launch the postwar recovery, spokespersons for science tried to convince Congress that such practical goals were best achieved through theoretical and experimental development of basic knowledge, conducted by scientists pursuing truth first and applications by-the-by, with minimal meddling from politicians.

Those who sought the inclusion of social science in this "new social contract with science"[13] faced the same rhetorical challenge: to argue persuasively that social science also works, and especially if it is left to develop its own knowledge base without immediate political and military interference. But this utility was not at all apparent to Congress and the public: after all, physicists had built the bomb that ended the war. Worse, the programs and policies developed to address unemployment and fiscal crises during the Depression—designed with the help of prominent economists and sociologists—were recast by some in the increasingly

12. On postwar images of "the bomb," see Weart, *Nuclear Fear;* Kevles, *Physicists.*
13. Guston and Keniston, "Social Contract."

conservative postwar mood as an embarrassing and dangerous flirtation with socialism. What physicists and chemists did to win the war needed little rehearsal; the wartime accomplishments of social scientists needed to be spelled out—as a warm-up for assertions about the general utility of their disciplines.

During the month-long Hearings on Science Legislation in October 1945 (two months after Hiroshima and Nagasaki), several witnesses indicated that social scientists had worked effectively and productively alongside natural scientists during the war. Brigadier General John Magruder, who served as director of the Office of Strategic Services, vouchsafed the loyalty of social scientists—an important reminder at a time when the boundary between sociology and socialism was often fuzzed up: "Scholars in social science . . . loyally served their country and contributed critically to minimizing the war's extravagances in human life and material resources."[14] The psychobiologist Robert M. Yerkes, who worked in Washington during the war as director of the Research Information Service of the National Research Council (where he culled results from the huge Army intelligence program), noted that "in World War II, psychological methods in aid of classification and assignments tremendously increased our military might and correspondingly lessened our costs in lives and materiel."[15] Harold Smith, who as director of the Bureau of the Budget steered Truman's vision of the NSF toward the Kilgore model, put in a good word for "the economist[s] who helped guide the organization of our industrial resources [but nevertheless] are hardly likely to get the public acclaim that they so well deserve."[16]

All eyes those days seemed to be looking forward to the peacetime recovery, and several witnesses suggested that the greatest utility of social science would be its role in figuring out how to keep that peace. The triumph of the bomb carried with it fears about how the awesome force could be used next—by whom, and for what ends. The permanent secretary of the American Association for the Advancement of Science, F. R. Moulton, suggested that social science could leash the monster: "The

14. Senate Subcommittee of the Committee on Military Affairs, *Hearings on Science Legislation,* 79th Cong., 1st and 2nd sess., October 31, 1945, part 1, 901 (hereafter, *Hearings on Science Legislation*).

15. *Hearings on Science Legislation,* October 29, 1945, part 4, 753. On psychologists' role in the war effort, cf. Capshew, *Psychologists on the March;* Herman, *Romance of American Psychology,* chap. 4; Buck, "Adjusting to Military Life."

16. *Hearings on Science Legislation,* October 10, 1945, part 1, 96. On Smith's tenure as director of the Bureau of the Budget, see Kevles, *Physicists,* 356.

further development of the social sciences may well determine whether the new and terrible forces which man has discovered through the natural and physical sciences become man's servant for enhancing his welfare or the terrible instruments for his destruction."[17] Many witnesses constructed a complementary division of labor in which material things (like bombs) become the domain of physical science, while the human uses of such material things become the domain of social science (such ontological gerrymandering is a leitmotif running throughout the case for including social science in the NSF).[18] Sociologist W. F. Ogburn argued that "it is the function of the physical scientist to make the bomb, but it is the function of the social scientist to tell you what its social effects [will] be."[19] Although the "Atoms for Peace" slogan would not appear for several years, these witnesses and others believed that social science should be included in the new science foundation because its developing knowledge would be essential to guide the use of scientific power toward productive ends.

The cartographic trope "physical science develops things/social science figures out what to do with them" was generalized from the specific case of the atomic bomb and applied to every technical invention. During the July 1946 floor debates culminating in Senator Hart's amendment creating the "permissive formula," Senator Thomas of Utah spoke against the amendment and for inclusion: "No invention, no patent, no scientific

17. *Hearings on Science Legislation,* October 9, 1945, part 1, 78. Secretary of Commerce Henry Wallace makes the same point: "It is only by pursuing the field of the social sciences comprehensively and understandingly that we can bring humanity abreast of these great natural forces which have been unleashed and if we don't do that we make for bigger and worse wars" (*Hearings on Science Legislation,* October 11, 1945, part 1, 144). Historian Paul Boyer notes the irony in such arguments, saying of both W. F. Ogburn's research on "urban dispersal strategies" and the work of the Social Science Research Council's Committee on Social Aspects of Atomic Energy in 1946: "Neither, however, bore out the oft-repeated claim that social science had profound and original insights to offer a society terrified of the bomb" ("Social Scientists and the Bomb," 36). This supports my contention that a resolution to Marx's hypothesis (one science) will come not from comparisons of the immediate research practices and products of natural and social scientists, but from analyses of how those practices and products are represented latterly and cartographically in circumstances where palpable resources—here, funding and scientific legitimacy for social science—hang in the balance.

18. Woolgar and Pawluch, "Ontological Gerrymandering."

19. *Hearings on Science Legislation,* October 29, 1945, part 4, 771, also 766. Captain Harry Malisoff, representing the Disabled American Veterans, elaborated on much the same point: "As our knowledge of atomic power is developed to the point where such power can be used for productive as well as destructive purposes, it is essential that we examine and analyze the economic and social implications of such development" (*Hearings on Science Legislation,* October 12, 1945, part 1, 190).

development amounts to anything for the benefit of the people anywhere unless it has its social aspect."[20] Russell Smith of the National Farmers Union told the subcommittee during the October 1945 hearings that "man must learn as much about the social world as he has learned about the physical, [and] that unless he brings to the use of the inventions he has discovered at least as much ingenuity as he has expended on their discovery, he faces an extremely gloomy future."[21] The technological fruits of physical science will go unrealized without a social scientific understanding of their uses, an argument that logically establishes an independent but vital utility for sociology, economics, and the rest—so said Secretary of the Interior Harold Ickes, a staunch supporter of inclusion: "The results of purely physical research unimplemented by the talents and ingenuity of others, expert in the fields of human relations and behavior, would be both barren and futile."[22] The matter was perhaps only slightly overstated by a physician from Massachusetts General Hospital: "One should not forget that the value of all science seems to be in the hands of social science."[23]

Moreover, proponents suggested, social science can prevent or fix the inevitable liabilities that would flow from the new material technologies emerging out of natural science research— in effect, natural science itself becomes a social problem. This was Ogburn's position exactly: "For every important mechanical invention that physical scientists make there is created a new social problem on which social scientists should work. . . . If the United States Government supports research in physical science, then, in my judgment, it is logically and morally obligated to support research in social science. For the discovery in physical science creates a problem in social science."[24] As late as March 1947, when it was politically improbable that social science would be explicitly included in the science legislation, the chairman of the National Research Council, Detlev W. Bronk, created an image of social science clearing up the

20. *Congressional Record,* July 3, 1946, 8232.

21. *Hearings on Science Legislation,* October 10, 1945, part 1, 127.

22. *Hearings on Science Legislation,* October 18, 1945, part 2, 341. Pro-inclusionist Harlow Shapley from Harvard agreed: "It is clear that we know as yet far too little about human behavior, individually and in groups, for the derivation of the full benefit now accruing from our knowledge and control of plants and animals" (*Hearings on Science Legislation,* October 9, 1945, part 1, 51). Shapley directed the Harvard University Observatory and worked with Bowman, Bush, and others to secure an autonomous postwar science as free as possible from immediate political interference (see Kevles, *Physicists,* 364).

23. Allan Butler, *Hearings on Science Legislation,* October 22, 1945, part 3, 488.

24. *Hearings on Science Legislation,* October 29, 1945, part 4, 765.

messes left behind by physical science: "I cannot think of any field of research in physical science which does not ultimately lead, and usually very promptly, to new social problems. . . . It is important, therefore, that competent social scientists should work hand in hand with the natural scientists, so that problems may be solved as they arise, and so that many of them may not arise in the first instance."[25] The hand-in-hand metaphor was picked up by other witnesses who saw a variety of social problems requiring joint solution by natural and social scientists: "Suppose we are planning a coordinated attack on malaria. We would probably start with such natural science techniques as study of the mosquito, study of the germ, study of insecticides and drugs, but we would eventually get to such socio-economic problems as the ownership of mosquito-breeding waters. . . . In a problem of this kind, it would be extremely shortsighted to insulate one science from another in a neat set of pigeonholes."[26] Other witnesses mentioned stream pollution and agricultural development as national problems requiring joint attention of natural and social scientists.[27]

But the "present state" of social scientific knowledge was felt to be insufficient to provide all this badly needed help—which became another argument for placing these disciplines in the middle of the proposed NSF. Ogburn's famous concept of "cultural lag"—the understanding and control of natural and physical phenomena has leaped far ahead of the understanding and control of the social consequences of technological change—was used to good effect by those favoring inclusion. Ogburn used the language of "maladaptation" to describe the disjointed alignment of fast-paced material changes and relatively slower changes in culture and social practices.[28] At the AAAS, Moulton had come to much the same conclusion: "The present lack of balance in the development of the physical and social sciences is one of the important reasons for including provision for the social sciences in the science bill."[29] Og-

25. House Committee on Interstate and Foreign Commerce, *Hearings: National Science Foundation*, 80th Cong., 1st sess., March 6, 1947, 38.

26. Watson B. Miller, National Security Administration, *Hearings on Science Legislation*, October 29, 1945, part 4, 798.

27. Harlow Shapley, *Hearings on Science Legislation*, October 9, 1945, part 1, 52; Abel Wolman, professor of sanitary engineering at Johns Hopkins, *Hearings on Science Legislation*, October 25, 1945, 666.

28. Ogburn, *Social Change*, 199–213, on "cultural lag."

29. *Hearings on Science Legislation*, October 9, 1945, part 1, 78. A metaphor of "bridging the gap" between natural and social science is introduced by anthropologist John M. Cooper of Catholic University: "Our management of our own lives and of our society has not kept pace

burn warned, however, that too much could not be expected of the social sciences too soon: just as physicists could not have split the atom "unless you had the discovery of the roentgen rays," investments in basic social science research would be imperative to realize the disciplines' full potential.[30] Still, he felt, "social inventions" were on the horizon, such as better actuarial tables and aptitude tests, and although these are "usually not recognized as inventions . . . because they are not as tangible as material objects and they are seldom patented and sold for profit," their development "follows about the same course as mechanical invention": basic research, then practical applications.[31]

A Common Methodology

Evidently some congressmen believed that "basic social science" was oxymoronic—and so an important task facing those in favor of inclusion was to get sociology and economics and anthropology under the tent of "pure research." An effort was made to show that social scientists relied upon the same methods as natural scientists, and that the findings of all were not just useful but objective. Howard A. Meyerhoff (executive secretary of the AAAS) told the subcommittee that "social scientists, too, feel they are a basic science," and Donald Scates (education professor from Duke University and president of the American Educational Research Association) provided the rationale: "Techniques have been developed and are widely known for attacking human problems with the same degree of scientific rigor that characterizes physics, chemistry and biology. I want to emphasize that human science is not basically different from physical science; it simply deals with different phenomena." Moreover, Scates continues, "we would deny any *scientific* validity to the distinction," thus denying legitimacy to anybody who might try to separate natural from social science.[32]

Methodological parallels are trotted out: "statistics" is twice men-

---

with the advances that have been made by the natural sciences" (*Hearings on Science Legislation,* October 29, 1945, part 4, 776–77).

30. *Hearings on Science Legislation,* October 29, 1945, part 4, 769.

31. Watson B. Miller, *Hearings on Science Legislation,* October 29, 1945, part 4, 796; Walter Rautenstraucht, School of Engineering, Columbia University, *Hearings on Science Lesgislation,* October 31, 1945, part 4, 896.

32. *Hearings on Science Legislation,* October 9, 1945, part 1, 90. Meyerhoff refers at this point to a AAAS poll which showed that 67 percent of the natural scientists surveyed favored inclusion of social science in a science foundation. See also later testimony by Scates, in *Hearings: National Science Foundation,* March 7, 1947, 267–68.

tioned as a "principal ingredient" of social scientific methods; anthropology is described as an "empirical" science; social scientists have been successful in "amassing . . . substantial factual data and their careful interpretation over a period of many years."[33] If physicists appear to have a more secure purchase on atoms and molecules than sociologists have on behavior and social institutions, this is not because the reliable methods of the former are not applicable to objects studied by the latter. It is only that "research is more difficult in social science because of the larger number of variables than are found in problems of the physical sciences." So, as Ogburn puts it, "prediction in social science must be in terms of probabilities," but this does not preclude "many very creditable achievements, which have proved reliable and trustworthy."[34] Or maybe the perceived differences between physics and sociology rest on inflated estimates of natural science: "They exaggerate the amount of exactness that there is in the physical and biological sciences [but] they underestimate the amount of evidential value that social science techniques can get out of raw data from economic and social fields."[35] At any rate, says Yerkes, "the fact that we [in social science] are not well along toward the solution of many vital social problems has nothing to do with the applicability of scientific procedure."[36]

Such declarations of a methodological communion between natural and social sciences came sometimes in response to the charge that social scientific claims lacked objectivity, that social scientists lacked credibility—thus, neither could be trusted. If social science was ever to get inside the NSF, its supporters needed to demonstrate that the same methodologies that permitted natural scientists to exclude considerations of values, ethics, politics, and economic interests from their conclusions would also permit objectivity among social scientists. Economist Wesley C. Mitchell, veteran New Deal adviser (like Ogburn) and now representing the Social Science Research Council, had no doubt at all: "The ear-

33. Shapley, *Hearings on Science Legislation,* October 9, 1945, part 1, 52; Senator J. William Fulbright of Arkansas, *Congressional Record,* July 1, 1946, 8050; John M. Cooper of Catholic University, *Hearings on Science Legislation,* October 29, 1945, part 4, 778; John M. Gaus, president, American Political Science Association, *Hearings on Science Legislation,* October 29, 1945, part 4, 748.

34. Ogburn, *Hearings on Science Legislation,* October 29, 1945, part 4, 767. Ogburn lists sampling techniques in statistics, quality control in industrial production, and predicting population trends and parole violations as illustrations of these achievements.

35. E. G. Nourse of the Brookings Institute, *Hearings on Science Legislation,* October 29, 1945, part 4, 763.

36. *Hearings: National Science Foundation,* March 6, 1947, 184.

nest and objective investigation of problems of human relations can produce results of inestimable practical value when properly trained research workers imbued with scientific detachment and integrity are given opportunity to apply themselves with adequate resources."[37] The claims of social science were variously described as "reliable and trustworthy" and as "objective and fairly precise."[38]

Pro-inclusion advocates did not shy away from admitting that social science touched on controversial issues—as if natural science did not. Senator J. William Fulbright celebrated such controversy, and turned it deftly into an argument for inclusion: "The fact that it is controversial is one good reason why we ought to attempt to find out something about this field of knowledge, which happens to be fundamental to the existence of a democratic society."[39] Indeed, the arguments went, controversy cannot compromise the objectivity of social scientific claims, resulting as they do from the same detached and reliable methods that are used by physical and biological scientists.

The Unity of Scientific Knowledge

The folk epistemology deployed by those seeking to get social science inside the NSF sounds every bit like classical positivist presumptions of "the unity of science." Scientific knowledge is indivisible, it is said: efforts to draw lines between territories of scientific knowledge are doomed to failure and—if attempted—fraught with liabilities. Senator Kilgore "recogniz[ed] the inherent unity of all science," a view that Yerkes echoed two years later: "Knowledge is a continuance, sir, and an attempt to divide [it] up into sciences is pretty futile."[40] Few witnesses or legislators, however, endorsed the disciplinary imperialism expressed, curiously, by an engineer: "All science is social science."[41]

Two specific arguments were advanced. First, some emphasized that

37. *Hearings on Science Legislation,* October 29, 1945, part 4, 741.

38. Ogburn, *Hearings on Science Legislation,* October 29, 1945, part 4, 765; Miller, *Hearings on Science Legislation,* October 29, 1945, part 4, 797. John M. Cooper argued: "Anthropology is an empirical science, not a normative one. It doesn't tell us what we should do. . . . They do not in the name of their science sponsor any specific philosophy of life" (*Hearings on Science Legislation,* October 29, 1945, part 4, 778).

39. *Congressional Record,* May 20, 1947, 5511.

40. Kilgore in Senate Committee on Military Affairs, Subcommittee on War Mobilization, *Preliminary Report on Science Legislation: National Science Foundation,* December 21, 1945, 21; Yerkes in *Hearings: National Science Foundation,* March 6–7, 1947, 184.

41. Rautenstrauch, *Hearings on Science Legislation,* October 31, 1945, part 4, 895.

intellectual developments in one region stimulated or redirected inquiries in the other. Russell Smith described as "important" the "impact of the natural upon the social sciences, and of social and economic developments on the natural sciences," a point later recalled by Fulbright on the Senate floor: "They all impinge on one another. Both the physical sciences and the social sciences are important. They react and affect one another, so it is difficult to compartmentalize science in that way."[42] Moreover, divisions of science into natural and social domains would impede the progress of both. Congressman Emmanuel Cellar of New York told a House committee that the immediate functioning of the new science foundation would be jeopardized by drawing lines through science: "I understand that most scientists believe that the Foundation would not function effectively if an attempt were made to draw sharp lines between the different sciences and to exclude the Foundation from supporting social sciences."[43]

Second, others emphasized that reality itself does not come neatly demarcated into disciplinary territories; if ontologically the universe is one, then our efforts to understand it scientifically must be one as well. Yerkes described how his search for scientific understanding frequently forced him to cross the supposed line between social and natural science—effortlessly, and without even noticing: "I happen to be a psychobiologist myself . . . a hybrid between a psychologist and a biologist, a psychologist being a person who is concerned with behavior and experience primarily, and a biologist, a person concerned with life in general. I am midway between."[44] Pro-inclusionists made the boundary between social and natural science "highly artificial" and "arbitrary."[45] Some called attention to the historical variability and thus relativity of boundaries drawn between the sciences. Yerkes suggested that such demarcations had outlived their usefulness: "Though necessary in the pioneer stages of scientific development and though they are of continuing usefulness for certain operative purposes, such divisions tend to become more shadowy and unimportant as scientific knowledge expands. They have no place in the thinking of the national legislature when it attacks the problem of utilizing the re-

42. *Hearings on Science Legislation,* October 10, 1945, part 1, 126; *Congressional Record,* July 1, 1946, 8051.

43. *Hearings: National Science Foundation,* March 6–7, 1947, 26.

44. *Hearings: National Science Foundation,* March 6–7, 1947, 185.

45. Yerkes, *Hearings on Science Legislation,* October 29, 1945, part 4, 755, 758; Miller, *Hearings on Science Legislation,* October 29, 1945, 798; Yerkes, *Hearings: National Science Foundation,* March 6–7, 1947, 189.

sources of science as fully as possible."[46] On the floor of Congress, just after Hart had introduced his amendment that would, in minutes, seal the fate of the social sciences, Senator Thomas of Utah dissented: "Sciences have been classified in one way today, and will be classified in another way tomorrow. This will always be the case in growing institutions of knowledge. We cannot have a tight, little, compact understanding and knowledge, and assume that we can fasten it in a box and label it one way or another."[47] He was not convincing. But before considering the more persuasive countermappings that would put social science *outside* science in the 1940s, I fast-forward about two decades to a moment when representations of the "unity of science" in fact carried the day.

REPRISE: KEEPING SOCIAL SCIENCE IN THE NSF, THE 1960s

Arguments in the 1960s *against* the creation of a new foundation specifically for the social sciences will sound almost exactly like what we have just heard: social science is *just like* natural science—in its utilities, methodology, and epistemology. A convincing case for just one science rendered pointless and even dangerous the creation of two separate federal agencies, each charged with funding different flavors of basic research. This "rhetoric of similarity" was given voice in the 1960s by members of the (natural) scientific establishment—principally NSF director Leland Haworth—who wanted to keep the NSF inclusive. Interestingly, the equivalent scientific establishment back in the 1940s—Vannevar Bush, for example—found a "rhetoric of difference" more useful to keep social science out of the foundation (as we shall see). Mappings and interests need not consistently align: the same goals may be pursued—in different structural or organizational settings—with completely inverted cultural cartographies.

Complementary Utility

The Great Society era assigned top priority to a set of national problems that would seem tailor-made for social scientific remedies. Between 1965 and 1968, as Congress reviewed the performance of the NSF and considered the possibility of a National Social Science Foundation, witnesses

46. *Hearings on Science Legislation,* October 29, 1945, 758.
47. *Congressional Record,* July 3, 1946, 8231.

drew up a long list of broken bits of society that the social sciences might help to mend: poverty, housing, pollution, overpopulation, mass transportation, public health (for example, smoking), urban growth, welfare, space, human mortality, agricultural productivity, malnutrition, and obesity. This catalog of societal woes seemed prima facie evidence of the need for a new NSSF: these obviously social problems cried out for an agency whose sole purpose was to enlarge the basic understanding of social processes. And yet every item on this list was specifically mentioned by somebody who actually wanted the social sciences to remain *inside* the NSF. Those opposed to a new NSSF argued that social science could be useful in solving pressing national problems only if it would sit tight with physics and chemistry in the NSF as presently constituted. The key was to show that all of those social problems required the conjoint attention and expertise of both social and natural scientists.

Haworth took the lead in defining this "complementary utility" as the mutual interdependence of all disciplines: "Many of the most crucial problems that the country faces, in fact, the world faces, are going to depend very very substantially on the social sciences and on their integration with natural sciences and engineering."[48] In preparation of a staff study on the place of social science research in the federal government, letters were solicited from many experts, including economist David D. Moore of the Battelle Memorial Institute, whose argument anticipates Haworth's cultural cartography: "Many, if not most, of the significant problems facing our society require inputs from both the social and physical sciences, and it is highly important that mechanisms be developed so that there is a better communication and understanding among the many disciplines involved."[49] An explicitly cartographic metaphor is em-

48. Senate Committee on Government Operations, Subcommittee on Government Research, *Hearings: National Foundation for Social Science,* 90th Cong., 1st sess., February 7, 1967, part 1, 64 (hereafter, *Hearings: NFSS*). Haworth later reiterated the same idea, now in the context of how the NSF might itself be rearranged to give greater attention to social science: "Of even greater importance are the needs of society for far more understanding of man and of the nature and behavior of social institutions, than are available to us today. More and more, solutions to the problems of society are becoming dependent on science as a whole, and more and more, the social and the natural sciences are becoming interdependent" (Senate Committee on Labor and Public Welfare, Special Subcommittee on Science, *Hearings: National Science Foundation Act Amendments of 1968,* 90th. Cong., 1st sess., November 15–16, 1968, 78; hereafter, *Hearings: NSF Act Amendments*).

49. Senate Committee on Government Operations, Subcommittee on Research and Technical Programs, *Staff Study: The Use of Social Research in Federal Domestic Programs,* part 3, "The Relation of Private Social Scientists to Federal Programs on National Social Problems," April 1967, 138 (hereafter, *Staff Study: Use of Social Research*).

ployed by Gerhard Colm of the National Planning Association: "The strict separation between physical and social sciences in itself is undesirable. Many of our most urgent problems lie on the borderline between the two fields and require for their solution . . . teams of people with experience in both kinds of disciplines."[50]

The list of national problems requiring the genuinely combined efforts of natural and social scientists strains credibility at times (such rhetoric is to be judged not for its accuracy but for its utility). Robert A. Levine of the Office of Economic Opportunity does not exactly spell out how physicists might aid the war on poverty, but he still sees their input as essential, as he also will in regard to housing ills: "As we have found in poverty research, there are increasing numbers of problem areas which require a combined natural science and social science input. Creation of a separate social science body might hinder the growth of these interdisciplinary developments."[51] Haworth returns to more plausible terrain when he argues that "a combination of social science, engineering, and natural sciences [is] required to attack problems such as those arising from water and air pollution, population pressures and the inadequacies of mass transportation systems."[52] Smoking and obesity were two other problems seen as lending themselves to solutions dependent upon both natural and social science.[53] And, as in the 1940s, the case for keeping

50. *Staff Study: Use of Social Research,* April 1967, part 3, 63.

51. *Hearings: NFSS,* February 7, 1967, part 1, 23, 27, 31.

52. *Hearings: NFSS,* February 7, 1967, part 1, 99. The specific problem of pollution is also mentioned by Paul Miller of the Department of Health, Education, and Welfare: "Researchers themselves are increasingly finding the boundaries between the two realms arbitrary and constraining. The problems which attract the attention of today's investigators do not easily divide into physical, biological or social components. . . . For example, air pollution is as much a concern of political scientists as it is of chemists, as much a concern of sociologists as it is of physicists" (*Hearings: NFSS,* February 16, 1967, part 1, 157).

53. Donald R. Young, who chaired the Committee on Government Programs in the Behavioral Sciences for the National Academy of Sciences–National Research Council, cited obesity: "I am quite convinced that all the sciences must work together. . . . We have one man who is studying the problem of gross obesity. This is obviously both a problem of metabolism, and also a problem of values, of habits, patterns of life requiring collaboration of biological and social scientists working in close association with each other" (Senate Committee on Government Operations, Subcommittee on Government Research, *Hearings: Federal Support of International Social Science and Behavioral Research,* 89th Cong., 2d sess., June 28, 1966, 136; hereafter, *Hearings: Federal Support*). Smoking is mentioned by Pendleton Herring of the Social Science Research Council: "There is a growing contact, let us say, in the field of biology with the behavioral sciences. It is a sharing, if not of common ground, at least common interest in what the other field is doing. . . . Very clearly, if you turn to a field such as public health, the problems there are as much social as medical. You are faced in the field of public health with problems of motivation, how to get people to come in and have their chests X-rayed, or what

the NSF intact was aided by assertions that social science could effectively deal with the problems sure to arise from (natural) scientific breakthroughs and the brand-new technologies they spawn.[54]

One new rhetorical wrinkle was the claim that social scientists—like their peers who study nature—have skills and expertise that make them better than ordinary folk in comprehending or solving urgent national problems. Boundary-work took place on two fronts: as the line between natural and social science was erased, an invidious demarcation was created between expert scientists and various outsiders less able to get the job done. Herbert A. Simon's testimony recalls Francis Bacon's line that "my system levels men's wits" by pointing out that the "powerful tools" developed by scientists in all fields gives them a leg up: "The actual or potential usefulness of social scientists to society does not and cannot depend on their being brighter or having more common sense than journalists, or lawyers, or ministers, or the man on the street. If they have something to contribute to the solution of social problems, it is because they, like other scientists, are concerned with developing and using tools of inquiry to help the unaided mind."[55] One such transdisciplinary tool was systems theory, which was in the 1960s the buzzword for associations between scientific research and government policy. John R. Borchert employed familiar cartographic imagery to fill up the spaces where natural and social science intersect: "The increasing application of systems theory to problems of resource utilization and environmental quality, especially bringing together diverse disciplines in the physical and social sciences, are demanding that gulfs between them be bridged, and providing new opportunities for disciplines that lie in the boundary zone between these two classical divisions of the sciences."[56] In this view, social science *is* more useful than mere common sense, in part because of its reliance on methods developed arm in arm with natural scientists.

---

do you do about cigarette smoking, and factors that really involve human response rather than the medical problems of research that are already pretty well under control. . . . We should have in mind this closer contact between the natural and social sciences as we try to improve governmental arrangements that would support social scientific research" (*Hearings: Federal Support,* June 28, 1966, 209).

54. A 1966 subcommittee report mentions that "anticipating and dealing with the social results of new applications of science and technology is part of the sphere of the social scientists" (House Committee on Science and Astronautics, Subcommittee on Science Research and Development, *Report: The National Science Foundation—Its Present and Future,* 89th Cong., 1st sess., 1966, 49; hereafter, *Report: Present and Future*).

55. *Hearings: NFSS,* June 7, 1967, part 2, 395.

56. *Hearings: NFSS,* letter dated March 22, 1967, part 3, 794. For a contemporary critique of the dangers of systems theory, cf. Boguslaw, *New Utopians.*

Common Methodology

Those opposed to the splintering of the NSF argued that obvious ontological differences between bugs and natives did not imply that entomologists and anthropologists used different methods of observation and analysis—moreover, everybody can get objective results. Paul Miller of HEW told Congress exactly that: "The fundamental processes of objective inquiry [do] not differ whether the focus be upon people or nature. Science is a set of procedures, a frame of mind, if you like; as such, it is neither more nor less applicable to different fields of investigation."[57] For Henry W. Riecken of the Social Science Research Council, these processes included "that common core of logical organization, experimental procedures, mathematical and quantitative techniques, and conceptual cumulations which constitute the unity of science and insure the objectivity of its findings."[58] Much was said about the altogether good statistical turn in the social sciences, and its increasing reliance upon computerized analysis of large data sets. Pendleton Herring told the subcommittee: "Certainly the spirit of the social scientist today is to be detached, analytical, objective and to quantify when he can quantify."[59] Indeed, Herring worried that social scientists face the risk of going too "hard": "Gad, if you could see the interest in using computers, you would sometimes think that maybe some of our brethren had gone too far in quantifying."[60] Sociologist James S. Coleman linked these quantitative

57. *Hearings: NFSS,* February 16, 1967, part 1, 156. Political scientist Aaron Wildavsky from Berkeley was among the several witnesses who also described the common methods that united social and natural science: "The subject matter of social science is different from natural science, but the method, I think—the spirit of inquiry—is quite similar" (*Hearings: NFSS,* June 2, 1967, 288).

58. *Hearings: Federal Support,* July 20, 1966, 269.

59. House Committee on Science and Astronautics, Subcommittee on Science, Research, and Development, *Hearings: Government and Science—Review of the National Science Foundation,* 89th Cong., 1st sess., July 20, 1965, 1:439 (hereafter, *Hearings: Government and Science*). Carl Pfaffman of Rockefeller University adds: "In addition to the increasingly empirical character of study is the trend toward quantification, the measurement and analysis of data" (*Hearings: Federal Support,* June 28, 1966, 120).

60. *Hearings: Government and Science,* July 20, 1965, 1:439. Paul Miller adds that "the coming in recent years of the computer, the exploration of mathematic models of research—all of that area . . . is always the never-ending need of all science" (*Hearings: NFSS,* February 16, 1967, 179).

"I hope that the rigorous and hard-nosed methods of the physical sciences can slowly extend themselves more and more through the biological and into the social sciences." The metaphor of a "soft social science" train heading toward or into the "hard-nosed natural science" station

methods to increased objectivity in social science research: "One very important value of highly quantitative research lies here: the researcher's own biases cannot allow him to ignore results quite so easily."[61]

Social science was said to be *like* natural science because useful applications of both are equivalently dependent upon prior basic research—another argument for keeping them administratively wedded: "The social sciences do not differ from the natural sciences in the long lead time required for basic research to prove fruitful and in the necessity for maturing and testing conclusions."[62] Social scientists too rely upon "the peer group review and evaluation procedure that is so commonplace in science."[63] Social scientists too demand increasingly complex and expensive research infrastructures: "They require laboratory resources akin to those now provided in the physical sciences" and "data archives . . . are,

---

is one still used, as it was by Gardner Murphy of the Menninger Foundation three decades ago: "It means, on the contrary, gradual movement of social science in the direction in which the natural sciences are making their tremendous progress. As in all science it means the retention of the human element, but the strengthening of the human element by disciplined observation of facts in light of well-defined theory. . . . It makes no difference to me whether we say that the social sciences are sciences already in the full sense, or whether we say that they are pre-sciences, not yet solid enough to be fully relied upon. This is like asking whether the train is in Grand Central Station, or is headed east from Niagara Falls. To the train dispatcher this is all important, but the main thing for the rest of us is that the train is headed in the right direction" (*Hearings: NFSS,* June 28, 1967, 647, 652). Ten years later, Anthony Giddens uses the same metaphor to effect a completely different image of social science: "Many . . . have relinquished the belief, for various reasons, that social science, in the near future will be able to match the precision or the explanatory scope of even the less advanced natural sciences. . . . Those who still wait for a Newton are not only waiting for a train that won't arrive, they're in the wrong station altogether" (Giddens, *New Rules,* 13).

61. *Staff Study: Use of Social Research,* April 1967, part 3, 56.

62. Haworth, *Hearings: NFSS,* February 7, 1967, part 1, 96.

63. Arthur Brayfield, representing the American Psychological Association, *Hearings: Federal Support,* June 27, 1966, 59. He later adds that the peer review system contributes mightily to the progressive growth of knowledge in social science: "You would much prefer to have your peers look at your work—and of course this is the way science is advanced, by having your critical colleague looking over your shoulder and telling you that you are wrong, your procedures are wrong, and you really do not know how to go about it, and you do not have a significant problem" (65). Procedures for judging claims "operate in the same way and by the same principles and criteria as do the natural science programs" (Father Theodore Hesburgh, president of Notre Dame, *Hearings: Government and Science,* July 21, 1965, 1:474).

Harvard political scientist and science policy adviser Don K. Price told the committee that bias or error is "the same problem . . . in all science. The natural scientists reduce it by exposing their experimental findings to checking by their colleagues and competitors. It is less possible in the social sciences to verify findings by experimental procedures but free publication and free debate seem to me to be the best guarantee of reducing bias and establishing a measure of objectivity" (*Staff Study: Use of Social Research,* April 1967, part 3, 159).

in a sense, the big tools of the social scientist, analogous to the accelerators, telescopes, and so on, of the physical scientist."[64] Social scientists too examine "sensitive" issues, and do so with the same cool as their natural science buddies: "When values conflict (the cancer-smoking issue, fluoridation, battery additives, supersonic aircraft, automobile safety, smog abatement), natural science and technology becomes as value-laden as social science."[65] Social scientists too will not tread on political turf: "A distinction has to be made between the objective views as a scientist and the personal and political views as a citizen, but this distinction can be made by the social scientist as well as by the physical scientist or biologist."[66] One method, one science, one foundation.

And then the coup de grâce: moving the social sciences to their own foundation would make it more difficult for them to comprehend society—or to nurse it to health. The rhetoric thus created a "stepchild" status for these disciplines: by hanging around with their mature kin in the natural sciences, immature social scientists would "grow up" to objective, systematic, and precise methods. It was pointed out that social scientists had had "little more than one-quarter century of experience in the application of methods to human social behavior," and that this "early stage of development [made] it especially important that they be helped by association with the more advanced physical and biological sciences."[67]

64. Charles Y. Glock, sociologist from the University of California at Berkeley, *Staff Study: Use of Social Research,* April 1967, part 3, 86; Haworth, *Hearings: NFSS,* February 7, 1967, part 1, 70. The various meanings of "big science" are explored by Capshew and Rader, "Big Science."

65. Herbert Simon, *Staff Study: Use of Social Research,* April 1967, part 3, 186. "These social science disciplines are by their very nature, by using the word 'discipline,' . . . endeavors to be analytical. And while they sometimes fail, the important thing is that these are deliberate rational efforts of men trying to think analytically about problems which as you say by their very nature are at times highly controversial" (Herring, *Hearings: Government and Science,* July 20, 1965, 1:442).

66. Ernest Hilgard, Stanford psychologist, *Staff Study: Use of Social Research,* April 1967, part 3, 108. "The social scientist may be able to provide an objective statement of what the consequences of a particular action will be. However, the question of whether these consequences are desirable or undesirable on balance is not one which is a matter of scientific determination, but rather one of political evaluation" (Carl Kaysen, director of the Institute for Advanced Study at Princeton, *Hearings: NFSS,* June 7, 1967, part 2, 373). On the tension between politically neutral facts and politically useful information with respect to environmental controversies, cf. Zehr, *Acid Rain.* Zehr calls this the "dilemma of objectivity versus utility."

67. Riecken, *Hearings: Federal Support,* July 20, 1966, 269; Donald R. Young, *Hearings: Federal Support,* June 28, 1966, 132. "Encouraging interaction between social and natural scientists, and encouraging social scientists to apply natural science standards of objectivity and technical ingenuity to their own research, are important ways of moving toward . . . levels of effectiveness like those that have been attained by the natural sciences" (Simon, *Staff Study: Use of Social Research,* April 1967, part 3, 191). Donald F. Hornig, director of the Executive

As orphans, though, the social sciences would face an even harsher world: "I fear that a separate Social Science Foundation may be [more] vulnerable to the winds of political controversy than would a single national foundation covering the whole range of science—natural and social."[68] If social scientists stayed inside the NSF family compound, they would bask in the "superior prestige" of the natural sciences; they " [might] be able to attract more governmental support in partnership with the sciences than they could independently; but "they would be unlikely to be consulted as often or as seriously . . . if separate bodies for the social sciences existed."[69] With prestige, money, credibility, authority at stake—the very stuff of boundary-work—witnesses opposing the NFSS could not afford to let social science appear "as something less than science."[70]

Unity of Knowledge

Positivist (like Marxist) dreams of a unified science in the postwar period were reenacted in Washington twenty years later, this time by those hoping to keep the NSF intact. In hearings testimony, witnesses variously point to "science as a unified enterprise," to "the unity of all the sciences," and to the "unity of knowledge."[71] An image of the unity of science is

Office of Science and Technology, made the same point: "The social sciences have not yet evolved the kind of basic laws and principles that the national [natural?] sciences have. They are only learning to become quantitative and predictive. And I think they are making great progress in those directions. But I think that the interaction of the social sciences with the more experimental, more quantitative approaches of the national [natural] sciences give strength to them" (*Hearings: NSF Act Amendments,* November 15–16, 1967, 64).

68. Carl Kaysen, *Hearings: NFSS,* June 7, 1967, part 2, 374. F. Max Millikan, director of MIT's Center for International Studies, did not want to "risk exposing this tender plant to the full blast of direct and possibly hostile congressional scrutiny by giving it separate legislation of its own. . . . Social sciences are much more controversial than natural sciences, and it will, I suspect, be a good deal harder for the Congress to resist the temptation to try by legislative history or otherwise influence the substance or the focus of the research" (*Hearings: NFSS,* June 27, 1967, part 3, 606).

Such fears of political intrusion were probably acute in the wake of Project Camelot, which may have prompted Leland Haworth to worry that "the Foundation you propose [would be] thought of as in part, at least, a cover agency, [and] that this would tend, not in actuality, but in the eyes of the critics abroad, to 'contaminate' the new agency" (*Hearings: NFSS,* February 7, 1967, part 1, 74).

69. Don K. Price, *Hearings: NFSS,* June 20, 1967, part 2, 415; Henry Field Haviland, Jr., of the Brookings Institution, *Hearings: Federal Support,* July 19, 1966, 154; Simon, *Hearings: NFSS,* June 7, 1967, part 2, 406.

70. Miller, *Hearings: NFSS,* February 16, 1967, part 1, 156.

71. Kaysen, *Hearings: NFSS,* June 7, 1967, part 2, 374; Murphy, *Hearings: NFSS,* June 28, 1967, part 3, 648; Frederick Burkhardt, president of the American Council of Learned Societies, *Staff Study: Use of Social Research,* April 1967, part 4, 113.

sometimes achieved cartographically by constructing a cultural boundary, not between social science and natural science, but between social science and the humanities. Thus, although Miller of HEW allows that the social sciences have had a "historical relationship . . . to the humanistic disciplines," he points out that this association has become distant recently as social scientists have "perfected additional instrumentations of observation" more typical of those in the natural sciences.[72] The humanities function as a foil, useful rhetorically for making social sciences look as if they fit far better with natural science. Haworth says flatly that "the Foundation . . . does focus primarily on those social sciences that are approached from the same viewpoint as the other sciences are approached, as distinguished from what one might call the more humanistic approach to the social sciences."[73]

As on the 1940s map "One Science," witnesses argued that disciplinary demarcations (reified in the proposed NFSS) are an arbitrary and pernicious way to carve up the whole ontological enchilada. "One feature of the current study of behavior," says Carl Pfaffman, "is the recognition of the complementary contribution of genetic factors on the one hand, and environmental influences on the other. The interrelationships between the physical, biological and social sciences are so extensive and fundamentally significant that it would be an unfortunate error to fail to take advantage of the opportunities for their improved coordination already so well initiated in the administration of the National Science Foundation."[74] I am (understandably) rather enchanted by Brayfield's observation that, regarding science, "you cannot find a really good demarcation." His purpose was to show that "all these disciplinary distinctions are really kind of artificial."[75] Other witnesses describe the boundary between hard and soft science as itself rather soft: Thomas L. Hughes, director of Intelligence and Research at the State Department, talks about "shadings all across the board." Anthropologist Ralph L. Beals puts cartographic language to perfect use: "There are core areas which are rather unique to a particular discipline, but around the edges, where

72. *Hearings: NFSS,* February 16, 1967, part 1, 163.

73. *Hearings: NFSS,* February 7, 1967, part 1, 68.

74. *Hearings: Federal Support,* June 28, 1966, 120, 132. Simon gives an example from another boundary-straddling field: "The boundaries between science disciplines are becoming less and less definite, and the communications across them increasingly important. To cite an admittedly extreme example, computer list-processing languages, invented for certain uses in psychology, have been applied to the modelling of RNA reproduction in biological cells" (*Hearings: NFSS,* June 7, 1967, part 2, 394).

75. *Hearings: Federal Support,* June 27, 1966, 69–70.

you come in contact with the other disciplines, it is very fuzzy.[76] If these boundaries are so arbitrary, then there must be danger in taking them for real. "Fragmentation and departmentalization" is described as "the plague of the universities." Another witness evokes distant memories of "natural philosophy, which was really all science," recalling how "it tended to splinter as time went on and there got to be rather narrow specialization [so that] many scholars were interested only in their own fields. I believe we must pull them back together."[77] Social scientists of the day must have been comforted to hear how much the NSF loved them!

INVIDIOUS DIFFERENCES: PUTTING SOCIAL SCIENCE OUTSIDE, THE 1940s

The case for excluding social science from the NSF-to-be would seem to have been the easier one to make. In May 1946, Vannevar Bush told a congressional subcommittee that President Roosevelt had not specifically mentioned social science in his request for an assessment of postwar scientific research—and so it never made it into *The Endless Frontier.* And with good cause, Bush continued: "There was no word in the call concerning the social sciences. That was quite appropriate."[78] But why was social science so inappropriate in these discussions about the future of government-financed basic research? Why was social science so "out of place?" Why did social science fit only awkwardly into the cultural space labeled "science," just "as hair and butter should be kept apart"? On what grounds was it reasonable to suggest that "the social sciences differ from the physical and biological sciences more than the latter differ from each other."[79] To judge from the testimonies, it was not enough

76. *Hearings: Federal Support,* June 27, 1966, 14, 91. For Kalman Silvert, professor of government at Dartmouth: "The distinction which is being made between the 'hard' or the physical sciences, and the 'soft' or the social sciences may not persist much longer. A better differentiation might perhaps be between those sciences which study recurrent phenomena, and those sciences which study historical, and thus necessarily unique, phenomena" (*Hearings. . .,* July 20, 1966, 214).

77. Physicist Robert W. Krueger, *Staff Study: Use of Social Research,* April 1967, part 3, 120; Haworth, *Hearings: NFSS,* February 7, 1967, part 1, 72. In chapter 5, we will see how Albert Howard also takes up cudgels against fragmentation and specialization in science—a task that *is* boundary-work.

78. House Subcommittee of the Committee on Interstate and Foreign Commerce, *Hearings: National Science Foundation Act,* 79th Cong., 2d sess., May 28, 1946, 53 (hereafter, *Hearings: NSF Act*).

79. R. J. Dearborn, representing the National Association of Manufacturers, *Hearings on Science Legislation,* October 12, 1945, part 1, 185; Col. Bradley Dewey, president of the Ameri-

to demonstrate difference: social science had to become less than science—not worthy of inclusion in a National Science Foundation. An image of real (natural) science was selectively but effectively concocted, then used to measure the inadequacies of social science. It was said that those non-sciences were dangerously unbounded, inseparable from politics and ethics, weakened by methods lacking objective and predictive force, and—to boot—no better at all than common sense. This cartographic representation of the social and natural sciences may be no more accurate than one mapping their common or overlapping utilities, methodologies, and ontologies, but eventual passage of the Hart Amendment (deferring scientification of the social sciences for several years) attests to its persuasiveness on that occasion.

Dangerously Unbounded

A physicist from Princeton admits that he is "not a social scientist," but hesitates not a moment to tell a hearings subcommittee that "it is very hard to say what shouldn't come under that, whereas it is fairly easy to define what you mean by research activity in the physical sciences." Vannevar Bush himself testifies that the term "social science" is "very vaguely defined," and then he invokes a (for me, delicious) cartographic metaphor: "It has imperfect boundaries."[80] James B. Conant of Harvard reports that social science is "very difficult to define," and another witness that it "has ill-defined limits."[81] Bowman says that "there is the widest difference of opinion as to what constitutes research in the social sciences," while Senator Warren Magnuson recalls that as the science legislation was promulgated "we could not quite put our finger on exactly where to begin and where to end it." Worse, Magnuson continues, "social scientists themselves can't quite agree on what is the program."[82]

Such unboundedness creates three difficulties. First, if the boundaries

---

can Chemical Society, *Hearings on Science Legislation,* October 30, 1945, part 4, 818; House Committee on Interstate and Foreign Commerce, *Hearings: National Science Foundation,* 80th Cong., 2nd. sess., June 1, 1948, 133.

80. Henry DeW. Smyth, *Hearings on Science Legislation,* October 25, 1945, part 3, 652–53; Bush, House Committee on Interstate and Foreign Commerce, *Hearings: National Science Foundation,* March 6–7, 1947, 240.

81. *Hearings: National Science Foundation,* March 6–7, 1947, 166; Alfred Winslow Jones, *Hearings: National Science Foundation,* June 1, 1948, 133. Jones's article originally appeared in the June 1948 issue of *Scientific American.*

82. *Hearings on Science Legislation,* October 8, 1945, part 1, 23; *Hearings on Science Legislation,* October 19, 1945, part 2, 453.

are so vague, or if they cannot be agreed upon easily, then who knows what strange beasts might qualify for government patronage? MIT president Karl Compton testifies: "Practically, I don't know where you would stop, because everything is social science, really, everything that human beings are interested in."[83] Second, the potentially limitless character of social science will dilute the quality of research done in those fields—there will be too many worthy problems and too few social scientists to examine them properly. Senator Radcliffe continues his harangue against inclusion: "The general field of social relations . . . is so exceedingly vague, and covers such an enormous field [that] we are likely to run into difficulties, into confusion and cause the research to be much too thin." Finally, ambiguous boundaries threaten a government bureaucracy that demands tight boxes. Senator Hart justifies his "permissive formula" compromise by alerting everybody to the unmanageability of social science: "No agreement has been reached with reference to what social science really means. . . . I believe . . . that the inclusion of social sciences in this bill is very apt to result in doubling the complexity of governmental bureaucracy, and quite likely, the expense of the Government."[84]

In the end, even Harley Kilgore—steadfast supporter of government-financed social science—capitulates. He tells the Senate in June 1948: "Social science has ill-defined limits."[85] Because of this putative ambiguity, legislators turn over the boundary-work to the National Science Board, to be completed at some future moment—never once asking where physics stops or biology begins.

Controversial Politics

Now we find out what awful things lurk about the murky borders of social science—matters of politics, race, religion, ideology. Such politically contentious matters become the social scientific hair that does not belong in

83. *Hearings on Science Legislation,* October 25, 1945, part 3, 631. Radcliffe, on the Senate floor, creates a specter: "Social sciences cover or touch upon almost everything one can think of in the way of human associations, conduct and relationships. I can remember, years ago when some sociologists used to insist that sociology included practically every form of human endeavor; that it was really the parent of economics and politics, and that even the natural sciences sprang from social science" (*Congressional Record,* July 1, 1946, 8049). The senator may or may not have been recalling the testimony of a certain physician from Boston, obviously pro-inclusion, who had declared that "all science seems to be in the hands of the social sciences" (see note 23 above).

84. *Congressional Record,* July 1, 1946, 8049; July 3, 1946, 8230.

85. *Congressional Record,* June 2, 1948, A3531.

the butter of detached and objective natural science. From the 1940s until almost the end of the 1950s, the very subjects of race, religion, politics, and ideology (sexuality was not mentioned, but could have been) were inherently so controversial that it was all but impossible to secure state patronage for research on such things.[86] Though Harlow Shapley spoke out for inclusion, he worried, for "social scientists have to deal with subjects that are close to politics at times."[87] For others, the rhetoric was no-holds-barred: "There is not anything that leads more readily to isms and quackeries than so-called studies in social science, unless there is eternal vigilance."[88] Social science was collapsed into "social philosophy," where "so much of human prejudice and tendency . . . enter into the study of social phenomena." On this score, as Cornell president Edmund E. Day drew the line, "social science research [was] basically *different from* natural science research."[89] That difference arose because "the natural sciences, of course, deal with the immutable laws of nature and politics doesn't enter into that. The social sciences, of course, on the contrary, deal with changing relations between men." Consequently, it was feared, there would be "a lot of pressure from pressure groups."[90]

From a logical point of view, social science could either be dangerously powerful or plainly ineffectual—but not both. Cultural cartography is not necessarily governed by rules of logic, and those hoping to keep social science out of the NSF played both tunes with little apparent worry about contradiction. On the one hand, social science was seen as so powerful that it could manipulate public opinion on controversial issues and preempt democratic deliberation: "[There is] great danger of the use of so-called research in the social sciences for political purposes and to influence legislation."[91] The case for inclusion was hardly helped by repre-

86. Those opposing inclusion lined up to list the topics that had no place in a detached and objective science: "all the racial questions, all kinds of economics, including political economics, literature, perhaps religion, and various kinds of ideology"; also "morality" and, again, because it was emphasized, "even racial and religious matters" (Senator Hart, *Congressional Record,* July 3, 1946, 8230; John Milton Potter, president of Hobart and William Smith Colleges, *Hearings on Science Legislation,* November 1, 1945, part 5, 943; Hart, *Congressional Record,* July 2, 1946, 8098).

87. *Hearings on Science Legislation,* October 9, 1945, part 1, 52.

88. Senator Radcliffe, *Congressional Record,* July 1, 1946, 8049.

89. Bowman, *Hearings on Science Legislation,* October 8, 1945, part 1, 23; Day, *Hearings on Science Legislation,* October 29, 1945, part 4, 790.

90. Boris A. Bakhmeteff, representing the Engineers Joint Council, *Hearings on Science Legislation,* October 26, 1945, part 3, 715.

91. Homer W. Smith, physiologist at New York University, *Hearings on Science Legislation,* October 23, 1945, part 3, 496.

sentations of social science as "one individual or group of individuals telling another group how they should live."[92] The specter of Comtean technocracy raised by one witness must have given the politicians pause (though they were reassured that their jobs were not threatened by physicists or chemists): "Social science comes very close to the fundamental political questions which are the questions of the day, and I begin to see the possibility of a Government's building up a certain body of opinion, a certain direction of thinking through that, whereas in the physical sciences I am not afraid of that simply because it is quite objective."[93] On the other hand, social scientists were presented as so inept that they could never be dangerous: "I do not believe a scientist will learn about society by the study of social science as a science. Still less will he learn about it by the establishment of a research project in social science as a science."[94] Cornell's Day concurred with this good-for-nothing image of social science: "Education can turn the trick of remolding public opinion [but] no amount of findings out of social science is going to serve that purpose. . . . As yet social science has relatively little to offer."[95] Either way—dangerously powerful as arbiter of contentious values or incapable of understanding much of anything—this framing of social science made it look thoroughly out of place in the proposed NSF.

Worse, opponents argued, to include those controversial social scientists in the fledgling foundation might well compromise the perceived ethical neutrality and taken-for-granted disengagement of natural scientists. Recall that pro-inclusionists had argued that social scientists would benefit from rubbing shoulders with natural scientists: they would learn the sophisticated methods and logic of science more quickly, and their claims would be shielded from attack by the cultural authority of science

92. Surprisingly, this was the opinion of Senator J. William Fulbright (*Congressional Record,* July 1, 1946, 8048), who apparently could argue both sides of the "inclusion" question (elsewhere, he is a staunch advocate of government-funded social science).

93. Columbia University physicist I. I. Rabi, *Hearings on Science Legislation,* November 2, 1945, part 5, 999. Rabi here circumvents a cartographic problem that plagued George Combe (chapter 3): politicians are inclined to protect their own authority over their own cultural space—and to look unfavorably upon maps that show technoscientists creeping into deliberations of public policy. Vannevar Bush also got the politicians thinking about their own cultural jurisdiction: "In the field of social sciences we are dealing with people. And we therefore come immediately upon questions which are the questions with which you gentlemen deal all the time, some of which are very difficult, and some of which I think would be entered upon by a Federal agency with some proper trepidation" (*Hearings: National Science Foundation,* March 6–7, 1947, 242).

94. Potter, *Hearings on Science Legislation,* November 1, 1945, part 5, 943.

95. *Hearings on Science Legislation,* October 29, 1945, part 4, 790.

tout court. In anti-inclusionist arguments, the direction of influence is neatly reversed: the presence of social science in the NSF will bring natural scientists down into the messy politics of their lesser "peers." Guilt by association was implied: "Many of the things which a social scientist has to say are controversial in nature. . . . I think hitching these two together, you might find . . . that the work of social science would become unpopular and would therefore reflect on the whole job. It seems to me that the two fields are sufficiently different that one wouldn't quite do that. It would not be wise to have them sink or swim together."[96] When inclusion is the goal, natural science *père* cleans up social science *fils;* but for exclusion, social science becomes the more powerful (and evil) force—polluting the objectivity of natural science with its controversial politics.

Flawed Methods

Karl Compton justified the boundary between natural and social science quite simply: "The methods are so different."[97] Typically, this meant that proven (natural) scientific procedures could not be carried out on human beings or social institutions—laboratory experimentation was frequently singled out as the sine qua non of real science and the unique source of credible knowledge. The key to experimentation was said to be manipulation and control, which is difficult if humans are under the lens. As Senator Fulbright noted: "We [in political science] can't control things like you can in a laboratory; if you have a test tube and there is a certain bug in there, you can make him do things, can't you, and you can't do that in the political field." Knowledge claims in social science are thus decided not by experimental evidence, but are "purely a matter of persuasion."[98] The lack of experimental "control" compromises the ability of social scientists to "prove things," as does the relative lack of mathematics and statistics.[99] Henry Allen Moe of the Guggenheim Foundation carefully lays out a scientific method with essential features designed to highlight their conspicuous absence from social scientific practice: "First, the

96. Rabi, *Hearings on Science Legislation,* November 2, 1945, part 5, 998.

97. Compton, *Hearings on Science Legislation,* October 25, 1945, part 3, 631.

98. Fulbright's remarks came during an exchange with a remarkably guarded J. Robert Oppenheimer of the Manhattan Project and H. J. Curtis of Oak Ridge (*Hearings on Science Legislation,* October 17, 1945, part 2, 327).

99. Day, *Hearings on Science Legislation,* October 29, 1945, part 4, 790; Rabi, *Hearings on Science Legislation,* November 2, 1945, part 5, 998, 999; Dewey, *Hearings on Science Legislation,* October 30, 1945, part 4, 818; C. E. MacQuigg, representing the Engineering College Research Association, *Hearings: NSF Act,* May 28, 1946, 33.

controlled observation of data, preferably under conditions which can be precisely and frequently repeated; second, rigorous measurement, if possible to an infinitesimal margin of error; third, rigorous reasoning, especially mathematical reasoning, upon the basis of data so observed and so measured." Such qualities are best approximated in physics, and they degenerate completely as "the possibility of control becomes less, the number of unknown variables increases, and measurement . . . becomes a matter of . . . gross approximation." That describes the social sciences science well, evidently—a place where scientific method gives way to "statistical acrimony and finally . . . open debate."[100]

The contrast between abstract theorizing and concrete empiricism is here deployed—as it was by Tyndall (Chapter 1) to play off science against its very different rivals (religion and mechanics). This time round, "concrete" is the science to be—and so the abstractness pinned to the social sciences makes them less than science. Even a witness generally favorable to inclusion of the social sciences is forced to admit that "the whole field of research in social science doesn't lend itself to the concreteness that it does in the natural sciences."[101] Senator Hart reproduces the claim in his successful bid to prevent legislatively mandated inclusion: "There is no connection between the social sciences, a very abstract field, and the concrete field which constitutes the other subjects to be dealt with by the proposed science foundation."[102] Some trace the "more complex and higher abstractions" of social science back to its "many areas where measurement is impossible." But whatever the source of this abstraction, its results have no merit: an "insecurity of findings that you don't get in laboratory science," and "all sorts of hare-brained studies about things not capable of objective study, things that in the end have to be determined by individual opinion."[103]

Paradoxically—demonstrating once again that logic does not rule cultural cartography—other anti-inclusion arguments drop the other shoe: social science *lacks* the powerful theoretical abstractions that give natural science its predictive force: "The complete lack of any fundamental laws . . . places this subject in the field of the 'humanities.' It is not science in

100. *Hearings on Science Legislation,* November 1, 1945, part 5, 939.

101. Abel Wolman, *Hearings on Science Legislation,* October 25, 1945, part 3, 670. His pro-inclusion views are cited above.

102. *Congressional Record,* July 3, 1946, 8230.

103. Jones, *Hearings: National Science Foundation,* June 1, 1948, 133; Day, *Hearings on Science Legislation,* October 29, 1945, part 4, 790; Bowman, *Hearings: NSF Act,* May 28, 1946, 14.

the sense that the term 'science' is to be interpreted in the bill."[104] Damned if you do . . .

No Better than Common Sense

The line between scientific understanding and common sense has long been a site for boundary-work. In most instances, the border between scientific understanding and common sense is drawn precisely and demonstrably, as scientists seek to show that their cognitive authority is justified by their better comprehension of reality than that of hoi polloi. In rare moments, the boundary between science and common sense is blurred or at least made porous—as in Combe's effort to enlist the masses of phrenological converts to his cause (chapter 3), or Howard's preference for testimonials from composting enthusiasts over experimental data manufactured by his scientific adversaries (chapter 5). As Congress debated the science legislation in the 1940s, a field of inquiry described as no better than "what everybody knew" stood a dim chance of inclusion in the NSF. And that is how social science was made to appear.

Fulbright suggests during the 1945 hearings that "in social science, everybody is an expert," an opinion with which John F. Victory will later concur: "Most American people are more or less social scientists." As evidence for his own status as a social scientist, Victory—what a name for 1945!—describes himself as "past chairman of the Committee on Charities and Corrections of the Washington Board of Trade, and the present secretary of the Community Chest of Washington and member of its Executive Committee."[105] He makes no mention of technical training and skills, or of professional experience in either research or teaching. Obviously, if everybody is an expert, nobody is—which is what Oppenheimer hints at in his testimony: "I am aware of difficulties of establishing in [social science] fields rigorous criteria of competence and qualification."[106] During the testimony of Johns Hopkins's Isaiah Bowman, Congressman Clarence J. Brown of Ohio makes a telling comment: "Everyone else thinks he's a social scientist. I'm sure that I am not, but I think everyone else seems to believe that he has some particular God-given

104. Roger Adams, chairman of the board, American Chemical Society, *Hearings on Science Legislation,* October 30, 1945, part 4, 827.

105. Fulbright, *Hearings on Science Legislation,* October 29, 1945, part 4, 762; Victory, *Hearings: NSF Act,* May 28, 1946, 62.

106. *Hearings on Science Legislation,* October 17, 1945, part 2, 301.

right to decide what other people ought to do, and what form of behavior they should follow and so on." For colorful emphasis, he identifies social scientists as "a lot of short-haired women and long-haired men messing into everybody's personal affairs and lives, inquiring whether they love their wives or do not love them."[107] Here gender-coded norms about proper hair length are skillfully used to move social scientists into yet another cultural space altogether: deviance (which is certainly where snoops belong, more so than in the NSF). Potter adds that we are better off, not with the latest scientific understandings of society, but with "social teaching based on the whole ages-old experience of mankind."[108]

Once more, cultural cartography defies conventions of logic: others argue that social science should not be included in the proposed NSF because it is too difficult for mere physicists or chemists to understand. Forget that other witnesses have equated the knowledge of social science to what anybody knows: now the specialized skills and techniques of social science exceed the competence of foundation administrators who—it is clear to all—will not be drawn from the ranks of sociologists, anthropologists, or economists. Because the "type of training" in social and natural science is actually very different," there are those who argue that "no board, no administrative organization . . . could . . . possibly be adequately qualified to administer such policies and carry on work in two fields so absolutely diverse." Inevitably, "there will be questions the answers to which can be derived only by expert statistical analysis and comparisons made by men trained in the social sciences."[109] Bush in effect ends the matter this way: "If [Roosevelt had] wished to have advice as to the position of the social sciences in the postwar world, he would naturally have turned to men who are professionally social scientists."[110]

FLATTERING DIFFERENCES: A NEW HOME FOR SOCIAL SCIENCE?—THE 1960s

The case for a National Foundation for Social Science required its backers to put a boundary—and, ideally, some distance—between natural and social science. Arguments for the new agency emphasized difference—but not the same difference that was used to exclude social sci-

107. *Hearings: NSF Act,* May 28, 1946, 11–13.

108. *Hearings on Science Legislation,* November 1, 1945, part 5, 939, 946.

109. Rabi, *Hearings on Science Legislation,* November 2, 1945, part 5, 998; Hart, *Congressional Record,* July 3, 1946, 8230; Bowman, *Hearings: NSF Act,* May 28, 1946, 10.

110. *Hearings: NSF Act,* May 28, 1946, 53.

ence from the NSF at its creation two decades before. Then, the key was to present social science as less than science, as not worthy of inclusion. Now, the key was to present social science as more than science, as requiring a separate funding agency to address the distinctive problems and promises of these disciplines. In the 1940s, the boundary was drawn invidiously as social science came up short on a yardstick calibrated by qualities attributed to natural science. By the 1960s, two yardsticks were needed to measure distinctive methods, goals, politics, and accomplishments of the two kinds of science.

Alternative Methodology

The case begins with the assertion that "social science activities should not be weighed by or against physical science criteria."[111] Judgments of the quality of social science research should be made "according to the internal logic and methodologies of the various disciplines, rather than in terms of standards derived from an abstract idea of what does and what does not constitute 'science.'"[112] The stepchild social scientist—who would benefit methodologically from rubbing shoulders with the mature natural scientists at the NSF—is in fact no stepchild at all: "Natural scientists often regard the social sciences as 'underdeveloped'; some refuse to accord the social sciences the status of science. The latter attitude is justified only if the success of a science is measured exclusively by the degree of prediction and/or control which it confers on its practitioners."[113] That old yardstick is just useless because social scientists have a raft of methodological attitudes and procedures that have no place anywhere in natural science: witnesses variously emphasize the "normative," "humanistic," or "impressionistic" elements of social science research; a more "speculative" or "historical and morally oriented" approach; and (specifically) "anthropological" investigations to "provide concrete

111. Silvert, *Hearings: Federal Support,* July 20, 1966, 233. Silvert has earlier been cited as arguing that the distinction between hard and soft sciences was rapidly disappearing—implying that he favored leaving the social sciences in the present NSF. Actually, he equivocates on this policy option (232), which is why he is also cited here as arguing that the methodologies of the different science are sufficiently distinct that no common yardstick can be used to measure them all.

112. *Staff Study: The Use of Social Research,* April 1967, part 4, 96.

113. Anatol Rapaport, mathematical biologist at the University of Michigan, *Staff Study: Use of Social Research,* April 1967, part 3, 164.

and detailed pictures of various areas of social life on the basis of direct and close-up observation."[114] The message: soft is good.

Some proponents of the NFSS developed the "argument from ontology": people are not rocks, and so the methods of their study must be different. The anthropologist Margaret Mead emphasized the great complexity of human phenomena, which precludes use of the reductionistic techniques of natural scientists: "Where the natural sciences began by simplifying the phenomena of nature, in order to probe it in its total complexity, the social sciences began with the realization of the complexity of human culture and human society." The complexity inherent in social life requires that social scientists "build into our research plans the very uncertainty and indeterminacy which is characteristic of human life itself." Mead wrote to Congress that the complexity of human phenomena is inevitably "mirrored in the way these problems are studied."[115] Others justified the new foundation in terms of the *Wertbeziehung* (or value relevance) of the social sciences: "Social science differs from other sciences in that it more obviously and by its nature becomes involved in values of people and the policies and values of their institutions."[116]

Congress in effect learned about the "double hermeneutic": not only do social scientists confront a social world already meaningfully understood by ordinary folk, but the "theories and findings in the social sciences are likely to have practical (and political) consequences" for those folks.[117] It came out this way in 1967, from Margaret Mead:

> Methods of research in the social sciences differ in many other ways from those employed in the natural sciences. We work with living human beings, and we must incorporate their welfare as individuals, as groups, as members of nations—our own and others—into research plans. To take a small amount

114. Thomas L. Hughes, *Hearings: NFSS,* February 8, 1967, part 1, 134; Gabriel Almond of Stanford, then serving as president of the American Political Science Association, *Hearings: Federal Support,* June 28, 1966, 111, 112; Alvin W. Gouldner, sociologist from Washington University, *Staff Study: Use of Social Research,* April 1967, part 3, 102. Hughes is speaking as a representative of the State Department, which "has no strong preference" for a new NFSS or for an NSF with a beefed-up social science division. His testimony takes the form of an equivocal "for-and-against" listing, and so extracts from it appear as evidence both for a separation and for an identity of natural and social science.

115. *Hearings: NFSS,* June 1967, part 3, 784.

116. Theodore Vallance, American University, *Hearings: Federal Support,* July 19, 1966, 174.

117. Giddens, *Constitution of Society,* xxxv.

> of blood from a healthy man, in order to perform a battery of tests, leaves him psychologically as the investigator found him. But to treat him as if he were inanimate matter, or a creature incapable of understanding what is happening to him, or to his group, while statistics are assembled, is quite another matter. We have had to learn to work with those whom we wish to understand, as collaborators, not as mere objects of research.[118]

It is precisely because social scientific results have such immediate personal and political consequences that some methods common to natural scientists are not available for social inquiry. "The limitation on the use of experimental methods, for example, is not due to the social scientist's ignorance of experimental method; it is due to the fact that human beings do not like being treated like guinea pigs." Sociologist Kingsley Davis of Berkeley responds only a bit defensively to the stepchild image of social science: "The key to the problems of the social disciplines, and the key to their distinctiveness, does not lie in 'youthfulness,' 'complexity,' or 'ignorance of the scientific method.' It lies rather in the fact that social science gets entangled with the control mechanisms of society. The operation of the society depends upon its members carrying in their heads notions of what people ought or ought not to do."[119]

The bottom line for those demanding a separate social science foundation was that—whether disciplinary differences are traced to ontology or politics—standard methods of natural science cannot succeed in the study of human and social phenomena: "I do not feel that the solution of our great social problems of today are necessarily found through the rats-in-the-maze type of laboratory research." Bluntly, "I am convinced that the natural-scientific paradigm is simply not sufficient for a fruitful methodology of social science."[120] Something methodologically *different* is required: "We are going to have to be experimental and innovative and original, and, therefore, controversial."[121]

118. Mead, *Hearings: NFSS,* July 1967, 784.

119. *Hearings: NFSS,* June 2, 1967, part 2, 267.

120. Stewart P. Blake, Stanford Research Institute, *Hearings: NFSS,* June 21, 1967, part 2, 453; Anatol Rapaport, *Staff Study: Use of Social Research,* April 1967, part 3, 165. Rapaport continues: "The social scientist must deal with concepts which have no analogues in natural science. Premature definitions of these concepts in terms acceptable to the natural scientists (e.g., operational definitions) often deflect the attention of the social scientist from the problems which he really wants to study to other more tractable, but also more superficial, problems. In some cases, this has led to trivialization of social science, especially in our country."

121. Senator Fred Harris, *Hearings: NFSS,* June 21, 1967, part 2, 476.

Celebrating Controversy

The celebration of the controversial aspects of social science by proponents of a separate NFSS is a case where the boundary between science and politics is dissolved—on so many other occasions they are meticulously kept apart—in order to legitimate extension of scientific authority and resources to an "iffy" set of disciplines. Social scientific research is said to be innately controversial—for two reasons. First, because the subject matter of these fields is "another human being or a social organization," as well as "the subjectivities and activities, the perspectives and operations of people," it follows that (as Senator Harris put it): "Almost anything you do in the social sciences has some rather immediate and direct applications in the minds of a great many people as it relates to their lives, to the Government, to policymaking."[122]

Second, politics enters social science through its investigators, who inevitably bring into the research process values and interests that often are intimately connected to the people and processes under study. Alex Inkeles uses Project Camelot to illustrate the folly of assuming that social scientists enter the research process with empty hands: "The social scientist is placed in conflict between the objectives of his sponsor and the self-defined interests of his subjects."[123] Fellow sociologist Kingsley Davis summarizes both the reflexive and recursive character of social science, which gives it its unavoidably controversial and political edge: "Any attempt to study social behavior from a strictly disinterested, analytical point of view thus suffers from two handicaps that do not afflict the study, say, of the geology of the earth's mantle. First, since the social scientist himself shares and conducts his life with reference to the basic sentiments, attitudes and political issues of his social milieu, he finds it

122. Sociologist Alex Inkeles of Harvard, *Hearings: Federal Support,* July 19, 1966, 183; Myres S. McDougal, Yale Law School, *Hearings: NFSS,* June 27, 1967, part 3, 514; Harris, *Hearings: NFSS,* July 13, 1967, part 3, 775. Harris uses this point to draw a boundary between the different sciences: "And that is one reason why social science research is often more controversial than that being done in the natural and physical sciences." Paul Bohannan, anthropologist from Northwestern University, makes the same point about social science: "It changes people's attitudes instead of changing people's ideas. Study physics, you change your ideas about the structure of the atom. . . . However, since social scientists deal directly in morally charged subject matter, they are judged by different canons from those used to judge the natural sciences. . . . Natural science deals primarily in the area of things we *know.* Social science on the other hand, deals primarily in subjects which the ordinary citizen *feels*" (*Hearings: NFSS,* July 12, 1967, part 3, 686, 689).

123. *Hearings: Federal Support,* July 19, 1966, 183.

exceptionally difficult to view them dispassionately. . . . Second, insofar as he adopts the role of the scientific observer toward other members of his society, he risks arousing their hostility."[124]

But it is *all right* that political controversy is an inescapable companion of social science—indeed, debate is the raison d'être of social science, its life blood. A federal agency charged with the patronage of pure science might be expected to "take care that the money would go only to 'safe' and sober men who were not to come up with unsettling conclusions or heterodox ideas," but "would that be the kind of research worth supporting in the first place?"[125] Emphatically not: "To convert the meaning of pure social science into an operational code book for noncontroversial social science would be self-defeating and, even in my opinion, suicidal. . . . What should be encouraged by congressional legislation is precisely research into dangerous areas."[126] Political or value significance is what makes a problem worth studying: "If a social problem is wholly neutral in its significance, I really doubt that anyone with much curiosity or intellectual enthusiasm would want to study it." Geoffrey C. Hazard Jr. of the American Bar Foundation seeks to expose the watertight demarcation of science and politics as mere contrivance, itself an accomplishment of interest-laden discourse rather than grounded in the actual practices of social scientists: "Until we are willing to be more candid and mature about the value-laden implications of most social science research, we will probably continue to engage in the sort of *rhetorical charade* that now goes on."[127] In this view, social science is better for this value relevance, not a lesser science because of it—and no methodological sleight of hand could or should extrude political controversy from its inquiries.

Several witnesses fashion a different role for social scientists, not as observers of social life but as critics of its flaws and participants in its improvement. "Many social scientists in fact construe their role as that of 'critics' of society" as they "reject the idea of disinterested scientific analysis." The contrast to natural scientists is heightened, and perhaps dangerously exaggerated, by Kingsley Davis: "The purpose is not to un-

124. *Hearings: NFSS,* June 2, 1967, part 2, 267.

125. Although this question was asked by Haworth, Senator Harris instantly turned it into an argument for an NSSF: "It seems to me it is a good reason for having a Social Science Foundation." That is surely not what Haworth wished to imply (*Hearings: NFSS,* February 7, 1967, part 1, 88).

126. Washington University sociologist Irving Louis Horowitz, *Hearings: Federal Support,* July 20, 1966, 241–42.

127. *Hearings: NFSS,* June 2, 1967, part 2, 310 (my emphasis).

derstand society, as a natural scientist would understand a termite nest or a buffalo herd, but to prove what is morally right or morally wrong about some form of behavior or projected policy."[128] Although the buzzword of the day was not revolution but "relevance," my choice to begin this chapter with Marx is thus better than I had imagined: "The philosophers have only *interpreted* the world, in various ways; the point, however, is to *change* it."[129]

Disunity of Science

The unity of science—so vital to the efforts to keep the NSF whole—falls apart in the testimonies of those seeking a separate social science foundation. Yale law professor McDougal dismisses the idea of unity as "mysterious." Indeed, he argues, "some presently existing 'unity,' or some future achievable 'integration' of the natural and social sciences would appear to border on, if not pass beyond, mysticism." Moreover, the hope that natural and social scientists will routinely and beneficially exchange substantive ideas or methodological tips becomes a chimera: "The contribution of the physical and biological sciences to most social science is relatively small, since most of the social sciences are relatively unrelated to new developments or new advances in the physical and biological sciences." But there is another reason for this lack of cooperation: natural scientists "have tended to look down their noses at the social sciences as not really being sciences at all."[130]

Also shattered in the NFSS debates is the plan for natural and social scientists to work shoulder to shoulder on urgent national problems. "The principal problems facing our country are not scientific and industrial and are not related to the production of goods. The principal problems are related to the optimum distribution of the products of science and industry"—by implication, a task inside social science.[131] In these arguments, the practical applications of natural and social science are not presented as *complementary* (as they were by those defending the extant NSF) but as *sequential:* science and technology raise practical problems

128. *Hearings: NFSS,* June 2, 1967, part 2, 268.

129. Marx, "Theses on Feuerbach," 145; *Hearings: Federal Support,* July 20, 1966, 258.

130. McDougal, *Hearings: NFSS,* June 27, 1967, part 3, 514–16; Launor F. Carter, manager at a computer software company, *Hearings: NFSS,* June 2, 1967, part 2, 298–99; Senator Harris, *Hearings: NSF Act Amendments,* November 15–16, 1968, 143.

131. James C. Charlesworth, president of the American Academy of Political and Social Science, *Staff Study: Use of Social Research,* April 1967, part 3, 54.

that natural scientists cannot solve, but social scientists can. "It is, after all, largely to the social and behavioral sciences that we must look for help in anticipating and coping with the vast changes being induced by the natural sciences and their technologies."[132] Moreover, practical results will not be as predictable as they might be in applications of natural science: "Unlike the scientists, we social scientists cannot guarantee that in the year 1970 or 1980 or 1990 we will make a significant advance that will land us on the moon, that will bring us truly perpetual motion, or that will even guarantee a better mousetrap or a tie that will resist gravy spots."[133]

Advocates of an NFSS were no better at avoiding logical inconsistency than their "predecessors" who sought to keep social science out of the NSF twenty years before. The same bad logic but good rhetoric is heard in their testimony: social scientific knowledge overlaps common sense, but natural scientists are not competent to evaluate it. Anthropologist Paul Bohannan says: "Unlike physics, which deals with factual material, social science deals with the kind of material that everybody else has experience in. We talk about government, about economies, about families. Everybody knows about families and everybody knows something about government."[134] Evidently, "everybody" does not include natural scientists: "Although both natural scientists and social scientists pursue the same basic goals of scientific integrity, the problems with which each must contend both in terms of topic and in terms of carrying out projects differ. Experience in one of the two areas does not provide full appreciation of the needs and problems of the other." Even more bluntly, Kingsley Davis argues that "the only people who can implement and judge such progress are those who themselves are in the forefront of development. It is for this reason that I see the creation of a new foundation, established *for* social scientists and *controlled by them,* to be eminently desirable."[135]

132. Hughes, *Hearings: NFSS,* February 8, 1967, part 1, 131. Alexander Archibald of the Arms Control and Disarmament Agency adds: "The world has not yet found out how to tame the many genii which the natural scientists have created, and which seem to have a natural tendency to escape from their bottle. . . . One would look to the social sciences to explore and propose how man shall master these genii" (*Hearings: NFSS,* February 7, 1967, part 1, 62).

133. Historian Joe B. Franz of the University of Texas, *Hearings: NFSS,* July 12, 1967, part 3, 694.

134. *Hearings: NFSS,* July 12, 1967, part 3, 685.

135. Gouldner, *Staff Study: Use of Social Research,* April 1967, part 3, 97; Davis, *Hearings: NFSS,* June 2, 1967, part 2, 269–70 (my emphasis).

No Longer the Stepchild

Throughout this chapter (and throughout the book), I have argued that identifiable professional or political interests drive the occasion, form, and content of boundary-work—yet "interests" are every bit as discursively pliable as "science." Sometimes the most effective way to create a boundary between science and something else is to represent the interests of each side as irreducibly different. That move turns up in the rhetoric of those seeking an NFSS in the 1960s: whatever methodological or ontological differences may distinguish social from natural science, it is now said that the two domains of knowledge have separate agendas that not only do not coincide, but compete. In short, natural scientists have become an impediment to the success of social science.

Put another way, the supposed infantilism of the social sciences is not inherent in their constitution but results from an "arrested development" forced upon them by natural scientists, who are said to gain from such a bi-generational image. The rallying cry becomes, in effect, "stepchild no longer." Bohannan tells Congress that he no longer wishes to have his discipline seen as a "little brother asking for a lift." Senator Mundt baits Paul Miller (who opposes the NFSS) by suggesting that "the social and behavioral scientists can stand on their own two feet and justify their existence without constantly insisting that we have the same degree of 'science' as a physical scientist." Senator Montoya later attributes to Miller (who denies it, of course) the conclusion that social science has been relegated by the NSF to "a second-rate, stepchild position." Sociologist Irving Lewis Horowitz, never a shrinking violet, adopts the infantilizing metaphor: "Social scientists, even though intellectually they may not be entitled to see themselves on a par with the physical scientists, do consider themselves on a par psychologically, ethically and in every other way. Social scientists would feel [if they stayed in the NSF] like stepchildren to a parent organization."[136]

Horowitz's statement evinces a lingering insecurity, a worry that social scientists are not quite the intellectual equals of natural scientists. But if there has been such a lag in the progress of social science, the blame should fall squarely on natural scientists, who—especially by their organizational power within the NSF—have thwarted social science's full

136. Bohannan, *Hearings: NFSS,* July 12, 1967, part 3, 689; Mundt and Montoya, *Hearings: NFSS,* February 16, 1967, part 1, 168, 174; Horowitz, *Hearings: Federal Support,* July 20, 1966, 250.

bloom: "Our reluctance to advance our case as vigorously as we might stems from that early rejection experience, of not right at the onset being welcomed into the scientific community through its major basic research arm."[137] Those old wounds continue to fester, as social scientists sacrifice their unique identity in a vain attempt to belong: "They imitate what they believe are the characteristics that enable the physical and biological scientists to obtain large grants. They sell out their own science and masquerade or try to masquerade as 'real' scientists."[138] Such misdirected emulation has taken its toll—a kind of arrested development. By using a single yardstick to measure all research, the range of social science projects even attempted has been unwisely truncated: "The Social Science Division of the NSF is bound to support the kind of social science that is defined by and approved by the physical and biological scientists. It is they who will ultimately decide what is scientific and what is not, for NSF support."[139] The result? "[The] NSF has been concerned about supporting research on such sensitive topics as civil rights, sex, religion, federal-state relations."[140] In the end, it comes down to money—and which science it is spent on: "The social sciences have lagged behind the natural sciences. Their development has suffered for a number of compelling reasons, one of which may be the low level of funds available for social research."[141]

The authority and prestige of social science have also suffered under these overbearing natural science parents: "Instead of feeding the scientists hemlock or one thing or other, we suddenly elevated them to such

137. Brayfield, *Hearings: Federal Support,* June 27, 1966, 63. Brayfield also provided testimony that seemed to urge that the social sciences remain in the NSF, though with an enlarged role and budget. His ambivalence may have been the result of his speaking on behalf of the American Psychological Association, which had not taken an official position on a new NFSS (60, 63). In the passage above, Brayfield is chiding the NSF administration for its relative inattention to the needs and wants of behavioral scientists. He tells Senator Harris that "psychologists tend to plump for diversity of support"—that is, for both an NFSS and an expanded place in the NSF.

138. Anthropologist John Buettner-Janusch of Duke University, *Hearings: NFSS,* June 6, 1967, part 2, 352.

139. Austin Ranney, political scientist from the University of Wisconsin, *Hearings: NFSS,* July 12, 1967, part 3, 662. "Research and other activities in these fields would be most likely to receive support insofar as it essentially deals with highly quantifiable things, measurable things, and much of the most thoughtful research going on in these fields is not really of that character" (Richard H. Sullivan, president of Reed College, *Hearings: Government and Science,* August 5, 1965, 1:749).

140. James A. Robinson, political scientist from Ohio State, *Hearings: NFSS,* July 12, 1967, part 3, 719.

141. Senator Harris, opening remarks, *Hearings: NFSS,* February 7, 1967, part 1, 6.

a pedestal that we are perfectly willing to recognize that they know a lot more about this are than we ourselves can possibly know. . . . Now, we cannot do that as far as the social scientists are concerned."[142] This credibility lag can be closed up if the social sciences can get out from under the NSF: "A separate foundation will bring prestige and status to the social science[s]." Other witnesses agree that there are "powerful organizational reasons" for creating a new agency "unfettered by control . . . primarily oriented toward the natural and physical sciences."[143] The most evocative declaration of competing interests that divide social from natural science refers to the paramount societal division of those days (and possibly still). The following text from the journal the *American Behavioral Scientist* was entered into a congressional staff study:

> Social scientists continue to resemble the "official Negroes" of NSF grants policy. . . . From experience with earlier NSF programs one can predict that for the next year NSF will toss a sop to behavioral science by adding two or so psychologists, then cautiously establish an unwritten quota of two or three others to be drawn from anthropology, economics, and sociology. Principle: If you wear the proper scientific garb, avoid political science and all controversial topics, and are properly respectful and grateful, one or two of you may be invited to an NSF cocktail party.[144]

From little brother to stepchild to official Negro—the social sciences need their own space to grow up in, it is said, where they can enjoy the freedom of self-determination.

CONCLUSION: A CONTENTIOUS BOUNDARY BROUGHT UP TO THE MINUTE

Marx was both right and wrong to predict the convergence of natural and social science. Whether the boundary goes between two sciences or around one depends little on what physicists or sociologists actually do when they observe reality, calibrate measures, crunch data, or write up results. The elusive boundaries instead get drawn downstream from all of that, in ever changing contexts where maps of cultural spaces—science social or natural—have diverse consequences for the interests of diverse people. Marx's "one science" is not a hypothesis at all, but only a

142. W. Willard Wirtz, Secretary of Labor, *Hearings: NFSS,* February 7, 1967, part 1, 10.

143. Blake, *Hearings: NFSS,* June 21, 1967, part 2, 453; Horowitz, *Hearings: Federal Support,* July 20, 1966, 249; Senator Harris, *Hearings: NFSS,* February 8, 1967, part 1, 139.

144. *Staff Study: Use of Social Research,* April 1967, 70 (from an article orriginally published in the September 1963 issue of the *American Behavioral Scientist,* 70).

more or less useful strategy, to be judged pragmatically by actors rather than logically or empirically by analysts. The choice between a "rhetoric of similarity" and a "rhetoric of difference" is permanently available for those disciplines seeking patronage, credibility, or epistemic authority—and for those seeking to deny such advantages to others.

Quite by coincidence, as I sought to put this chapter to bed in 1995, Congress yet again took up the problem of whether social science really belonged in the National Science Foundation. On May 11, Congressman Robert Walker (R-Pennsylvania), who chairs the House Science Committee and also serves as vice-chair of the Budget Committee, scared the bejabers out of my colleagues by suggesting in a news conference that the Social, Behavioral, and Economic Sciences Program at the NSF would be excluded from the 3 percent increase in the foundation's proposed budget, and in response to a question added: "In large part, we think [social science is] an *area* [that] the National Science Foundation has largely *wandered into* . . . in recent years, [and this] was a kind of a politically correct decision. . . . And that is [one] *place* where the science budgets can be rescoped. We think that the concentration ought to be in [the] *areas* of the physical sciences."[145] Electronic bulletin boards lit up: we social scientists were implored to write Washington immediately, giving good argument for why social science funding should be restored, if not augmented. Not even a flurry of debate over whether Walker really "meant it" could prevent the subsequent avalanche of letters urging government officials to sustain the social sciences' right to federal patronage through the NSF.

What should I do? I could have sent this chapter to my elected officials—all 49 pages of it—in order to set the record straight about when and how social science came to be included in the NSF. But with no executive summary to append and unable to break it up into sound bites, I had doubts that it would get read or appreciated. Worse, its message is not necessarily well tailored to save our funding programs: after all, those testimonies from hearings long ago equivocate on the central question of whether sociology and kindred disciplines belong with physics, chemistry, and biology. With steady dispassion, I have reported here as many reasons why social science is *not* science as why it *is*. Clearly, what was needed was not an analysis of boundary-work, but more boundary-work.

145. I just had to highlight Walker's spatial metaphors. Print discussions of his press conference and its wake include Curt Suplee, *Washington Post*, May 12, 1995, A11; *Science*, May 19, 1995; Andrew Lawler, "Scientists Mobilize to Fight Cuts," *Science*, May 26, 1995, 1120; Colleen Cordes, *Chronicle of Higher Education*, May 26, 1995, A28.

Marx would help me make social science appear at one with natural science. I wrote to Congressman John T. Myers and to Senators Dan Coats and Richard Lugar on May 18, 1995:

> I urge you to do everything you can to sustain funding for the social, behavioral and economic sciences through the National Science Foundation.
>
> Yesterday was the 41st anniversary of the landmark Supreme Court decision *Brown vs. Board of Education,* which enabled the racial desegregation of our public schools. Those who argued in favor of school desegregation drew on sociological and psychological data produced by the wartime "American Soldier" studies, which demonstrated scientifically that racial tolerance increased among soldiers assigned to mixed race fighting units.
>
> The contributions of the social, behavioral and economic sciences to the health and wealth of this nation are documented in the new exhibit at the Smithsonian's National Museum of American History on "Science in American Life." Visitors will read about the "American Soldier" studies, and about James R. Murie's pioneering anthropological research on the Pawnee Indians, and about psychologist B. F. Skinner's attempts to program pigeons to guide missiles during WWII.
>
> Visitors to "Science in American Life" will also learn about the history of intelligence testing in this country, a scientific issue that has recently reached public attention with the publication of Herrnstein and Murray's *The Bell Curve: Intelligence and Class Structure in American Life.* There is much debate within the social scientific community about the meaning of IQ and its relationship to race or poverty, just as there will be much debate in Washington over the book's policy implications. *The Bell Curve* could not have been written without data gathered by the National Longitudinal Survey of Youth and other long-term social science research projects—whose existence is jeopardized by proposals to eliminate NSF funding for the social, behavioral and economic sciences.
>
> Whether the issue is racial desegregation of schools or the effects of IQ on inequality, the social, behavioral and economic sciences stand ready to provide systematic data and objective analyses that will enable informed and rational public debate of these important matters.
>
> The social, behavioral and economic sciences have found a place in the National Science Foundation not because of "political correctness," but because they provide reliable descriptions and interpretations of American society and its people. Our basic research allows the public and its political leaders to better understand the many social problems we now face, and to more accurately anticipate the consequences of policies and programs we choose to implement.

Two months later, Myers wrote back to say that the 1996 NSF budget would be cut, but that the "reduction is to be taken without prejudice toward any academic discipline." It is difficult to say whether my cartographic efforts—along with the cultural maps rhetorically drawn by my fellow social scientists—were persuasive or decisive. I am slightly more certain of this: continued NSF patronage for the social sciences required some interpretative justification beyond "show me the money." Absent our arguments that social science was just as useful, reliable, accurate, objective, and systematic as physics or chemistry, the case for leaving my team out in the cold would have been far easier to make.

Chapter Three

MAY THE BEST SCIENCE WIN: COMPETITION FOR THE CHAIR OF LOGIC AND METAPHYSICS AT THE UNIVERSITY OF EDINBURGH, 1836

In 1815, the liberal-minded *Edinburgh Review* described phrenology as "a mixture of gross errors, extravagant absurdities," displaying "real ignorance, real hypocrisy"; in short, it was nothing more than "trash, despicable trumpery" propagated by "two men calling themselves scientific inquirers."[1] So what's a poor phrenologist to do with a panning like this? That problem faced George Combe of Edinburgh, who staked his career as scientist and as reformer on the truth of phrenology. Twenty-one years after those kind words were published, the undaunted Combe sought to ride phrenology into the prestigious chair of logic and metaphysics at the University of Edinburgh. He failed: the town council gave the chair instead to Sir William Hamilton, an Oxford-trained defender of traditional Scottish "common sense" philosophy.

The outcome may be less interesting for us than Combe's rhetorical tactics: rebuffed by the intellectual elite of Edinburgh who posed as guardians of real science, Combe redrew the culturescape with a different place for science—new borders, new territories, new landmarks all around—that put phrenology inside its authoritative space. The "science" on Combe's cultural map had little resemblance to its place and

In translation, the epigraph reads: "Take good heed concerning the treading down of the boundaries of the fields lest horrible calamities be brought upon thee" (from *The Teaching of Amen-em-apt,* ed. and trans. Wallis Budge [London: Martin, Hopkinson, 1924], 151).

1. [John Gordon], "The Doctrines of Gall and Spurzheim," *Edinburgh Review* 25 (1815): 268.

shape on maps drawn and used by Hamilton and his friends at the *Edinburgh Review.* A radically rearranged culturescape enabled Combe to attach the growing cultural authority of science to his phrenology, and more consequentially, to enlist allies from other quarters of Edinburgh and beyond. But when those supporters were asked in 1836 to submit testimonials to the town council—rehearsing their faith in the truth of phrenology and their trust in its staunchest advocate—the redrawn maps of science became the undoing of their favorite candidate.

I begin this chapter by describing the substance of phrenology, its roots and competing legacies, giving special attention to the dramatically different opinions it elicited in Edinburgh during the early nineteenth century. A brief prosopography of its detractors and enthusiasts moves us toward a sociological understanding of this divided opinion: identifiable social, political, and economic interests of some were advanced by a belief in the truth of phrenology, just as interests of others were advanced by its falsity. I shall say something about the individual histories of the protagonists Hamilton and Combe, for their lives also tell much about why separate groups and interests were attracted to their different versions of science, truth, and the good society.

"Interests" alone, however, do not explain Combe's defeat and the sad fate of phrenology, for those with the power to advance their interests—whether the formal power of the Edinburgh Town Council or the garden-variety power enjoyed by the city's elite—needed maps to tell them where to go and what to avoid. Not just any maps, but fresh ones capturing and momentarily stabilizing the undulating cultural and social terrain of a society on the verge of bourgeois liberal industrialization and professionalization. Indeed, the rhetorical cartography of Hamilton, Combe, and their supporters provided more than one such map. In what world of meanings did Combe's phrenology seem plausible, truthful, useful, scientific, and chair-worthy? How did meanings meld with interests to yield choices—a testimonial for Combe, or a vote for Hamilton?

To set the stage for the 1836 competition for the chair of logic and metaphysics, I discuss the cultural life of early-nineteenth-century Edinburgh, focusing on recent changes in the organization of the town council and its role as patron for the university. The centerpiece of this episode of cultural cartography is an analysis of testimonials published by Combe and by Hamilton, each consisting of formulaic effusions by scholars, scientists, and other professionals seeking to make a persuasive case for the selection of their man. Read side by side, the two sets of testimonials present dramatically different maps of science and its spatial relationship to other cultural fields, such as the mundane knowledge of Everyman; philosophical studies of mental functioning; professional domains of medicine, law, education, and penology; and religion. The map in Hamilton's testimonials reproduced in rhetorical form the cityscape of Edinburgh, preserving distinctions between cultural spheres and social institutions geographically separated in their brick-and-mortar form. The map for Combe messed up these tidy spatial and institutional classifications: lines between science and politics were erased or blurred, as were lines between physiology and mental philosophy, science and religion, expert knowledge and common sense. But through it all, phrenology *was* science (albeit "science" with a novel spin) and Combe *was* a scientist (but not like Hamilton), for those who used his cultural map to chart their travels through life and work and who decided that the best logic and metaphysics could be found in the bumps of the skull. Unfortunately for Combe, only three of thirty-three members of the Edinburgh Town Council had that map in mind as they voted for Professor David Ritchie's successor on July 15, 1836.

MIND, BRAIN, AND SKULL IN EARLY-NINETEENTH-CENTURY EDINBURGH

The nuts and bolts of phrenological beliefs are simply put.[2] First, the brain is the physical embodiment of the mind. Second, the brain is made up of separate organs, each corresponding to distinct mental faculties. Third, the size of a particular organ is a measure of the power of the associated mental faculty. Fourth, the relative size of organs can be read in a craniological examination from the pattern of bumps on a person's skull (a protuberance indicates a large organ underneath that spot and, thus, an exceptional capability for some mental activity). Fifth, behavioral and dispositional differences are determined at birth by the relative size of the mental organs, but through phrenological examination and exercise of various faculties, anyone can discover their own capabilities or limits and exploit them efficiently. The phrenological list of mental facul-

2. Those who wish to dip into the voluminous writings by phrenologists—or into the slightly less voluminous writings about them—will be ably assisted by Cooter, *Phrenology in the British Isles.* On the content of phrenological beliefs, see especially Shapin, "Politics of Observation."

ties included *sentiments* such as combativeness, self-esteem, benevolence, and veneration, and intellectual *talents* such as imitation, order, time, number, tune, and wit.[3] An individual with a large organ for "amativeness"—ascertained by a big bump at the correct place on the outside skull—was expected to have a large appetite for "feelings of physical love."

Evidence for phrenologists' claims typically came not from dissection or other anatomical explorations of the brain itself, but rather from educated readings of heads of the living, skulls of the dead (and even of animals)—with bumps then correlated to "known" behavior and mental dispositions of the subject. (One could easily guess the size of the region for amativeness in rabbits.) Phrenology began in the late eighteenth century with the Viennese-born, Parisian-trained anatomist and physician Franz Joseph Gall (1758–1828); it was carried to Edinburgh at the turn of the next century by Gall's student Johann Spurzheim (1776–1832), who passed it along to the initially skeptical but then converted George Combe (1788–1858). Its influence throughout society was democratic: Queen Victoria and Prince Albert twice asked Combe to examine the heads of their children, but the British working class also flocked to phrenological demonstrations and classes.

Phrenology's claims destined it to be a cause célèbre. It teetered on almost every philosophical seesaw of the nineteenth century—mind/body, nature/nurture, biological determinism/free will, socialism/capitalism, collectivism/individualism, materialism/spiritualism, good/evil, radical politics/reform, heterosexuality/homosexuality—crossing the great divides central for this chapter: science/religion, science/politics, science/pseudoscience.[4] Phrenologists moved easily from brain to

3. The number of organs and, thus, of mental faculties was a much debated matter among phrenologists: Gall believed in twenty-seven, Spurzheim upped that to thirty-five, but later proponents argued for as many as ninety-two. The "most popular figure" seems to be thirty-nine faculties (Walsh, "Is Phrenology Foolish?," 359).

4. Some phrenologists considered homosexuality the result of a prodigious organ for "adhesiveness" (see Lynch, "Here Is Adhesiveness"). That particular claim is strikingly modern: an AP wire report in the *Bloomington (Ind.) Herald-Times* (August 1, 1992) reported that scientists at the UCLA School of Medicine had determined from autopsies that the anterior commissure—a structure of nerve cells in the brain, near the back of the skull—is 34 percent larger among homosexual than heterosexual males. Cooter offers the reminder that all these boundaries were matters for our nineteenth-century Edinburghian subjects to negotiate: "Because phrenology is poised between the intellectual boundaries that have come to be erected between science and pseudo-science, it provides the incentive to expose and evaluate those boundaries. Phrenology is doubly illuminating because phrenologists themselves actually popularized those

mind to skull to society, turning the ever contentious question "How is man constituted?" into referenda on (say) the scientific truth of the claim that the brain is the physical organ of mind or on the political desirability of using craniological exams to decide the likelihood of recidivism among convicts. Facing those many philosophical conundrums, Combe and his fellow travelers did not come down on one side or the other in consistent or decisive fashion. For example, although phrenology argued that an individual's capabilities and limitations were determined at birth by the relative size of organs and faculties, the doctrine denied the fatalism that such biological determinism sometimes warrants. Room was made for "environmental" factors in human development: individuals could "exercise" their faculties to move toward a level of "perfection" that was constrained only in its outer bound by physiology.

Phrenology also crosscut political cleavages separating those who favored individualist or collectivist strategies of social reform. Its message was reformist: individuals could improve themselves—and by aggregation improve society (no class struggle here)—by learning about their talents and limits via a phrenological rendering of their skulls. The doctrine's emphasis on individual hereditary differences drew the wrath of utopian socialist Robert Owen, whose collectivist experiment at New Lanark presumed that environmental improvements (in education and working conditions, for example) were key to reforming the sorry lot of early-nineteenth-century laborers. Yet George Combe's brother Abram established an Owen-like utopian community at Orbiston in 1827, based in part on phrenological principles of social reform (that Abram lost much money in its quick failure might have had something to do with brother George's growing dissatisfaction with Owen and his ilk).[5]

The long-haul legacy of phrenology is a double one, as this cultural artifact suggests:

---

demarcations"—and so did their opponents, as we shall see (Cooter, *Cultural Meaning of Popular Science,* 8). Cooter's superb book, incidentally, should be the first thing read by those interested in the reception of phrenology among diverse British audiences, or in the function of phrenological beliefs for legitimating the emerging new world order of industrial capitalism.

5. On the filiation, and then separation, of Combe's ideas and Owen's, see Grant, "New Light." Details on Abram Combe's experiment come from brother George's entry in the *Dictionary of National Biography,* 4 (1917): 883–85.

"Tattoo." Bernard Schoenbaum © 1991 from The New Yorker Collection. All rights reserved. Reprinted by permission.

Phrenology has moved from science to sideshow legerdemain, from the hands of scientists and reformers to those of hucksters and tatoo artists. But the man having his bald head mapped out into phrenological regions hardly fits the image of carnival lowlife: chiseled features, formally dressed in suit and tie.[6] Why . . . he might even be a physiological psychologist, for that is the other legacy of phrenological beliefs: some of its claims are now authoritatively encapsulated in scientific facts about, for

6. The cartoon echoes an earlier lampoon of phrenology, *Travels in Phrenologasto,* by John Trotter. Writing in 1825 under the name Don Jose Balscopo, Trotter describes his ascent in a balloon and discovery of a "race of people whose heads are shaved, painted white, and 'chalked out by black lines into a variety of little fields and enclosures, very much in the same style as we see a Gentleman's estate in England laid out on a map'" (Parssinen, "Popular Science and Society," 10). Evoking the same imagery, Davies situates phrenology in the twentieth century as "harmless quackery practiced upon the gullible at county fairs and the Coney Island boardwalk" (*Phrenology,* ix). On one occasion notable for its exhortations of the triumphs of science—the Century of Progress exposition at Chicago in 1933—the "Temple of Phrenology" is located in the Midway area with seventy-five other curious attractions, such as "Milne's Handwriting Analysis," "Trained Fleas," the "Two-Headed Baby," "Wilson's Snake Show," and "Bozo the Clown." All of these exhibitions were comfortably removed from the real Hall of Science (*Official Guide Book of the Fair,* 1933).

example, cerebral localization. Phrenology has become both science and pseudoscience, both truth and error—contradictory accomplishments of two centuries of boundary-work, through which the whole of phrenological doctrine and practice gets torn up into bits scattered among diverse cultural fields—some granting authority to its claims and praising its achievements as science, others mocking it into "a scientific miscarriage" and "psychology's great faux pas."[7]

Boring's encyclopedic *History of Experimental Psychology* measures how wrong phrenologists are now thought to be: "It came to occupy the position of psychic research today, looked at askance by most men of science because unproven, using unscientific methods and indulging in propaganda." For historian Robert Young, "phrenology, of course, is nonsense. . . . To read about it . . . one must look in [Martin] Gardner's *Fads and Fallacies in the Name of Science,* where it shares a chapter with the pseudo-sciences of physiognomy, palmistry and graphology."[8] Most difficult to reconcile with contemporary physiological truth is the presumed congruence in size between regions of the brain and bumps on the outside of the skull. Boring summarizes today's knowledge: "The thickness of the skull varies greatly and apparently adventitiously" (that is, not as a simple function of brain matter pushing it outward in spots). But the truth of a correlation between brain and skull was a sine qua non for phrenology, for this was a theory not just of brain but of its mental and behavioral manifestations, and it was a theory designed not just for explanation but for accurate prediction of character, temperament, and talents (the doctrine hinged on being able to infer inside causes from outside measures).

The now-presumed fallacies of phrenology were not only empirical—like this one—but methodological and logical as well. Boring castigates phrenologists of the nineteenth century, still hoping to be scientific, for committing elementary statistical blunders such as selecting cases to fit the theory by examining subjects known in advance to be criminal or insane, rather than randomly sampling from the entire population and then examining them "blind." He faults them logically for their ad hoc evasion of falsification: when a specifically identified protrusion did not match "what everybody knew" about the mental and behavioral dispositions of the subject, the phrenological examiner would suggest that other, even more dominant faculties were suppressing or altering those indi-

7. Winkler and Bromberg, *Mind-Explorers,* 8; Flugel, *Hundred Years of Psychology,* 44.

8. Boring, *History of Experimental Psychology,* 57; Young, *Mind, Brain, and Adaptation,* 9–10.

cated by the first big bump. In this account of the legacy of phrenology, little has changed from Adam Sedgwick's judgment in the 1840s: "that sinkhole of human folly and prating coxcombry."[9]

But this is only half the legacy of phrenology, the side with tattooists. Boring modulates his denunciation of the obvious pseudoscience by suggesting that "phrenology was wrong only in detail and in the enthusiasm of its supporters," and that some of its claims are "yet still not absolutely disproven."[10] Another legacy makes phrenology salutary for science, first, by advancing claims about physiology that ended up true and that stimulated anatomical research on the brain and its functioning, and second, by advancing the domain of science into questions of mind, morals, and politics once thought to be distant from the scientific ken.

There is little question that, as Shapin puts it, "the phrenologists are frequently admitted to have 'got it right.'"[11] Gall's idea that different parts of the brain are responsible for different physical, behavioral, and mental functions has been legitimated since as the real-scientific theory of "cerebral localization." Right-brain/left-brain distinctions have become common sense. The claim that the brain is the organ of mind was phrenologists' monopoly in the early nineteenth century, though today it is fact for everybody. If phrenologists "got it wrong" by correlating the size of brain regions with cranial bumps, their other claims pushed science forward by moving the question of mental functioning from metaphysics and epistemology to biology, anatomy, and physiological psychology. Once mind could be observed in nature as the brain, the nascent discipline of psychology moved from "the domain of the speculative philosopher . . . to the special study of the naturalist and physiologist."[12] Bruce suggests that phrenology "advanced science chiefly by driving opponents to serious research," thus reconstituting Gall and Co. as something less than serious. Boring adds that Flourens—who identified different functions for the cerebrum, cerebellum, the medulla, and the cord later in the nineteenth century—found his positions "much strengthened because he could appear as a conservative correcting the pseudo-science of Gall and Spurzheim."[13] Phrenology has found it difficult to get respect.

Phrenology's reconstructed contributions to the advancement of sci-

9. In Young, *Mind, Brain and Adaptation,* 10.

10. Boring, *History of Experimental Psychology,* 58, 57.

11. Shapin, "Politics of Observation," 147.

12. Young, *Mind, Brain, and Adaptation,* 16, also 3, 10, 12; cf. Boring, *History of Experimental Psychology,* 57.

13. Bruce, *Launching of Modern American Science,* 118; Boring, *History of Experimental Psychology,* 61; cf. Grant, "Combe and the 1836 Election," 184.

ence extend beyond its several assertions about brain and mind that—with serious scientific attention—have come true. George Combe in particular is described as paving the way for the reception of Darwin, by making it increasingly acceptable for Victorian Britain to think about a wide range of moral, political, and social issues in naturalistic and scientific terms. Combe's entry in the *Dictionary of National Biography* is nice boundary-work: "Though his theories have fallen into complete discredit, he did something . . . to excite an interest in science and a belief in the importance of applying scientific method in moral questions."[14] Phrenology announced to a world still dominated by supernaturalism that universal laws of nature rule the affairs of people—from individual minds to the body politic. Human behavior was here a matter for science to grasp in the form of patterned protuberances on the head. Wrobel suggests that "by the time phrenology arrived on [American] shores in the early 1830s . . . it [had come] to resemble a social science . . . promising a rationalistic means for describing man's place in society and his relation to nature's laws."[15] Moreover, phrenological soap-boxers like Combe in Britain or the Fowlers in America carried the message of scientific naturalism to the middle and working classes, creating the assumption far and wide that "science could become significant in daily life," and spurring on a "powerful movement for the spread of scientific education that is associated with [John] Tyndall" (whose own boundary-work we mapped in chapter 1). Phrenology had almost everybody "nibbling at the teats of science."[16] The implication is that *The Origin of Species* would have faced even frostier audiences had not Combe and other phrenologists earlier brought problems of human nature and social order within the compass of science.

The parsing of phrenology into science and pseudoscience, truth and error, physiology and politics, anatomy and morals, began with its mixed reception in Edinburgh at the turn of the nineteenth century. The struggle was on to locate phrenology somewhere on a cultural map, and those opposed to the truth of its claims or to the message of its politics found space for phrenology in socially constructed categories of pseudoscience or blasphemy.[17] Evidence abounds for successful efforts by bor-

14. *Dictionary of National Biography,* 4 (1917): 885.

15. Wrobel, "Phrenology as Political Science," 124, also 126–7; Cooter, *Cultural Meaning of Popular Science,* 271; Cooter, "Phrenology," 216; Davies, *Phrenology,* 171.

16. Henry Cockburn, quoted in Saunders, *Scottish Democracy,* 93; Bakan, "Influence of Phrenology," 203; Temkin, "Gall and the Phrenological Movement."

17. "The science-pseudoscience split was the active creation of intellectuals in a specific socio-economic context" (Cooter, *Cultural Meaning of Popular Science,* 19).

der guards to keep phrenology outside the gates of science. In 1834, Combe asked the British Association for the Advancement of Science to create a phrenological section, a move that no doubt would have done much to establish his program within the panoply of science. The request was denied, for reasons attributed to Adam Sedgwick by a pro-phrenology observer of the meeting: the British Association should "confine their researchers to dead matter, without entering into any speculations on the relations of intellectual beings; and [Sedgwick] would brand as a traitor that person who would dare overstep the prescribed boundaries of the institution. . . . It was feared that moral or political discussion might be introduced."[18] For the British Association, phrenology belonged in politics and morals and, for this reason, could not also find a place in true science. That kind of thinking no doubt led the Royal Society of Edinburgh to prevent any phrenological advocate from addressing the scientific society (at least before 1830), and nurtured a move to prevent purchase of George Combe's masterwork *The Constitution of Man* by the Edinburgh public libraries. Combe was also denied use of rooms at the University of Edinburgh for public presentation of his science.[19]

Edinburghians who argued that phrenology was not just wrong but not "science" included both anatomists from the university's medical college and its moral philosophers. Anatomists grounded their attack on what they knew best: the structure of the brain. John Gordon in 1815, John Barclay in 1822, and Francis Jeffrey in 1826 questioned, for example, the reality of phrenologists' thirty-six distinct regions of the brain. Their anatomical dissections were said to reveal no such discrete and bounded sections, and as Barclay writes: "It seems to require no small share of creative fancy to see anything more than a number of almost similar convolutions . . . all exhibiting little difference in their form and structure as the convolutions of the intestine." Jeffrey concluded that, anatomically, the theory was a "radical absurdity."[20]

18. From an account published in the *Phrenological Journal* 9 (1834): 121. On the British Association's rejection of phrenology, and Combe's subsequent creation of the "alternative" Phrenological Association, cf. Morrell and Thackray, *Gentlemen of Science,* 276–81; Shapin, "Phrenological Knowledge"; DeGiustino, *Conquest of Mind,* 50. Other analysts have argued (as I shall) that the constructed *interestedness* of phrenology—as politics and morals, as distinguished from its claims about the brain or mind—was sufficient grounds for its exclusion from real science, a move which did much to promote science as value-free, objective, and disinterested. Cf. Cooter, *Cultural Meaning of Popular Science,* 27; Shapin, "Politics of Observation," 147.

19. Shapin, "The Politics of Observation," 231.

20. Ibid., 154; Cantor, "Edinburgh Phrenology Debate," 212. Cantor's is the best account of the substance of anti-phrenological arguments.

Moral philosophers grounded their attack on what *they* knew best: the functioning of mind. Here, phrenology ran headlong into the distinctively Scottish brand of moral philosophy, the cherished "common sense" metaphysics (at least north of the border) of Thomas Reid and Dugald Stewart. There is more irony in this intellectual battle, for Gall initially borrowed the idea of separate mental faculties from the Scottish tradition. But defenders of Reid and Stewart argued that Gall and subsequent phrenologists were seriously misdirected in their use of the idea of faculties. John Gordon, an M. D. who later held the chair of moral philosophy at the University of Edinburgh, argued in the *Edinburgh Review* in 1803 that phrenology unjustifiably reified the faculties of mind by locating them physically in distinctive places in the brain. For "common sense" philosophers, faculties such as "goodness" or "collecting and retaining facts" were reflective categories or processes of the mind at work rather than places in the brain. For them, the brain was an indivisible and homogeneous organ capable of different kinds of mental activity, though these different activities were not thought to occur in distinctive physical regions. Moral philosophers also rejected phrenologists' empirical efforts to study mind by looking at its presumed physiological correlates. The Scottish tradition preferred a different sort of empiricism known as "reflection," in effect, an introspective, systematic examination of one's own mind at work. Mental processes could be understood only by the philosopher looking inward; nothing outside could unlock its secrets.[21]

Churchmen grounded their attack on what *they* knew best—God, the Bible, and ethics—and located Combe's *The Constitution of Man* in the realm of blasphemy, the theological equivalent of pseudoscience. In 1836 the *Scottish Guardian,* a voice of the Kirk, denounced Combe, "whose opinions so widely differ[ed] from the doctrinal standards of the Church," and pronounced even his candidacy for the logic chair that year "one of the bad signs of the times."[22] In Combe's doctrine, it seems, there was too much about the laws of nature and too little about the laws of God. According to phrenology, they preached, natural law rather than divine intervention and God's will governed not just beast, but now man. Combe was vilified as an atheist for his evisceration of morality: good and

21. Cooter, *Cultural Meaning of Popular Science,* 25; Cantor, "Edinburgh Phrenology Debate," 198, 206.

22. Grant, "Combe and the 1836 Election," 175. On religious objections to phrenology, cf. Cantor, "Edinburgh Phrenology Debate," 203–4; Cooter, *Cultural Meaning of Popular Science,* 129; Parssinen, "Popular Science and Society," 10.

evil were no longer matters of free will and choice, but were fatalistically determined by the "wiring" of one's brain. Where was moral responsibility in phrenology, and what happens to the soul if the brain is defined as the organ of mind? Evil was not sin but a poorly developed faculty, hardly a matter helped by church attendance every Sunday. Combe was even said to deny the Fall: perhaps his choice of the word "perfectible" was the wrong one in a cultural context where religious authority in moral affairs was unequaled. In his system, the fallibilities of men and women could be corrected—in the direction of perfection—not by faith but by a good phrenological diagnosis and proper exercise of less-developed faculties.

Supporters made a different space for phrenology: inside science (as *they* drew its boundaries). Evidence abounds for the welcome acceptance of phrenological doctrines by diverse audiences whose social structural characteristics changed significantly during the four decades of active debate. At the start, phrenology drew most of its support from members of the medical profession, typically younger practitioners not affiliated with the university's medical school. The attraction is obvious: phrenology literally opened up potential new markets by opening up the mind as a neurological and anatomical object now well within the physician's trained capacity. The many phrenological societies that sprang up throughout Britain were, even until 1840, predominantly attended by practically minded medical men. A phrenological lecture in the Royal Medical Society in 1823 drew more than three hundred listeners, and the doctrine received early support from the *Scotsman,* a newspaper that appealed to the emerging middle classes with its criticisms of old wealth and privilege.[23] After the 1830s, the complexion of phrenological audiences began to change. Now members of the skilled working and lower-middle classes flocked to phrenological demonstrations and lectures. Combe's *The Constitution of Man* was a mass market best-seller: Harriet Martineau (herself an enthusiast) claimed that the book's circulation of 100,000 copies sold (by 1855) ranked it just behind the Bible, *Pilgrim's Progress,* and *Robinson Crusoe.* Phrenology was a popular course in mutual instruction societies and workingmen's institutes throughout Britain. Combe and other promoters promised self-improvement—getting in touch with your underappreciated talents and constraints—and pro-

23. Cooter, *Cultural Meaning of Popular Science,* 28, 32; Cantor, "Edinburgh Phrenology Debate," 201; Shapin, "Phrenological Knowledge," 224.

grams of social reform based on science and skills rather than on privilege and power.[24]

How can the different receptions of phrenology in early-nineteenth-century Edinburgh best be interpreted? Why was Combe's doctrine dismissed as error or absurdity by anatomists and mental philosophers in the university, and loathed as blasphemy by the Kirk? Why was the "same" doctrine accepted at first as the one true science of man by practicing physicians and other professionals on the make, and later by working men and women seeking self-help? The eventual truth and falsity of phrenological claims offers no answer.[25] Present-day scientific facts say nothing about why scientists and others believed what they did in earlier times, and moreover, the long-haul legacy of phrenology finds as much truth in its program as error. Shapin and Cooter introduce the needed interpretative wedge: look at the social characteristics and thus the interests of those repelled by or attracted to phrenology, and ask *cui bono?* What do these different strata gain or lose from the promotion or denunciation of claims that the brain is the organ of mind, that brain and mind are divided into mental faculties determining an individual's character and that these can be scientifically read from the pattern of swellings on the skull?

Briefly put, Shapin and Cooter suggest that critics of phrenology were typically those with an interest in preserving the status quo and in particular its allocation of cultural authority and attendant bases of power and privilege, while supporters had a stake in changing all that.[26] Anti-phrenologists were insiders: inside traditional cultural institutions of power such as the university and the Kirk. Generally speaking, opponents were drawn from the literary elite whose privilege and authority were based on inherited agricultural wealth and title. Phrenology had some unnerving messages for them: its programs of utilitarian social and politi-

24. On the changing social characteristics of those attracted to phrenology, cf. Cooter, *Cultural Meaning of Popular Science,* 120, 141, 163; Shapin, "Phrenological Knowledge," 224, 227. The Martineau attribution comes from Parssinen, "Popular Science and Society," 9.

25. As Cooter recognizes, "hindsight permits us to see that neither side had, as it were, the 'correct' information on the nature of the brain." Indeed, it is clear that "when it comes to the Truth about nature and human nature, there can be no touchstones other than those that exist in the minds of participants in history. . . . The task before the historian of science . . . is to determine how and why some conceptions of reality acquire the mantle of objective scientific truth and enter into the domain of common sense while others come to be regarded as arrant nonsense" (Cooter, *Cultural Meaning,* 34–35).

26. Shapin, "Phrenological Knowledge"; Shapin, "Politics of Observation"; Cooter, *Cultural Meaning of Popular Science.*

cal reform challenged aristocratic control, just as its claims about mind and brain challenged the truth of established "common sense" philosophies and approved anatomical models. Most important, by drawing attention to individual differences that started in biology and ended up in behavior, phrenology exposed what the traditional elite had reason to fear: the growing individualism and atomization of Edinburgh society, which challenged their preferred "one-ness"—in theory (for "common sense" philosophy, all members of society, within their assigned places, were equal in God's eyes) and in ideology (the social order was grounded in common values and interests). Differentiation of brain and mind was, for these critics, a call for a differentiation of the social fabric, and inevitably the undoing of old orders from which they benefited. Phrenology challenged entrenched social and intellectual dominance and, besides, what if the old minister was found to have only a modest bump for veneration?

By contrast, those who picked up phrenological ideas in the 1830s and carried them forward even into the twentieth century were decidedly plebeian. They were outside established structures of power and privilege, and generally came from lower social statuses with few associations to the university or other elite cultural institutions. Phrenology clamored for changes that would advance their social interests: its message of progressive individualism replaced aristocracy with meritocracy, seeming to open opportunities for those with (phrenologically ascertained) talents to get a-head. Combe invited broader participation in a growing scientific culture by promoting aggressive expansions of scientific education up and down the social ladder. By attending phrenology classes, working people were making facts and producing new truths, even as they learned useful things about themselves. Moreover, phrenology promised to do something about social ills that perhaps more often afflicted this social stratum than those higher up: enlightened treatment of the insane and workable efforts to help convicts were all to be desired. Phrenologists offered a science that would reorder the social world in naturalistic symbols, solve its disorders, and find a place for Everyman.[27]

27. Parssinen argues that phrenology's reformist message went to liberal but not radical ends. "Phrenology served the needs of those who advocated the reform of educational and penal institutions and insane asylums, since it claimed to be able to diagnose the exceptional talents or deficiencies and to show the road to their improvement. . . . The doctrine places practical limitations to the extent and rapidity of institutionally-sponsored social change. . . . Not only did it argue against levelling social tendencies from below, but also against domination from above by those who owed their position to inherited rank rather than merit. . . . This was a social philosophy that was made to order for those who were, or who felt they were, upwardly mobile

The diffracted response to phrenology by Edinburgh's professionals—mainly physicians, but also those in the law and clergy—lends further support to Shapin's and Cooter's interests-based explanation. As DeGiustino notes, "for every surgeon who rejected phrenology, there was one who embraced it."[28] In occupational terms, professionals for and against phrenology would seem to have few social interests capable of dividing them over the putative truth or falsity of its claims. But other interests pulled them apart: in 1830, physicians opposing phrenology were generally older, more secure in their careers and incomes, perhaps more comfortable with extant theories of mind and brain than those who endorsed it. Their biographies reveal other differences: critics often entered medicine after a gentlemanly education (sometimes at Oxford or Cambridge) enabled by the wealth and title of their families. Opponents were not only younger and less established professionally, but typically had struggled to pay for medical school and were more likely to have apprenticed themselves to an apothecary. Shapin concludes: "Phrenology furthered the symbolic expression of disaffected groups in Edinburgh, insofar as it could be seen as the 'not-x' to the 'x' of [resented] elites."[29]

Shapin and Cooter map the acceptance or rejection of phrenology onto the social interests of diverse strata in early-Victorian Edinburgh society, and show convincingly that the truth of the doctrine depended in large measure on what good it might do you—not just academically, but pragmatically, politically, and economically. It is quite another matter to investigate how each side made their case—how vocal opponents and advocates of phrenology sought to convince everyone else that their position was right—a question that brings us back to cultural cartography. Both sides drew rhetorical maps of the culturescape, each making a different space for science that put phrenology in a place likely to advance their separate interests. My argument is this: cartographic depictions of science in reconfigured culturescapes became the rhetorical means to make sense of phrenology and of its implications for diverse audiences in early-nineteenth-century Edinburgh. Before deciding whether phrenology was true or false, useful or useless, chair-worthy or not, town councilors and anyone else needed a cultural map on which they could

---

in early Victorian society. It reinforced their demand for admission to the ruling class and still resisted the extremists who would abolish all ranks in society" ("Popular Science and Society," 5–6).

28. DeGiustino, *Conquest of Mind,* 40.

29. Shapin, "Politics of Observation," 145; cf. Cooter, *Cultural Meaning of Popular Science,* 42–44.

locate George Combe and his claims. Mapping out the boundaries for science became a prerequisite for understanding the meaning of phrenology and its pragmatic implications for one's life and interests. So where was science in this episode? Where were its borders, who were its neighbors, what were its landmarks?

Critics and supporters of phrenology could agree that science was a valued space, capable of surrounding claims with cultural authority. But just who and what went inside? That depends on how one drew the boundaries. By 1836, George Combe faced an uphill battle to get inside science, following his consistent and unyielding rejection from the gatekeepers of "official" science (university-based anatomists and mental philosophers). But nothing required Combe to conform to his critics' definitions of science, so in effect he shaped his own place—with two important consequences. First, new bounds of science were created with criteria that justified the inclusion of phrenology within. Rather than cut phrenology to fit the pattern of science prepared by the intellectual and cultural old guard, Combe threw out the pattern and offered up a new science that would extend its cultural authority to his claims, doctrines, and programs. But if this reconfigured culturescape was to become something more durable than a pipe dream, and if the reality of phrenology as truthful and useful science was to become an enduring feature of social structure, Combe needed allies—obviously not to be found among Edinburgh's academic and cultural elite.

The second consequence of Combe's boundary-work was to extend the frontiers of science outward into once autonomous regions of politics, religion, metaphysics, and common sense, in order to get other audiences interested in his project. The old kind of science proffered by William Hamilton and the *Edinburgh Review* was so confined, so circumscribed, that huge chunks of Edinburgh society (and beyond) had no reason to be interested in it and no hope of actually entering its halls. Combe's map stretched out a vast terrain for science, capacious enough to hold the masses of people now given a reason to learn about *this* science and even to become a part of it.[30] But Hamilton and his friends had their good reasons as well for drawing in the boundaries of science tight enough not only to exclude (and thus delegitimate) phrenology, but also to preserve

30. Cooter says this well: "[Supporters] had come to believe that science could provide them with easy solutions to problems of existence. They went to phrenologists and they furtively studied the science because they were told that if they did so they might . . . be able to rise above the misery of their station. . . . Phrenology can so aptly demonstrate the meaning and significance attached to science and scientific movements" ("Phrenology," 228).

walls between their science and other cultural spheres of politics, religion, or public opinion. Testimonials prepared for the town council by supporters of Combe and Hamilton throw these geographic cultural differences into high relief, as we shall see.

This cartographic reading of the Edinburgh phrenology debates challenges earlier interpretations by Davies and Cantor. According to these historians, by 1836—when Hamilton and Combe took their long-running disagreements into the competition for the chair of logic and metaphysics—scientific debate over the facts of phrenology had more or less been settled, and what remained was political debate over proposed social reforms. Davies writes: "Like Darwinism, phrenology left the laboratory and thereafter its 'proof' lay in debating forums and public acceptance, not in scientific experiments. . . . Its enthusiasts, as well as its critics, insisted on discussing the discipline not as a science or a body of data but as a philosophy." Cantor dates the supposed transition: "With the publication of *The Constitution of Man* in 1828 [Combe's] interest in establishing the scientific basis of phrenology waned. Instead, he followed Spurzheim in concentrating on the social, political [and] religious implications. . . . To all intents and purposes, the Edinburgh debate over the scientific and philosophical basis of phrenology ended in 1828. Thereafter, the phrenologists progressively became more involved in examining the subject's social implications."[31]

For several reasons, Davies and Cantor's position is off the mark. If one takes seriously the idea that boundaries between science, politics, and religion are there for participants (not analysts) to draw and move, then evidence from testimonials prepared for the 1836 competition belies their empirical assertion. Combe's supporters routinely describe phrenology as science, and I would argue that their word choice is anything but incidental. It was vital for Combe's candidacy that phrenology be clothed in the cultural authority of science: the efficacy of its social programs hinged on the credibility and facticity of correlations between mental faculties, brain regions, and skull protrusions. Shapin agrees that even into the late 1830s, "its practitioners made phrenology out to be a science and offered proofs of its scientific character."[32] But why? I disagree with Shapin's conclusion that locating phrenology in science was "an attempt . . . to conceal the role of their social interests. . . . Credibility may be secured by the production of apparently naturalistic knowledge,

31. Davies, *Phrenology,* 162; Cantor, "Edinburgh Phrenology Debate," 202.
32. Shapin, "Homo Phrenologicus," 53.

in which it is impossible for opponents (or historians) to discern social interests."[33] That may be a distinctively modern view of how scientists often gain credibility for claims, insisting on their objectivity, neutrality, and givenness in nature. But Combe was offering a different kind of science, one where credibility was not severed from interests but grounded in them. It was precisely the incorporation of interests and values within a rescribed space for science that enabled Combe to keep his claims alive as truth by attracting allies and converts. "Interest-free science" was in fact to be found only on maps by supporters of William Hamilton, and they were criticized by pro-phrenologists for just that reason.

Davies and Cantor show themselves as essentialists on the demarcation question, a view that discourages sociological retrieval and analysis of the endlessly negotiated boundaries of science. Davies in effect takes "experiments in laboratories" as the criteria separating science from politics and other cultural realms. But why should sociologists privilege that particular definition of science—which, in this episode, happened to be part of the position taken by William Hamilton's supporters? Combe saw nothing unscientific about taking phrenology out of university rooms or laboratories to the masses gathered at theaters, workingmen's institutes, or even prisons. The Edinburgh phrenology controversy asked whether science was something only university anatomists or philosophers did on their home turf to advance knowledge, or something many others could do at other venues to advance pragmatic political or moral ends.[34] So much hung in the balance: nature, truth, social reform, and the chair of logic and metaphysics.

AN UNSETTLED SETTING

Before turning to the testimonials in detail, I pick up several disparate threads that come together at the moment of the chair's vacancy in 1836. The central message is this: the Edinburgh Town Council was asked to

33. Shapin, "Politics of Observation," 168. Cooter picks up the idea: "If the phrenological expression of human nature was to effectively serve bourgeois hegemony, it would have to gain the appearance of being more scientific than ideological" ("Deploying 'Pseudoscience,'" 251).

34. Cantor at least recognizes that more than one science was then afloat: "Opinion in Edinburgh appears to have been polarized between those who adhered to established norms in science, and those who subscribed to the new counter-culture which developed its own standards of scientific explanation" ("Edinburgh Phrenology Debate," 201). But he is careful to distinguish the early obviously scientific arguments over, for example, anatomical evidence for discrete regions of the brain, from later just as obviously non-scientific arguments over penal reform and educational policy. The distinction simply would not make sense for followers of Combe.

pick Ritchie's successor amid unsettled circumstances. Many historians have called attention to the upheavals in Scottish society and culture at this time—in politics and at the university—and I shall trace these out. It would be quite at odds with the theoretical spirit of this book, however, to see such "rapid social change" as the result of faceless structural forces, such as professionalization or industrialization. The unsettling of Edinburgh came at the hand of identifiable groups and individuals who had cause and means to shake things up. And few in 1836 were more unsettling than George Combe.

As it happens, both Combe and William Hamilton were born in 1788, but they share little else. Combe was one of seventeen children born to a lowland Scottish brewer. He was a sickly child, but managed to survive a cruel master during high school to enter the University of Edinburgh, where he studied for three years. In 1804, at age sixteen, George became a clerk in preparation for a career in the law. Eight years later, he established a legal practice, but close family ties required that he spend time running the brewery and helping his siblings through school (brother Andrew, George's closest friend, became a physician). Spurzheim visited Edinburgh in 1815 to defend phrenology against John Gordon's attacks, and initially Combe—upwardly mobile young lawyer—was as scornful as writers for the *Edinburgh Review.* But Combe was curious enough—perhaps stimulated by his brother's medical education—to order cast skulls from London, and he spent the next several years ascertaining for himself the validity of phrenology. Perhaps his 1817 visit to Spurzheim in Paris was the turning point, for Combe returned to Edinburgh hot to promote the science. There is no phrenologist like the convert! Combe published *Essays on Phrenology* in 1819, started the Phrenological Society the next year, and in 1823 (at age thirty-five) became founding editor of the *Phrenological Journal.*

Cooter is probably right to suggest that Combe's unwavering attraction to phrenology was psychobiographical in part, but it was also instrumental. His father's house was Calvinist, and this new complete science of man seemed a ready answer to his "struggle for personal identity and a search for certainty." But Cooter adds that when Combe went public with his newfound truths, he discovered "an audience as anxious as himself to understand the relationship of the skulls to the science of craniology and character." Perhaps by studying phrenology and promoting it, Combe felt that he might "achieve a little fame."[35] His convictions were

35. Cooter, *Cultural Meaning of Popular Science,* 107, 108; other quotations are from the entry for George Combe in the *Dictionary of National Biography.* Cf. Gibbon, *Life of George Combe.*

tested in 1825 by attacks from Francis Jeffrey and again in 1827–28 from William Hamilton, and each time Combe rose to the defense. *The Constitution of Man* was published in 1828, but its eventual popularity was slow to come, no doubt aided in 1832 by a sizable bequest from a grateful phrenological subject, which lowered its cover price. In 1833, he married Cecelia ("daughter of the famous Mrs. Siddons"), whose fortune of £50,000 eased Combe's financial pains for the rest of his days. His choice of mate may have been decided more on phrenological than pecuniary grounds: "Her anterior lobe was large, her Benevolence, Conscientiousness, Firmness, Self-Esteem and Love of Approbation amply developed; whilst her Veneration and Wonder were equally moderate with his own." Predictably, the marriage worked. Combe was forty-eight years old when he entered the competition for the chair of logic and metaphysics in 1836, happily married, financially secure, and extraordinarily visible from lectures, examinations, and sales of the *Constitution* then running near 2,500 copies per year.

William Hamilton took a different road to the 1836 competition. His father was a professor of anatomy at Glasgow, though he died when William was two. From the start, William was closely tied both to university life and to medicine (his initial career choice); many members of his extended family were either professors or physicians. He was an exuberant child, described during his undergraduate years as "strikingly handsome and [having] great athletic power." At age fourteen, he began three years of study at the University of Glasgow, moved to Edinburgh in 1806 to study medicine, and then on to Balliol College, Oxford, in 1807. For his superior academic performance at Glasgow, Hamilton's Oxford education was supported in part by a Snell Exhibition prize; he continued there his "prudent and temperate ways," and became known as the consummate Aristotelian scholar. Around 1811, Hamilton shifted his interests from medicine to law, took his B. A. degree from Oxford in 1811, and returned to Edinburgh in 1813 to become an advocate. At age twenty-nine, the young lawyer scoured his family history to determined that he was male heir to Sir Robert Hamilton of Preston; Sir William assumed the hereditary title not only for "social position" but also because of a strong sense of family.[36]

From all accounts, Hamilton was never much of a lawyer and lacked fluency as a speaker. It was said that he was more comfortable in the advocate's library than in Parliament House, preferring abstract thought

36. Veitch, *Memoir of Sir William Hamilton;* other quotations are from the entry for Hamilton in the *Dictionary of National Biography,* 8 (1917): 1111–16. Cf. Monck, *Sir William Hamilton;* Rasmussen, *Philosophy of Sir William Hamilton;* Stirling, *Sir William Hamilton.*

to the minutiae of the law. Evidently, Sir William read almost everything, and he quickly became known among the literary circles of the town, securing a position among the literary elite via articles in the *Edinburgh Review* (1829–36). With many of his young professional peers, Hamilton was a liberal Whig at a time when the Tories were ascendent, and those politics may have cost him a shot at the chair of moral philosophy in 1820 (John Wilson won it, with support from the conservatives). The following year at age thirty-three, Hamilton was elected to the professorship of civil history at Edinburgh. But the subject was not required for any degree, and even though Hamilton managed to attract twenty or thirty students to the course, the pay was poor to nonexistent.

Hamilton gave up lecturing about the time that he first made serious inquiries into phrenology in the early 1820s; perhaps he thought that his earlier medical training would enable him to judge for himself the anatomical claims of Gall and his epigone. These studies led Hamilton into his first bout with George Combe in 1827–28, although after its indecisive end, Hamilton engaged in no further anatomical or physiological research of any kind. Hamilton married his cousin Janet Marshall in 1828, who became his amanuensis and "cheered him through his long period of declining power" in later life. In 1832, Hamilton was appointed to the minor legal office of the solicitorship of the tiends, and by the time he successfully pursued the chair of logic and metaphysics four years later, he had secured his reputation as an intellectual-about-town.

Combe's and Hamilton's lives first crossed in 1827–28, in a battle over the truth of phrenological claims about mind, brain, and skull. Although Hamilton was in no sense a "professional" medical man, perhaps on the basis of his early medical training (and no doubt with help from physician friends), he carried out anatomical research on the head. His summary judgment differed little from what Gordon and Jeffrey had been writing for two decades. Phrenology is "the most tormented and misshapen of all the systems which have lately sprung from the teeming womb of impudence and conjecture." The gauntlet had been flung down: "[Phrenologists'] reasonings are incomplete and preposterous—their systems incongruous and subversive of itself—their nomenclatures confused and barbarous, and their classification of faculty radically absurd. . . . We know that the observations are false, and the observers themselves unworthy of belief."[37]

37. *Sir William Hamilton and Phrenology, An Exposition of Phrenology shewing the complete inefficacy of the objections lately advanced in the Royal Society, and the Real Grounds on which the System ought to be assailed* (Edinburgh: William Hunter, 1826), 4, 10.

Hamilton's objections were several, mainly consisting of assertions that phrenologists had got the brain and skull all wrong. Although there was little explicit mention of political programs (perhaps because Hamilton and Combe held similar views about social reform), the debate itself is a marvelous example of the politics of reality making. Hamilton challenged the idea that the brain was a differentiated organ: "It has never been shewn, and never can be shewn, that the brain is much more than a homogeneous substance . . . just the same sort of vessels and functions, the same pulpy and fibrous textures." The supposedly distinct regions of the brain were the phrenologists' own invention, "having chalked out a series of fanciful formations." The correlation between organ size and behavioral capacity was dismissed with a hypothetical question that answered itself: "Has it ever been discovered that the delicacy of taste might be presumed from the size of the palate or the tongue?" Moreover, some fundamental human functions—the elementary senses—had been left off phrenologists' roster of faculties: "The man of phrenology would not be able to hear, because the necessary organ had not been planted in his skull . . . but he would be an admirable judge of music because he had been freely gifted with the faculty of TUNE!"[38] Hamilton apparently falsified a phrenological prediction about gender differences in organs for sexual appetite that was grounded more in Victorian culture than in nature, offering evidence that the cerebellum—seat of sexual activity—was larger in female skulls than male, just the opposite of what phrenology claimed. But he chose to devote most of his anatomical critique to the question of frontal sinuses. Existence of these spaces between brain and skull, which Hamilton suspected to cover about one-third of the phrenological organs, denied the possibility that relative size inside would be mirrored by protuberances outside.[39]

Hamilton expressed his indictment of phrenology both privately and publicly. He wrote to George Combe in April 1827: "So long as phrenology is the comparison of two hypothetical quantities, a science of proportion without a deterministic standard and an acknowledged scale; so long as it can be maintained that its facts even if not affirmative constitute only a partial induction which can never represent the universality of nature; I deem it idle to dispute about the application of a law which defines no phenomena and truth of a hypothesis which has no legitimate

38. Ibid., 14, 15, 13, 25.

39. On the issue of frontal sinuses, see Shapin, "Politics of Observation," 149–50, 239; on Hamilton's attack in general, cf. Cantor, "Edinburgh Phrenology Debate"; Grant, "Combe and the 1836 Election."

conclusion."[40] The debate is here opened up to questions of methodology: phrenologists had always argued that the important fact was the relative size of faculties and organs, but Hamilton felt that without a proper measure for determining absolute size, the enterprise was something of a glass-bead game. He challenged as well the representativeness of the skulls on which phrenologists based their theories, suggesting that they might have oversampled old and deformed cases. Hamilton went somewhat public with his charges in presentations at the Royal Society of Edinburgh in 1827, and later in a public lecture at the university. But only "somewhat": the rules of the Royal Society prohibited open debate during their meetings, and Hamilton refused to publish under his own name the complete body of his researches. Moreover, Combe never really had a proper opportunity to defend his science at either venue. For example, although Hamilton invited Combe to prepare a rebuttal following his university lecture, the university senatus denied Combe the chance to speak—ostensibly because he was not a member.

But Combe's message did get out, privately in letters to Hamilton, and publicly through the *Phrenological Journal* and presentations to friendlier audiences. He challenged Hamilton's anatomical expertise (thus seeking to undercut the authority of his claims about an undifferentiated lump of brain), and threw back in Hamilton's face the charge of unrepresentativeness in the selection of skull samples. But the strongest message in Combe's reply was that Hamilton was not a disinterested and neutral judge of phrenology, and his partiality against the doctrine made it impossible for him to see what phrenologists routinely saw. The debate turned from nature to those who would be competent and unbiased enough to read it properly, a move often seen by sociologists in scientific controversies.[41] Combe wrote to Hamilton: "You would consider phrenologists as partial judges, while I would regard non-phrenologists as unqualified to decide, through deficiency of elementary information. How is this to be settled?" Combe wanted to put the evidence before as wide and large an audience as possible, so that ordinary people could decide

40. William Hamilton to George Combe, April 1827. All correspondences in this chapter may be found in the George Combe collection at the National Library of Scotland, Edinburgh. My thanks to Professor David Edge for assisting with my access to these and other materials.

41. Hamilton invoked an "empiricist" repertoire against Combe's "contingent" one: "I regret extremely that you will not consent to try the truth of the phrenological anatomy by an appeal to what must be considered as the authentic evidence of nature" (Hamilton to Combe, April 27, 1827). Hamilton wanted nature to decide, but Combe knew that people decided nature—and he had to get the "right" people as arbitrators. On repertoires, see Gilbert and Mulkay, *Opening Pandora's Box*.

for themselves the merits of each case: "The object of both of us ought to be to enlighten and convince the public on the subject in dispute; but how can this be accomplished by a private discussion before individual arbitrators. Their decision, whatever it might be, cannot carry conviction to the understandings of those to whom it might be proclaimed, and in a matter of philosophy, no one thinks of believing on mere authority, if fact and argument are within the reach of his own mind."[42] In the end, Hamilton won the battle over protocols, and the matter was set before three umpires, all from the medical faculty at Edinburgh. The outcome was a genuinely Scotch verdict: though the judges said little in favor of phrenological claims, they dismissed Hamilton's evidence as "insufficient to sustain his allegations."[43] "Not proven," and probably few minds were changed by the confrontation of claims and evidences.

It should not come as a surprise, perhaps, that facts of nature could not be settled in this court, for the society and culture in which the acid test was sought itself rested on shifting sand. *Edinburgh* was as unsettled as theories of brain and mind, which made it hard to see human nature clearly and decisively during the years leading up to the 1836 competition. What Cooter writes of the 1820s probably applies as easily to the next decade: "Virtually every cultural tissue that in one way or another had bonded preindustrial society was in a state of dissolution, while the new cultural tissues that would bind industrial society were only at the point of their formation."[44] The city's population had doubled between 1800 and 1840 (to 160,000 people), and the government fell into bankruptcy in the early 1830s. The most consequential changes affected reallocations of power and authority—and the criteria through which possession of those resources was legitimated. Saunders has described Edinburgh then as a "democratizing and professionalizing society," two movements that would be reflected in testimonials written for Combe and Hamilton as they pursued the logic chair. An ancien régime anchored in landownership or hereditary wealth and title was losing its grip on the powerful institutions of the city.

To call this "democratizing" is not to suggest that no new elite had arisen to replace them, but only that the grounds for entry into higher strata had changed. Class distinctions were ever so palpable, but now those nearer the top were likely to be professionals engaged in medicine, the church, law, education, banking, or the military. And there were more

42. Combe to Hamilton, April 28 and May 7, 1827.

43. Grant, "George Combe and the 1836 Election," 182.

44. Cooter, *Cultural Meaning and Popular Science,* 69.

than a few of them: in 1840, about 10 percent of the male Edinburgh population fell into the occupational category "capitalists, bankers, professionals and other liberally educated men," as contrasted with only about 2.6 percent of the equivalent population in the manufacturing hub of Glasgow.[45] At least in theory, individuals were now to be judged and rewarded by education and talent, not by accident of birth or material possessions. That meritocratic theory was in effect put into practice through liberal social reforms; the rising professionals demonstrated increases in their Whig-inclined political clout by initiating many programs of "civic patriotism." The perception that education and skill could now legitimate power and authority perhaps predictably lead to the increased value of education and training for all social strata, and to a growing confidence in scientific, technical, and "professional" solutions to social ills that were hard to overlook.

The 1830s have been described as a time of "popular enlightenment," as middle and working classes sought instruction in political economy and phrenology, from lectures at mechanics' institutes and from relatively cheap books promising self-improvement. In response to epidemics of cholera and other urban maladies, "the authority of technical knowledge, not common sense, was invoked. . . . Charity became organized but it also tended to become scientific, optimistic and catholic. . . . It spoke with particular authority in the activities and writings of the medical men." Cures for moral diseases, like biological ones, were zealously sought in the applied sciences of human behavior. The embrace of scientific authority was certainly not an overnight change: doctors struggled to convince others that their cures for cholera (or immorality) were more efficacious than the time-worn traditions they labeled superstitious. Still, "the majority of the population took their doctor's advice . . . and listened to their ministers."[46]

This popular enlightenment, the rise to power of a professional class, a faith in science and expertise to solve social problems brought the spirit of liberal reform to the doorstep of the university. Once counted among the leading universities in Europe, Edinburgh was apparently slipping—both in its intellectual stature and in the ability of its training to respond to changing conditions. Between 1820 and 1836, the number of students fell by more than 600, and critics suggested that "the teaching at Edin-

45. J. Stark, *Picture of Edinburgh, containing a description of the city and its environs,* 6th ed. (Edinburgh, 1838), 3.

46. Saunders, *Scottish Democracy,* 96, 163, 183. Cf. Youngson, *Making of Classical Edinburgh.*

burgh was ceasing to attract." Many problems stemmed from a struggle for control of the university between its patron the town council and the faculty senatus. The two often disagreed on what should be done to improve higher education in Edinburgh, and so by 1836 little real reform in university structure or procedures had been accomplished. Magistrates of the town council saw themselves in a position of "general and ultimate supervision," and indeed their College Committee was responsible for appointing the principal and some professors, as well as deciding curricular matters such as courses required for designated degrees. But questions raised by the faculty about the extent and legitimacy of the town council's control over university affairs brought to a head the contradiction between popular enlightenment and professional authority: Should the organization of the university in effect be placed in the hands of public opinion, as expressed through political bartering and party affiliations of the magistrates? Or are academic matters so arcane that mere politicians have no hope of grasping their nuances, something only expert academicians themselves can hope to do? Was university governance a matter for popular or professional control, for amateurs or experts? These questions had certainly not been decided by the time Combe and Hamilton presented their testimonials to the councilors in 1836.

Each side sought a different university. The magistrates, perhaps under pressure from powerful professional associations and from "enlightened" middle classes, wanted the curriculum to reflect the latest intellectual currents and practical methods. The challenge was to fit new areas of scholarship into increasingly outmoded structures of disciplines and chairs—made difficult by the fact that established faculty had strong financial reasons to oppose such changes. A university chair was in effect a monopoly awarded to a faculty member for the teaching of a subject, and though each carried a small stipend, most income was generated from class and examination fees paid per head. Any changes in the curriculum—say, creation of a new chair in a closely related field, or elimination of a subject required for degree—could threaten a faculty member's livelihood. And so "the recognized teachers in the faculties tended to form a conservative alliance, jealous of their prestige and legal position and concerned to defend the traditional disciplines and standards of their institutions."[47] The magistrates heard complaints that some subjects had simply grown too large or diverse to be covered by a single chair in a single course and, facing the recalcitrant faculty, provided support for

47. Saunders, *Scottish Democracy*, 331; cf. 317, 329.

extramural chairs outside the formal university structure to better provide training in those fields (a move predictably resented by the professors). With this struggle as backdrop, the magistrates—in deliberating a successor to David Ritchie—would be especially sensitive to candidates' descriptions of just what subject matter would be included in the logic and metaphysics course.

Crosscutting the town council's wish to keep Edinburgh abreast of recent intellectual developments was a desire—perhaps more salient—to restore or extend distinctive and traditional Scottish approaches to a subject. Nowhere was this latter feeling so high as in discussions of the glorious past and uncertain future of Scottish philosophy. On the strength of renowned works by Thomas Reid (1710–96), Thomas Brown (1778–1820), and Dugald Stewart (1753–1828), Scotland enjoyed in the late eighteenth century a philosophical golden age that had dulled considerably by the 1830s. The Scottish school creatively attacked philosophical and psychological problems of the first order, and their efforts were widely recognized as fundamental. Reid sought explanations for how raw sensations became perceptions of real objects, and Brown introduced the concept of "suggestion" (a precursor to "association") to interpret the process. Stewart is appreciated for consolidating the ideas of Reid and Brown, popularizing the message for lay audiences. Recall that this "common sense" philosophy developed the idea of distinctive mental faculties (later picked up by Gall and incorporated into phrenology) and argued that they could best be understood through systematic reflection on the operations of one's own mind (a method phrenologists dropped in favor of craniology). But the tradition was "noticeably waning" by the time Hamilton embraced it in the 1830s, in part perhaps because of the ineffectiveness of his predecessor in the chair of logic and metaphysics. David Ritchie, whose occupancy of the chair stretched from 1808 to 1836, was "a tall, big-boned, strong man, with a powerful rough voice and great energy, though little polish in his delivery [and] more illustrious on the curling pond than in the Professorial Chair."[48] How, then, to restore Scotland's philosophical claim to fame? The question was also very much on the minds of the Edinburgh Town Council in 1836, and historian G. E. Davie writes that "in these debates over appointments, a central and recurring issue was how far the reforming professor was to be a

48. In Grant, *Story of the University of Edinburgh,* 332. Cantor notes the waning of the "common sense" school ("Edinburgh Phrenology Debate," 218). On the history of this tradition, see Boring, *History of Experimental Psychology;* Davie, *Democratic Intellect.*

man who would keep alive the distinctive Scottish approach to the subject."[49]

The town council itself did not escape pressures for reform. Just a year before the competition for the chair of logic and metaphysics, the council was reconstituted as an elected body. Before 1835, nineteen of the thirty-three magistrates were chosen by their predecessors, and the others were chosen by incorporated trades and survivals of historic guilds. This selection procedure left virtually unrepresented merchants, advocates, the medical colleges, the university, and the clergy—that rising elite of professionals whose number and cultural authority had not yet translated into formal political power. That changed: after the municipal reform of 1835, for example, the majority of the College Committee were drawn from the liberal professions—four lawyers and four physicians among the eleven members. However, this shift to a popularly elected council fueled arguments that such a body was even less able to make decisions about university affairs, since now more than ever the magistrates would need to be responsive to the whims of public opinion. While these political reforms were under consideration in 1834, Sir William Hamilton argued in the *Edinburgh Review* for the autonomous control of internal academic affairs by the faculty, on grounds that the town council "is a collection of individuals—numerous, transitory and obscure, and the function [of electing chairs] was an appendage altogether incidental to their office. . . . Any ignorant religious party may calculate on the suffrages of its adherents in the Council in favor of any candidate whom it may propose."[50] Perhaps the council newly elected for 1836 chose not to remember this opinion when they selected Sir William for his chair.

Thus, the town council's consideration of David Ritchie's successor was, amid these unsettled circumstances, hardly a routine matter. It was full of significance for how Scottish philosophy, politics, and education would eventually get settled. According to an eyewitness account of the meeting in which the council cast its final vote, "not only were the eyes of the citizens fixed upon them, but the eyes of all England who were interested in the advancement of science—nay . . . the attention of the most distinguished philosophers both in Europe and America was fixed upon the decision of this day."[51] The procedures for election were certainly more straightforward than the criteria for judging the worthiness of any particular candidate. In all, fourteen men presented themselves

49. Davie, *Democratic Intellect*, 105.

50. In Saunders, *Scottish Democracy*, 321; cf. 94, 322–23.

51. *Phrenological Journal* 10 (July 1836): 227.

as candidates, but only four eventually received votes: Hamilton, Combe, Patrick Campbell MacDougal (later given the chair of moral philosophy at Edinburgh), and Isaac Taylor (who was publicly endorsed by the Kirk).

It was customary at the time for candidates to make testimonials available to the magistrates, securing letters from admirers attesting to their qualifications for the chair.[52] Combe and Hamilton took different tacks. The phrenologist assembled well over one hundred letters from supporters, including many from physicians, churchmen, officials at prisons and asylums, and philosophers—from throughout Europe and America. Combe published the testimonials as a book and presented the town council with a copy on May 10, 1836, at the start of the election process. He also gave copies to judges, ministers, and university professors, and by one account "there can be little doubt that their appearance caused some stir in Edinburgh, where Combe's wider, international reputation was little known."[53] Not surprisingly, the *Phrenological Journal* crowed that "the testimonials are so strong and numerous that we are not surprised to learn that they have astonished many who previously thought phrenology too absurd to merit serious attention."[54] The journal's defensiveness extended to the candidate himself, and may in part account for what might be considered overkill in the number and boring sameness of Combe's testimonials. The idea, apparently, was for Combe's testimonials to swamp the town council with evidence of the breadth and diversity of phrenological supporters—allies so numerous that critics writing for the *Edinburgh Review* would appear as a tiny minority opinion. Ironically, critics accused Combe of grandstanding and of trying to cash in on the popularity of the published testimonials. Combe wrote to Richard Carmichael shortly before the election: "The weight of the authority contained in them is operating as a rebuke of their own conduct, and they are loudly and bitterly abusing me for the 'exceeding bad taste' of 'selling' my testimonials and pushing them all over Britain." Combe confided to Carmichael that he was, in fact, making no money on their sale, and that his goal was changing minds—especially those of the town council—not bringing in money. To make the point, he distanced himself from his science, the truth of which should be the issue at hand: "[The testimonials] relate to a science much more than to me."[55]

52. For a historian's use of similar testimonials, cf. Crosland, "Assessment by Peers." On the general subject of attestation in science, see Turner, "Forms of Patronage."

53. Grant, "Combe and the 1836 Election," 180.

54. *Phrenological Journal* 10 (July 1836): 103.

55. In Gibbon, *Life of George Combe*, 322.

Hamilton pursued the chair self-confidently, from an obvious position of strength. His relatively few (twenty-one) testimonials came from Edinburgh's cultural elite (including professors from the university and ministers at its important churches), his teachers at Oxford, and three continental philosophers. Nine of these testimonials had been prepared for Hamilton's candidacy for the moral philosophy chair sixteen years before, including a short one from Dugald Stewart. Only twelve new testimonials were prepared for the 1836 competition: clearly, Hamilton preferred to rest his case not on the number of those who would support him but on their high standing and reputation within powerful institutions. His biographer recalls that "stronger evidence of capacity and attainments was never presented a body of electors."[56] Hamilton refused to canvass the magistrates one by one, trying to persuade each of his worthiness for the chair—though this was standard operating procedure. Indeed, his confidence in his candidacy borders on chutzpah: "I am assuredly most anxious to obtain this Chair; but I am ambitious of it not as a boon granted, but as a right recognized. I only ask—and I would only accept the appointment, on the grounds of superior qualification. To mendicate the votes of Patrons by the private solicitation of myself or friends . . . are proceedings which I not only scorn, but of which, as morally dishonest, I trust I am incapable."[57] Clearly, Hamilton believed that he had the strongest hand.

If Hamilton's confidence was justified, did Combe's candidacy ever have any chance of success? Boring suggests that "he was a strong candidacy for the chair," but Grant says that "Combe claimed little hope of success."[58] His letters during the several months leading up to the vote reveal a beleaguered man who felt misunderstood and unfairly attacked from all sides. "The clergy are bitterly hostile, because I wrote the *Constitution of Man,* the Tories are unfriendly because I am a Whig, and the Whigs hate phrenology because they have pitted their own reputation as philosophers against its truth; so that there is only truth and reason against a whole host of prejudices."[59] But Combe's hopes were soon raised by the widespread attention accorded his testimonials, as he wrote to his brother Andrew: "I hear that the interest excited by my testimonials is daily increasing, and I am told that if the election had been delayed

56. Veitch, *Memoir of Sir William Hamilton,* 185.

57. William Hamilton to the Edinburgh Town Council, 1836, 4.

58. Boring, *History of Experimental Psychology,* 56; Grant, "Combe and the 1836 Election," 180.

59. George Combe to Dr. Robertson (Paris), May 26, 1836.

for three months longer . . . I would certainly have succeeded and that even as it is, I have the chance equal to that of any other candidate." Combe got part of his wish when the election was postponed for three weeks, from June 28 to July 15, prompting him to write: "I am told that my chances for the Logic chair are rising, and that it is very uncertain who will succeed."[60] He also picked up the support of two papers, the *Scotsman* and the *Spectator.*

Thus, in his own mind, and no doubt in the minds of those who wrote testimonials on his behalf, Combe was a serious candidate—even if his final tally of votes fell far short of victory. Reality set in three days before the election, as Combe's reckoning of the magistrates' votes implied for him a thorough defeat—and for Hamilton as well! "I am informed now that Sir William Hamilton has a poor chance of success, and that Mr. Isaac Taylor will be the future professor." Combe does not miss the chance to dump on his old foe: "I shall be better pleased at his success than at Sir William's, because he has some originality and boldness, and in his address he stated distinctly that the old system is useless." Now certain of defeat, Combe expresses bitterness at his rejection by Whigs and just about everybody else who mattered: "With the exception of my very small circle of friends . . . I am left absolutely alone, to abide . . . the pitiless storm of bigotry and prejudice."[61]

The vote on July 15 finally brought Combe's deep mood swings to an end, though things did not turn out exactly as he had anticipated. In discussion accompanying the nomination of candidates, the magistrates reflected ambivalent views about their preparation for judging scholarly horseflesh.[62] Bailie Macfarlan, while nominating the Kirk's choice, Isaac Taylor, took the populist view that Everyman could judge the *character* of each candidate for themselves: "There are other testimonials within everyone's reach, and upon which every man of common sense was able to form a correct opinion. These were the public works of the individual, and these works they could judge of with as great freedom as of the testimonials of learned and scientific men." This was a slam not only on

60. George Combe to Andrew Combe, June 14, 1836; George Combe to James Stewart, June 24, 1836.

61. Combe letter, July 11, 1836.

62. The town council was not unerring in filling professorial chairs. It appointed Alexander Munro *tertius* to the anatomy chair in 1798. One student said of him: "He used to read his grandfather's lectures written about a century before"; and Charles Darwin added: "Dr. Munro made his lectures on human anatomy as dull as he was himself, and the subject disgusted me." Unfortunately for everybody, Munro *tertius* stayed in the chair for forty-eight years! No wonder there was clamor for reform of the university (Comrie, *History of Scottish Medicine,* 240).

Combe's many supporters, but also Hamilton's fewer but especially expert ones. Yet another councilor, Mr. Deuchar, justified his vote against Combe and for Taylor on grounds that the town council was not the proper body to adjudicate the truth of phrenology: "While the great majority of learned men are opposed to the doctrine, it would be great presumption in this council of thirty-three, and indeed a gross dereliction of duty, were they to suppose that by joining the minority they could turn the scale of opinion, and thereby establish phrenology."[63] No such hesitation is found in Dr. Neill's nomination of Combe: "I am deeply persuaded that logic will never make sure progress til it be taught on [phrenological] principles. . . . Ten years ago, [phrenologists] were treated with contempt and ridicule by the *Edinburgh Reviewers*, but now those learned gentlemen maintain a most respectful silence. The day for twitting about bumps has gone by." Convener Dick, a veterinary professor not previously expected to support Combe, offered a second: "While I cannot properly be called a phrenologist, I believe that the principles are founded in nature; and as Mr. Combe had proved himself a successful and popular teacher, and a sound physiologist and logician, I shall . . . vote for him." A Hamilton supporter responded: "For the sake of the university, for the credit of the Council, and to meet the expectations of all men of science, he trusted Sir William Hamilton would be the choice on this occasion."[64] On the first ballot, Neill and Dick went for Combe along with Thomas Milne; on the final ballot, with one member absent, Hamilton took eighteen votes to Taylor's fourteen.

How did Hamilton win, and Combe lose? It would be reductionistic to suggest that the election turned on religion or politics or science alone; rather, the outcome is a measure of how the town council sorted out the jumbled arrangements of all three cultural territories—using maps offered them in the candidates' testimonials. Indeed, the candidates were given a religious litmus test, and Hamilton at one point almost failed: "Hamilton was not known except from his connection with the *Edinburgh Review,* and it was well known that the theological philosophy of that journal was not in high repute." The winner was saved here by another councilor, who argued that he was tired of "making a stalking-horse of religion."[65] Party politics surely played a role as well, although on this score there was little to distinguish Hamilton from Combe—both

63. These eyewitness accounts of the council's meeting on July 15 come, with all expected prejudices, from the *Phrenological Journal* 10 (July 1836): 230–31.

64. Ibid., 227–28, 234, 227.

65. Ibid., 232.

Whigs and both supporting the same kind of liberal reforms then in favor with what had become the majority party. The real question is why the predominant Whigs went for Hamilton instead of Combe, and the answer brings together in the town council that day not just politics and religion, but also the popular enlightenment, the power of experts, the golden age of Scottish philosophy, penal reform, the future of science, mind, brain, skull.

The key to Hamilton's victory is found in his testimonials, just as, ironically, the key to Combe's defeat is found in testimonials written by those who thought they would help. What cultural map was drawn by Hamilton's testimonials? Everything in its place: the church is blocks away from the university, which is in turn far from the infamous Bridewell prison. Hamilton's testimonials in effect reproduced brick-and-mortar Edinburgh, preserving in cultural form geographic distances between the institutions of church and politics, preserving the space between different scientific disciplines at the university, preserving the separation of the Royal Society from the Canongate slums. Science keeps to its own place, in the university and the Royal Society, the preserve of a learned few. Hamilton's attestors did not want to rock any harder a boat that had, for several decades, been rocked furiously. But why would such a conservative map—leaving intact the well-bounded spaces of the town's fundamental social institutions—appeal to those Whig magistrates? After all, through the 1820s, these rising professionals pushed for liberal social reforms that sought to bring greater democracy and professionalism to Scottish life. Their support for a culturally conservative map is a measure of how far they had moved into positions of power: now the goal was to consolidate their recent gains, to stabilize and legitimate ideologically and structurally the grounds for their ascension to authority and privilege. With control in their hands, the map, the philosophy, and the candidate that preserved the emerging status quo appealed more than a rival who wanted to continue to shake things up.

That is just what Combe's maps did. Phrenology may have scared the Whig magistrates no end, or at least enough to vote against Combe—not because of the falsity of its claims, and not because its methods and logic were pseudoscientific. These modern meanings of phrenology had not yet been settled, though the council's eventual vote surely helped to move Combe's science in that direction. Testimonials for Combe again unsettled the Edinburgh culturescape, which was just beginning to come together in its new shape, with liberals at the helm, professionals in charge. The maps blurred distinctions between cultural institutions, which invited doubts about how the new social order might legitimate itself. Science, in particular, pushed out in every direction; phrenology forced reconsideration of the place of experts in deciding questions of fact—and of religiosity, morality, politics. Its enthusiasts harped on the uselessness of the once-gloried "common sense" tradition of Scottish philosophy. Their testimonials trashed the monopolistic jurisdiction of subject matters at the university. Whig liberals who a decade ago might have supported phrenology as part of the battery of social reforms destined to move them up the socioeconomic ladder and into power now stood to gain little from voting for its leading spokesperson. Cooter writes: "Far wiser, surely, to see the irrelevance of phrenology among former advocates . . . in terms of the accomplishments by the 1840s of what had yet to be secured by the liberal bourgeoisie in the 1820s, and what, with the most amazing social tensions and contradictions had been attained in the 1830s."[66] What once had been an ideological tool was now a threat.

How, then, did Combe come to employ such an ineffective cultural map before a town council that would decide his worthiness for a university chair? Remember how phrenology had first been received in Edinburgh some thirty years before: "ignorance" and "trumpery." Rejected by university men, established cultural elites, Kirk officials, and other arbiters of truth, Combe had sought allies elsewhere: among physicians or lawyers seeking new markets for their training, among superintendents at asylums and prisons seeking workable means to rehabilitate convicts and patients, among working men and women seeking self-awareness and self-improvement. But how to get them on board? By shaping phrenology into a science that interested *them*, that solved *their* problems better than what had up to then passed as a science of human behavior and mental functioning. And what a different science it had to become, one that no longer fit into the neat niches of Edinburgh society. It was a science as suitable for insane asylums and churches as for the Royal Society, a science that linked anatomy and mental processes, a science with truths that Everyman could easily see. Combe simply could not sell phrenology among those who had circumscribed science so narrowly, from

66. Cooter, *Cultural Meaning of Popular Science*, 97. Clearly, Cooter's view—like mine—challenges an early interpretation by Leahey and Leahey: "Phrenology's reforms, aiming as they did at the individual, presented no very deep threat to the established order, except insofar as some of the movement's partisans espoused materialism. . . . Phrenology was unconcerned with political reform and serious structural change" (*Psychology's Occult Double*, 95). On the contrary: phrenologists' redrawing of the boundaries of science demanded fundamental changes in institutional arrangements and allocations of authority.

their chairs at the university or from the *Edinburgh Review*—it took a different science to attract enthusiasts from elsewhere, a science pushing out into cultural domains (politics, religion, common sense) where it had not gone before, enrolling allies along the way. These allies, in their testimonials, then redrew for the town council a map of Edinburgh's culturescape that must have been unsettling enough to make a vote against Combe easily justified.[67]

SCIENCE SHOWS ITS VERSATILITY: THE TESTIMONIALS

So Hamilton's science won, and Combe's lost. Crucially, it was not the validity of their rival claims about mind, brain, and skull that was decisive here: lines on heads mattered less than lines in culture—preserved in recognizable form by Hamilton supporters, bent and squeezed into unnatural shapes by Combe supporters. The rhetorical cartography on display in these testimonials offer different verbal renderings of four pairs of cultural spaces: expert knowledge/lay knowledge, mental philosophy/physiological anatomy, pure science/useful applications, and science/religion.

Expert Knowledge/Common Sense

The two sets of testimonials allocate authority differently for adjudicating matters of fact: Hamilton's "elitism" contrasts sharply with Combe's "populism." For supporters of Hamilton, only the trained expert, the learned man, has the specialized skills to judge recondite matters of logic and metaphysics. Not so in Combe's cultural world: anyone can see and judge nature for themselves, and widespread acceptance of phrenology among the many is evidence of its truth (even if pointy-headed experts continue to treat it with scornful disbelief). Combe's cartography empowers ordinary people, inviting them into science and the process of fact making; Hamilton's map excludes those ordinary people as incapable of understanding ideas so abstruse that experts alone can handle them.

Though they were written in favor of the obvious front-runner, testimonials for Hamilton are remarkably defensive in two respects. First, as if he were sheepish about their paltry number compared to the flood written for Combe, Hamilton introduced his attestations by telling the town council that it is not the number that counts but their authority.

67. Combe's "translations" of the interests of potential allies into something phrenology could advance recalls Pasteur's maneuvers, as described in Latour, *Pasteurization of France*.

> But as there are now unfortunately in Great Britain few persons entitled to express an opinion on metaphysical [subjects], and as I shall neither tax the partiality of my friends for eulogy, nor found on vague generalities and incompetent assertion, I trust that the evidence I do adduce may be estimated not by the number, but by the authority, of the witnesses. . . .
>
> . . . In a department where so few are entitled to express an opinion, I have avoided . . . a multitude of witnesses." (*WH Testimonials*, 5, 7)[68]

In order to confirm the idea that Sir William had chosen his witnesses carefully, and on the basis of their demonstrated expertise in philosophical matters, the testifiers touted their own credentials. For example:

> My early, and for many years exclusive, devotion to that branch of Philosophy which includes Logic as one of its sections, and my acquaintance with the sentiments of the learned . . . will, I think, be viewed by yourself as enabling me, in some sort, to judge of the qualifications required for such an appointment. (34)

Another witness put himself in the place of spokesperson for a learned body of experts: "Nor am I now giving my own opinion merely, but that of every competent judge with whom I have conversed" (43). Hamilton is said to be a member in good standing of that small club: "He has, in fact, been hailed by some of the most celebrated philosophers of the day, as a welcome and worthy member of their illustrious family" (36).

The testimonials convey as well a defensiveness about Hamilton's putative capabilities as lecturer. As a lawyer, he had not distinguished himself by his oratorical skills in the courtroom, and indeed, in the course of the 1836 competition, critics called attention to the obscurity of his prose as evidence for his likely inability to convey thoughts clearly to his students. But witnesses turn this possibility into an asset, first by replying that the perceived obscurantism is not Sir William's fault but the result

68. The statements of Combe's supporters can be found in *Testimonials on Behalf of George Combe, As a Candidate for the Chair of Logic in the University of Edinburgh* (Edinburgh: John Anderson Jr., April 1836). Pagination continues from this book into a second volume, published about seven weeks later: *Additional Testimonials on Behalf of George Combe, As a Candidate for the Chair of Logic in the University of Edinburgh* (Edinburgh: Neill and Co., May 1836). Other testimonials for phrenology appear in *Documents Laid Before The Right Honorable Lord Glenelg, By Sir George S. MacKenzie, Bart., Relative to the Convicts Sent to New South Wales* (Edinburgh, April 1836). The other side is presented in *Testimonials in Support of Sir William Hamilton's Application for the Chair of Logic and Metaphysics, Vacant in the University of Edinburgh.* (Edinburgh: H. and J. Pillans, April 1836). All quotations from the various testimonials are cited in the text, initially by short title, thereafter by page. Names of witnesses are generally excluded; it is the substance of what they wrote that interests us here rather than their biography.

of untutored audiences, and second, by arguing that supposedly impenetrable prose is in fact a measure of his philosophical acumen. One testimonial closes the door on all but a few whose specialized training equips them to grasp what Hamilton is up to:

> I affirm that the article is so excellent, that there cannot be fifty people in England competent to understand it. . . .
>
> . . . The soundness of his views, the extensive acquaintance with philosophical systems, and the profoundness of thought which it exhibits, can only be appreciated by those who are of the metier; it is, in short, an article written for a few minds only throughout Europe, whilst to the multitude its very power and merit will render it obscure. (13, 15)[69]

George Combe made hay with the presumptuousness of this testimonial, writing in a personal letter: "Sir William Hamilton has printed a very foolish letter to the electors, assuring them that although he wrote articles in the *Edinburgh Review* so recondite that there were not fifty men in Britain who could understand them, yet that he can stoop to the humble capacity of students and that he is the man of the chair and no one else! The public are laughing at him!"[70] Evidently, Hamilton and his friends heard no ridicule, for their strategy was different from what Combe suggests—aggressive rather than apologetic. Several attestors implied that there was no need for Hamilton to talk down to the common level of understanding:

> But these were, avowedly, discussions of the most abstruse and recondite questions in psychology, and addressed only to those who were familiar with the language, and proficient in the methods and systems, of the science. They are necessarily obscure, therefore, to the uninitiated, just as a discussion between Laplace and Sir John Herschell upon some difficult question of astronomy would be. (31)

The testimonial aligns Hamilton's science of logic and metaphysics with more obvious examples such as astronomy (and below, with mathematics and philology), and the candidate is described as ready to take his place within the territory alongside other landmarks such as Herschel. Attention focused on Hamilton's papers in the *Edinburgh Review,* for these constituted the corpus of his published work. Were they difficult to un-

69. The obscure quality of Hamilton's prose is noted by Yeo (*Defining Science,* 43), who also provides this useful reminder about the 1830s: "The word 'science' had not entirely lost its earlier meaning of systematic knowledge, or *scientia*—for some people, logic, theology, and grammar were still 'sciences,' and the term was still used synonomously with 'philosophy'" (33).

70. George Combe to James Simpson, May 6, 1836.

derstand, lacking clarity? Hamilton himself told the councilors that he saw no reason to write for anyone other than the few who spoke his language:

> Nor is it easy to see why a metaphysician, more than a mathematician or philologer, should be required to bring down the highest problems of his science to the comprehension of the "general reader." . . .
>
> . . . A journal like the *Edinburgh Review* is not the place for elementary expiation . . . Its philosophical articles are addressed not to learners but to adepts. (2, 3)

But another testimonial softens the blatant elitism somewhat, by suggesting that the short length of articles in the *Review* did not allow Hamilton to fully unpack his ideas—something done perhaps more eloquently in a series of lectures at the university.

> Sir William Hamilton has not even the very slightest appearance of obscurity. His style is substantial and severe, but of perfect plainness for everyone acquainted with the subject, and not incapable of attention. . . . To be popularly clear, there is only wanting to Sir William Hamilton the space requisite fairly to develop his thoughts; and that space is not found in a Review—it is only fully obtained in a course of Lectures. (21)

The testimonials are consistent in drawing a line between what untrained audiences might see as obscurantism and the true profundity of Hamilton's ideas, recognized by all learned experts:

> No man deserving the name of philosopher or scholar, can read these papers without bearing testimony to the profundity of thinking, the metaphysical acumen, the vigorous tone of reasoning, and the stores of corroborative and illustrative learning, which they so richly display.
>
> I shall hail his appointment as one sure to be approved by all the learned of Europe in the department of philosophy. (37, 26)

But who cares about *them,* the typical testimonial for Combe would ask. His science places its truth-granting authority in the hands of ordinary folks rather than in those of experts.[71] And Combe's prose is so straightforward that it encourages more than their comprehension of its

71. "The phrenologists considered that factual evidence from nature had to be brought forward so that each individual could decide on its worth. . . .This emphasis on individual conviction implied a disregard for knowledge of experts. . . . Phrenologists resented the uncritical acceptance of evidence solely on the authority of an expert. To them, sense evidence predominated over authority" (Cantor, "Edinburgh Phrenology Debate," 215–16; cf. Cooter, *Cultural Meaning of Popular Science,* 72, 80, 174–75).

claims: "As a lecturer his language is forcible, yet plain and simple; his demonstrations are always clear and easily understood, and his arguments at once are logical and convincing" (*GC Testimonials,* 37). As Combe tells the council in an introduction to the first volume of his testimonials, the intelligibility of his discourse extends from experts to ordinary people:

> In addition to testimonials from persons of station and of philosophical eminence, I have procured several from individuals in various ranks and employments, with the view of shewing that the true philosophy of mind is calculated not exclusively to adorn palaces and academic halls, but also to recommend itself for its truth and utility to intelligent men of every grade. The pupils who attend the Logic class are the sons of such persons, and partake of their mental qualifications. (vii)

The last was no slur on Edinburgh's undergraduates, but rather an exultation of their sometimes plebeian origins—any occupant of the logic chair would have to be able to address those who had not grown up in rarified university air. Combe's witnesses echo the idea that his is a science for the streets:

> Phrenology . . . by presenting us with a true theory of mind, enables us for the first time to render Logic, as a science, useful, consistent and intelligible to all persons of ordinary capacities and attainments; which in my *humble* opinion, it can never become until it is taught on phrenological principles.
>
> I conceive that *all* classes of the community in the kingdom, and more especially the citizens of Edinburgh, have a very deep interest in the proper selection of a Professor to fill that Chair. (27, my emphasis; 121)

Evidence for the clarity of Combe's presentations is the widespread acceptance of his beliefs and methods. Data on the sales of Combe's books, and on attendance patterns for his talks, are included in the testimonials—this summary from Combe's own publisher:

> Phrenology was subjected to much ridicule, and there was little demand for works on the subject; but the state of matters for the last six or eight years has completely altered; the study has become popular; in this city, and elsewhere, courses of lectures have been attended by numerous auditories, especially of the young; and there is now a regular and increasing demand for your writings. (72)

Mention of the youthfulness of Combe's enamored audiences opens a theme common throughout the testimonials: phrenology is the new, up-and-coming philosophy certain to supplant the tired old tradition of

Reid, Brown, Stewart—and of Hamilton. The growth in popularity of Combe's talks and books attests, on the one hand, to his pedagogical skill (not altogether incidental to carrying out the duties of the incumbent chair): "Of your competency to teach the true philosophy of mind, no individual testimony can be needed, it being unequivocally attested by the unparalleled circulation of your works, and the high estimation in which you are held as a mental philosopher, not only in this country, but over a large portion of the globe" (4). But Combe's widespread acceptance attested to something even more significant than his teaching talent: from inside this populist epistemology, the more people who believed a claim, the more likely it was to be true.

> Now, the fact of a book on a subject so abstruse, running counter also to old opinions, and courting no prejudice, being so eagerly purchased by the humbler classes, seems to me to afford a strong presumption that its doctrines are in accordance with the laws of Nature, and therefore find a response in the common sense and common feelings of mankind. (58)

This witness's use of "common sense" was in no way an allusion to the old Scottish school—and indeed may even have heightened the contrast between it and the newer phrenological brand of mental philosophy. By announcing their acceptance of phrenology, witnesses were in effect ratifying the truth of its claims—they were participating *in* science—and it did not matter that they were not studied in logic or metaphysics. Still, one testimonial hedges its bets, and exempts the town council from ultimate authority to decide the credibility of phrenology:

> The appointment of Mr. Combe to the vacant chair would not necessarily imply a conviction on the part of each individual elector, that Phrenology is true; all that it would imply would be, that sufficient evidence, on the part of individuals of known intelligence and respectability who had studied it, along with the public fact of its wide diffusion and increasing popularity, had produced in their minds a reasonable presumption, on which, as public men, they are entitled to act, that the science is founded in nature, that it is about to become the standard philosophy of the age. (50)

Interestingly, immediate evidence from nature in support of phrenology—for example, craniological data, or correlations between bumps and character—is rare in the testimonials. It its place comes statistical evidence on a cultural phenomenon—growth in the number of phrenological believers: "I am now engaged in collecting statistical evidences to show the present state of phrenology and shall shortly lay before the public the most conclusive proofs that a steady conviction of the truth and value of phrenological science is now extensively and rapidly spreading through society" (6–7). At the same time these testimonials open science up to the many who accept the credibility of phrenology, witnesses also suggest that the exclusivity of Hamilton's elite science (open only to the métier) is itself a sign and cause of its obsolete uselessness. Combe taps into a widely held assumption that university faculty were conservative in their judgments of new thought systems—fearing perhaps a displacement of their students and thus their income. He drives a wedge between faculty jealously guarding their entrenched philosophies and the "enlightened" town council open to something fresh: "One of the greatest reproaches that has hitherto attached to the established Universities, is their pertinacious adherence to erroneous opinions after they have been abandoned by the general judgment of enlightened men" (v). Combe continues by empowering the town council to decide for themselves the truth of phrenology, just as he had so often empowered his audiences drawn from the humbler ranks.

> It is a characteristic feature in the constitution of the University of Edinburgh, that its Patrons [are] engaged in the active business of life; one which, although not boasting of a scientific character itself, possesses intelligence sufficient to appreciate the value and to understand the direction in which the currents of science are flowing, and which therefore is more open to the adoption of new truths than are those learned bodies, which cease to oppose improvements only when their individual members who have been educated in exploded opinions cease to exist. (v)

Those staid and protectionist philosophers have not even played fairly, misrepresenting phrenological claims, for example, through writings in the *Edinburgh Review:* "Some of them confessed that all their information on the subject was presented to them through the distorting medium of the *Edinburgh Review*" (79).

Laying bare the populist attack on the monopoly of the learned over science, several witnesses explicitly point out that the old mental philosophy—Hamilton wraps himself in its tradition—excluded everyone but the expert few:

> While the elder schools of mental science never consisted of any but a few learned persons, the new may be considered as tending, without the sanction of the learned, to embrace the great body of people.
>
> [Combe] has . . . fixed the attention of large audiences on disquisitions from which, however important and necessary to be thought of and understood, the public had by common consent been formerly debarred. (55, 24)

To be sure, such rhetoric was designed to remind magistrates that they were, after all, popularly elected. Why did it backfire? I believe the logic of this populist epistemology—majority rules on matters of nature, as in democratic politics—threatened the authority of professional knowledge, and so challenged the interests of the councilors themselves or of those who had their ear.

Combe's populism could be carried to excess: if virtually anyone can know the intricacies of mind, brain, and skull, then what could stop a self-proclaimed "phrenologist" from going into practice with no training from Combe and hence with no respect for "proper" phrenological delineations of faculties and organs? Every science invents its own kind of pseudoscience, and phrenology was no exception. One testimonial sought to distinguish Combe from "the wand of the conjurer," on grounds that the true phrenologist had no need to exaggerate his prowess: "At least he can enumerate the forces which are enlisted on either side [whether good or evil], though, being no charlatan, and not pretending that he is a *prophet,* he will not venture to predict what specific action . . . will result, under certain circumstances" (*Glenelg Documents,* 27, 26). Modesty is the way toward credibility, and away from pseudo-phrenological impostors. A staple in Combe's *Phrenological Journal* was a tricky sort of double boundary-work, in which the truths of phrenology were there for anyone to judge, although not anyone could do phrenology properly: "There are individuals who make it their business, have their shops, and receive pay for their manipulations, at so much per head! This practice not only degrades the science, but gives rise to superficial converts. . . . It turns a dignified science into a system of *legerdemain.*"[72] This is exactly how Hamilton and his allies would put it, although they never did so in the testimonials. Perhaps even to mention phrenology would give the science more legitimacy than they thought it deserved.

One other bothersome issue hung over the populist epistemology of Combe's testimonials. Why would one who denied the authority of experts over questions of natural truth bother to seek the authority inherent in a university chair—that haven of archaic thinking and self-interested

72. "Phrenological Quacks," *Phrenological Journal* 9 (May 1835): 517–19. "But perhaps most important for the creation of phrenology's scientific image was the identification and isolation of persons held to be advocating a 'corrupt' version of the knowledge—in part, mammonistic lower-class practical 'professors' of phrenology. . . . Through deriding 'bumpology' and 'phrenological palmistry' as vulgarizations, those seeking to monopolize the social resource of phrenology were subtly to elevate the idea that there actually was, at root, a 'pure' phrenology and that they, the respectable phrenologists, were its proper guardians" (Cooter, *Cultural Meaning of Popular Science,* 79, also 80).

rejection of the new and different? I believe that Combe's candidacy was only partly a matter of his seeking "social academic status," as Cooter implies.[73] It would also, as suggested in the letter from Combe mentioned above, establish once and for all the legitimacy of phrenology as the one true science of humankind—even after George Combe was no longer around to promote it. His students would be there to carry on.

Moral Philosophy/Anatomy

Is there a connection between mind and brain—and how does the skull come into play, if at all? Testimonials for Combe and Hamilton answer this crucial question in fundamentally different ways, and thus they establish clear distinctions between the two candidates in their ontologies, their methodologies, and in the implications of their philosophies for disciplinary jurisdictions. Hamilton is the candidate of preservation: he is attached to the traditional Scottish "common sense" philosophy of Reid, Brown, and Stewart, which examines mental processes through introspection and reflection on the workings of the philosopher's own mind. Mental faculties are processes of the mind at work, not physical places in the brain. A comfortable distance is sustained between the jurisdictions of anatomists (studying brains, physiologically but not functionally) and moral philosophers (studying the functions of mind with no reference to brain).

Combe's science disregards established boundaries between these disciplines. Phrenology is offered as the only complete science of mind precisely because it alone among philosophies of the day brings mental processes and anatomical structure together: the brain is the organ of mind, and for any individual, the size of a region or faculty in the brain is positively correlated with behavioral or moral traits and with protuberances on the skull's exterior. This view renders the philosophy of Reid and Stewart not just obsolete, but incomplete for its inattention to the physiological basis of logic and metaphysics. Moreover, as the testimonials make clear, phrenology replaces the "mere" speculations of the reflectionists with inductive empirical observations from nature: Combe could feel bumps and see the brain, but Hamilton (by implication) could only conjure the mental processes described by the "common sense" school.

The first testimonial for Hamilton suggests that the Scottish "common sense" philosophy is definitely worth sustaining (or, by 1836, resuscitat-

73. Ibid., 263.

ing), but the writer (Victor Cousin, the French philosopher) laments that no capable spokesperson has shouldered the responsibility:

> It is much to be regretted that no great writer has of late come forward as the champion of [the Scottish school], to make known to the scientific world . . . its good sense and searching analysis, its dislike of all extravagant theories, and its strict adherence to what appears to be within the limits of human reason. (*WH Testimonials*, 14)

Cousin's fears are unwarranted now that Sir William has offered himself for the chair: "I did not know that there was in Scotland so worthy a successor of Dugald Stewart" (13). It was important to establish Hamilton's fidelity to the philosophical tradition that had made Scotland famous in the century before:

> Sir W. Hamilton is the man who, before all Europe, has, in the *Edinburgh Review*, defended the Scottish philosophy, and posted himself as its representative. . . . It is Scotland herself who ought to honour by her suffrage him who, since Dugald Stewart, is her sole representative in Europe. . . . Sir W. Hamilton never deviates from the highway of Common Sense. (20–21)

How better to establish the continuity of philosophical tradition than by telling the town council that "Reid and Stewart, if they were electors, would choose Sir. W. Hamilton" (21).

In these testimonials, it is all to the good that Hamilton align himself with the "common sense" school—good for philosophy, for the university, and for Scotland's fragile self-esteem. It was a potent message that the magistrates would find hard to resist. Still, it was clear to everyone by then that the tradition had fallen on hard times since the glory days of Reid and Brown; David Ritchie is nowhere included among those leading lights. And so, at the risk of giving the appearance of an ambivalence toward Scotland's own moral philosophy, Hamilton's witnesses point out that he is more than a mere follower of Reid. Rather, Hamilton has familiarized himself with the latest works in continental philosophy:

> Reid, Stewart and Brown disliked the German systems as extravagant, and as going beyond the bounds of legitimate inquiry; and were therefore little acquainted with them. Sir William, on the other hand, has understood and studied profoundly the philosophy of the ancients, and also that of Germany and France. . . .
>
> . . . But the result has been only an increased attachment to the philosophy of Scotland. (14, 15)

The idea is to ground Hamilton's commitment to "common sense" philosophy not in mere ritualism or misplaced patriotism, but in the careful consideration of its rivals: "He has not declared himself an adherent of the Scottish school, until after a careful examination of other systems" (14). In short, he is pictured as a scholar who will gather useful ideas wherever they come from, rather than one who parrots Reid in a doctrinaire manner.

> Hamilton [is] the man eminently qualified, by the extent of his erudition, and the power of his rare abilities, to amplify the edifice of Scottish philosophy, by sinking deeper its foundations, by completing its majestic and simple elevation, and by prudently enriching it with new treasures drawn from foreign mines. (28)

Hamilton's biographer—many years later—reinforced this image of philosophical tolerance, but also noted his virulent reaction to phrenology as the exception that proves the rule: "So tolerant was Sir William of all opinions, that I may say that phrenology was the only doctrine that he could not tolerate." But then the biographer makes mention of an episode from Hamilton's life that curiously receives no attention whatsoever in the testimonials written for the 1836 competition: "He studied [phrenology] with care, and mastered very completely the anatomy of the brain. . . . The result was, he had come to look on phrenology as a mischievous humbug."[74] So what happened to Hamilton's anatomical researches of the late 1820s, when he went after Combe with evidence on the homogeneity of the brain and on the role of the frontal sinuses in preventing the brain from pushing out the skull? Why is this anatomical proficiency nowhere brought up in these testimonials written less than a decade after the much publicized battle with Combe? Their absence is evidence—in Hamilton's layout of the sciences—of the complete irrelevance of anatomy and physiology for the teaching of logic and metaphysics: the structure of the brain has nothing to do with the processes of mind. Reid argued that, and so does Hamilton. It is challenging, of course, to find textual evidence from the testimonials for subjects that had to be excluded. But the following attempt to bound the field of logic and metaphysics makes no place for anything biological, and stands in bold contrast to the "same" field drawn out by supporters of Combe.

> [Logic and Metaphysics] is placed at the very threshold of intellectual study,—forming the grand stepping stone between philology and philosophical inquiry—between those pursuits which exercise the attention and the

74. Veitch, *Memoir of Sir William Hamilton,* 116. These words are attributed by Veitch to George Moir.

> memory, and those higher and severer ones which employ the faculties of judging and reasoning; and it includes all those analytical and inductive investigations of the phenomena and laws of Thought, which constitute what has, in modern times, been called the Philosophy of Mind." (35)

Nothing anatomical going on here at all. Instead—as a way of defining the subject matter of the vacant chair—we have a recitation of Reid's mental faculties and an endorsement of his introspective methods. This solid border between anatomy and mental philosophy would please incumbents of several chairs around the University of Edinburgh, for, as Saunders reminds us, "the holder of a chair [had a] direct interest in asserting his legally recognized monopoly of instruction in his subject matter."[75]

Combe's testimonials present phrenology as a complete science (because it brought mind and brain within a single compass), a true science (because it was founded on observations in nature), and a new science (destined to sweep the Scottish tradition into the dustbin). It would, of necessity, force a reconsideration of how various "psychological" subjects (and the students who pay to learn about them) would be allocated around the university. The need to bring mind and brain together in order to understand mental philosophy is mentioned in most of the testimonials.

> It is truly a comprehensive system of mental philosophy.
>
> Phrenology . . . is the most certain and complete science of the faculties of man.
>
> It is my intimate conviction, that Phrenology is the true Science of Mind, and the only real Physiology of the Brain—that it is indispensable to a sound System of Logic, which is the art of applying the intellectual faculties to the discovery and communication of truth. (*GC Testimonials,* 38, 31, 89)

This combination of anatomical study and analysis of mental processes is presented as a necessary element of effective training in logic and metaphysics:

75. Saunders, *Scottish Democracy,* 329. Cantor also notes the mutually exclusive subject matters of anatomists and moral philosophers: "Barclay and John Gordon [extramural anatomists] showed great interest in the anatomy of the brain and nervous system, although they took care not to attribute specific non-physical functions to the brain." Indeed, "the Edinburgh moral philosophers had given little consideration to physical factors and instead concentrated on self-interrogation." ("Edinburgh Phrenology Debate," 199, 204).

> Having, for a series of years, taken great interest in the sciences of Metaphysics and Logic, and felt assured that the subjects could not be clearly elucidated, especially to students, without the aids which are afforded by the doctrines of Phrenology, I have ever been anxious that some of our Universities or high places of learning should associate them as kindred sciences. (11)

But it is not enough for the attestors to make phrenology different from the heretofore separate inquiries into anatomy and mental functioning: it must be shown to be better than each of them. Combe himself goes on the attack in his introduction to the first volume of his testimonials, "combatting the metaphysician with arguments, and the physiologist with facts" (ix). It is said that phrenology does a better job of rendering the structure of the brain than do the presumably university-based anatomists not informed by Gall's organology: "The relations subsisting between the brain and other organs have been unfolded by this science with uncommon clearness, and with a precision and accuracy hitherto undreamt of by physiologists" (16). But witnesses save their harshest criticism for the old and feeble Scottish school, which they deride as speculation accomplishing little. It is mere metaphysics because, unlike phrenology, it is not grounded on observations of nature. One letter spells out the liabilities of the old school:

> In studying works on mental philosophy by Dr. Reid, Mr. Dugald Stewart and Dr. Thomas Brown, who form the boast of Scotland in this department, the following observations strike a reflecting reader: . . . they make no inquiry into the organs of the faculties; [and] they give no account of the obvious fact of different individuals possessing the faculties in different degrees of endowment, which fit them for different pursuits. (127)

Such faults make the old logic and metaphysics a worthless enterprise, a word game rather than an empirical science:

> An old student myself in the Logic class, I can never look back but with regret on the barren path I then found myself compelled to tread; attempting, with faculties still boyish and immature, to grasp abstractions, and to gather positive knowledge out of a series of discussions of *names.* Familiar for many years past with the philosophy of mind which rests on the observation of nature at large, which has followed our better acquaintance in the present day with the physiology of the brain, I feel myself competent to declare how much of interest and importance attach to the study of the operations of our minds. . . . The old system of metaphysics explained nothing satisfactorily; and like all persons who attempted to arrive at definite results by its assistance, I experienced only mortification and disappointment. (14–15)

How bad would it be—how archaic, even embarrassingly so—for the council to condemn the chair of logic and metaphysics to the old Scottish school?

> The filling of the present vacancy with one who persists in describing the mind as consisting of memory, judgment and imagination, would appear to me as a solecism not less great than would the appointment to the Chair of Chemistry of one who continued to describe fire, earth, water and air as the elements of matter. (57)

Nature is frequently brought in as the ally of phrenology but the enemy of speculative metaphysicians, who think they can understand mind without looking at the brain:

> [I] regard the doctrines of Phrenology, not as visionary speculations, but as essentially in accordance with truth and nature, and, consequently, as affording a satisfactory basis for the study of the human mind.
>
> As a system of Metaphysics, it differs from all others, in being founded on fact and built up by observation. Physiological facts are substituted for metaphysical speculation.
>
> The principles of Phrenology are maintained with evidence and illustrations equally appropriate and conclusive, and with the dignity and strict accuracy of pure inductive science. (9, 20, 38)

Combe supporters could rightfully be accused of hubris, so confident are they in the validity of his doctrines. Their unbridled enthusiasm makes phrenology into an imperialistic philosophy with megalomaniacal tendencies: "[We] believ[e] that Phrenology forms the true basis of the science of mind, . . . and [are] also fully impressed with the conviction that it must eventually *supersede* every other system of Mental Philosophy" (8, my emphasis). This image might have been invidiously compared to Hamilton's more tolerant, *bricoleur* attitude toward others' systems of thought, but after all, how else would the town council know who was right?

If phrenology is really that good, why hasn't it already taken the university by storm, to say nothing of the *Edinburgh Review*? For Combe's case to be complete, the traditional Scottish mental philosophy (sans brain) had to be described not just as different, not just as old, not just as ineffective, but also as actively resistant to new thought and retarding the progress that phrenology promises. Combe himself clearly recognized the barriers facing phrenology: "[It is still] a new doctrine, which has not yet been admitted into any of the older Universities as science" (iii). The insistent juxtaposition between old and new puts the town council in the position of retarding progress, were they to vote for Hamilton over Combe. In doing so, they would be no different in their behavior than other academics and archaic philosophers who have blocked the inevitable rise of phrenology: "During the whole period of its history, it has been opposed, ridiculed, misrepresented, and condemned, by almost all the men whose intellectual reputations rested on the basis of the philosophy which it is extinguishing" (84).

But in the eyes of its supporters, the failure of phrenology to break through university walls has less to do with its intrinsic merits or faults, than with prejudiced professors interested in their reputations, fees, and legacies—and not in the truth. Pro-phrenologists seize the high moral ground, suggesting that those who resist their science are addled by self-interest. Combe says as much in a letter published in the *Phrenological Journal* two weeks before the election: "The history of all scientific discoveries establishes the melancholy fact, that philosophers educated in erroneous systems have in general pertinaciously adhered to them, in contempt equally of the dictates of observation, and of mathematical demonstration."[76] In his introduction to the testimonials, Combe puts Playfair—no doubts about his good standing in science—on his side, to evidence the resistance of some scientists to new ideas.

> Professor Playfair observes that . . . "the introduction of methods entirely new must often change the relative place of the men engaged in scientific pursuits, and must oblige many, after descending from the stations they formerly occupied, to take a lower position in the scale of intellectual improvement. The enmity of such men, if they be not animated by a spirit of real candour and the love of truth, is likely to be directed against methods by which their vanity is mortified, and their importance lessened." (iii–iv)

It is thus up to the Edinburgh Town Council, with a vote for Combe, "to burst the trammels of prejudice that have so long confined the public bodies in the United Kingdom" (12).

Several witnesses link Combe to landmarks in the territory of real science, men who suffered through the same scornful indignity as phrenologists, before their ideas were accepted as true. Combe takes his place among the scientific martyrs to prejudice—and hopes for the same vindication they eventually enjoyed: "I would respectfully beg leave to ask your Lordship, whether the Newtonian philosophy, for example, was less true, or less important, towards the latter part of the seventeenth century, when it was carped at and oppugned by many a mathematician and self-

76. *Phrenological Journal* 10 (July 1836): 223.

styled philosopher in Europe" (51). A better example of "the truth will out" may never be found:[77]

> There is no example of a truth, once fairly launched, having failed to make its way. It must, however, pay the usual tax of entry; some one must be put to inconvenience in its progress, and few persons are fond of being set aside. . . . The earth did not become immovable . . . because, three or four centuries ago, Galileo was forbidden to declare that it moved; and the circulation of the blood was not arrested by its being obstinately denied for many years subsequent to the labours of Harvey. (33)

Galileo, and less often Harvey, is a good friend to all who find their beliefs excluded by guardians at the frontiers of science.

One final issue: if the phrenologists' link of mind to brain was essential for distancing their philosophy from the "common sense" variety, was it also essential to hook in the skull? I noted above that phrenologists' craniological assumptions—their ability to read character and talent from an informed feel of bumps on the head—were vital for their ability to market their doctrine among those looking for self-improvement or for a scientific basis to help those obviously in need of improvement (the insane, the criminal). Not until magnetic resonance imaging has it been possible to examine the structure of the brain without cutting open the subject's head, a risky move then as now. But the craniology was also that portion of phrenological doctrine that had been found most lacking in logic and evidence (recall Hamilton's argument about the frontal sinuses). Was there any reason for Combe's witnesses to bring the contentious skull into the competition for David Ritchie's chair? Would it do more harm than good for their candidate's chances? One witness opens the possibility (even if minute) that phrenology might be wrong in some of its claims:

> It were indeed wonderful if an author so original and adventurous, should in all respects escape error; but even if errors should anywhere mingle with his speculations—and I know of none that are material—his contemporaries are bound, by what they owe to the first interests of society and of truth, to see that his labours and deservings be appreciated. (23)

Phrenology is thus a good thing to promote even if it is wrong here or there, an argument we shall see again in the discussion of another scorned belief: cold fusion (chapter 4). Another witness suggests that even if the link between mind and brain were shown to be without foun-

77. Gilbert and Mulkay, *Opening Pandora's Box,* 90–111.

dation in nature, phrenology would still be an improvement over what is presently taught in logic and metaphysics:

> I am convinced that even if all connections of the brain with mind were regarded not merely as doubtful, but as perfect chimera, still the treatises of many phrenological writers, and especially [Combe's], would be of great value, from their employing a metaphysical nomenclature far more logical, accurate and convenient as Locke, Stewart and others of their schools. (5)

A third writes that the truths of phrenology are more than skull-deep:

> None but those who are totally ignorant of Phrenology heard it as a means of merely discovering natural powers and dispositions by external signs. Those who have studied it know, indeed, that the natural powers and dispositions are, *caeteris paribus*, in conformity with the size of the various parts of the brains. (48)

Apart from these arguments that phrenology does not depend only on its craniology (or only on a correlation between organ size and talent or disposition), there are just a few detailed reports of actual phrenological findings, ones based on actual examinations:

> I have also been in the constant practice of examining skulls and casts from the heads of deceased persons, and comparing these with their known mental characteristics and their actions exhibited during life; and I have found a constant and uniform connexion between the talents and natural dispositions, and the form and size of the head.
>
> The form of head possessed by all dangerous and inveterate criminals is peculiar. There is an enormous mass of brain behind the ear, and a comparatively small portion in the frontal and coronal regions. (1, 35)

Far more common in the Combe testimonials is a general rehearsal of faith in the credo that mind, brain, and skull form another trinity worthy of respect and authority: "I believe Phrenology to be founded in truth,—that the brain is the organ of mind,—and that the character of an individual can be inferred with considerably certainty during life from the external form of the skull" (57). Despite scattered misgivings, then, the weight of opinion is on the side of the skull as an essential element of Combe's science, and of his candidacy.

Pure Science/Applications

The 1836 competition for David Ritchie's chair brought to the fore an enduring tension in efforts to define science, and to legitimate it—a tension played out in more than a few of these episodes of cultural cartography. Is science Aristotelian or Baconian, to be judged by its cumulative advance of truth or by its utilitarian accomplishments? Is science a contemplative activity best cloistered in universities or an engine of action easily transported to settings where practical decisions are made? Is the scientist a scholar, a teacher, and a knower—or a doer? Science becomes one or the other, depending upon which song is thought to play better in a given setting.

Hamilton (playing Aristotle) and Combe (playing Bacon) each had reason to believe that his was the right song for the Edinburgh Town Council. Scottish universities had a love-hate relationship to their ancient cousins in higher education across the border. On one hand, Edinburgh (along with St. Andrews and Glasgow) no doubt envied the scholarly reputations of Oxford and Cambridge, and the Scottish elite often sent their sons down south for the kind of genteel education (and prestige) not available to gentlemen back home. The intense desire in some quarters to polish up the once brilliant philosophical tradition of Reid and Stewart arose in part from insecurities about how well Edinburgh matched up against the scholarship and pedagogy at other leading European universities. On the other hand, Scottish universities prided themselves on the practicality of their university training, and there is no better evidence of this than the stellar reputation of Edinburgh's medical college. Scotland's higher education was closely linked to professional associations, who pushed the curriculum to include courses of immediate utility with up-to-date skills. Indeed, some scorned higher education across the border for being nothing but a finishing school for gentlemen sufficiently well-off that they did not need practical training in order to make a living. Testimonials for Hamilton and Combe evoked both of these images of the university—and of its science—and both could certainly appeal to the magistrates.

Hamilton's testimonials introduce him as the consummate scholar—widely read, steeped in many philosophical schools, survivor of Oxford examiners, a widely published intellectual, a teacher. The testimonials together base their recommendation of Hamilton for the chair on the depth and breadth of his knowledge—what he *knows:*

> It shows great power of thinking, great comprehension, and great acuteness, united with an extent, a depth and accuracy of erudition, seldom met together. . . . I now venture to say, that there is not in existence an individual who has so completely mastered the whole learning which related to the philosophy of mind, as Sir William Hamilton. (*WH Testimonials,* 29–30)

Some testimonials emphasize how deeply Hamilton had dug into tough philosophical systems: "For the perseverance and depth of research into any subject that has occupied his mind, as well as for ingenuity of conception, I have perhaps never met with any one that equalled, and certainly have never known any one that excelled him" (46). Others call attention to the many places he has dug—his intellectual range: "It is impossible for me to speak in too high terms of your comprehensive knowledge of the circle of the sciences. . . . I have never met with any individual whose knowledge was at the same time so extensive and so accurate, so varied and so minute, as yours, even in such fields of inquiry as lie most remote from your professional studies and employments" (55).

Evidently, all those hours in the advocates' library were now paying off, but Hamilton's reputation as a scholar was first established during his college days at Oxford. Here a witness explicitly informs the councilors of just what qualities they should be looking for, playing to Hamilton's longest and strongest suit: "Proficiency in the mental sciences can only be acquired by long and patient study, and . . . such proficiency forms the primary qualification for a Professorship of Logic" (40). Hamilton's Oxford accomplishments are paraded several times, as a sign that he had "held his own" with the possibly superior intellects there, and that he would bring—if selected—a solid reputation among Oxford dons to the University of Edinburgh:

> In the department, however, of science, his examination stood, and I believe, still stands alone; . . . in a University like Oxford, where the ancient philosophers are the peculiar objects of study and admiration, and the surest passports to academical distinction, his examination should not only remain unequaled for the number, but likewise for the difficulty of the authors. . . . It contained every original work of antiquity difficult or important. (49)

Hamilton's expertise in classical philosophy perhaps was as appealing to some strata in Edinburgh as the classical architecture chosen for the city's New Town—Greek thoughts, Greek temples, a certain symmetry. Another testimonial establishes Aristotle as Hamilton's scientific totem, as Bacon would be for Combe:

> Your splendid career at Oxford, and your known intimate acquaintance with the Dialectics and Metaphysics of Aristotle have been followed up by a devotion of your time and talents to the study of the Philosophy of Mind. . . . I shall never forget the sensation caused by the refusal of the Oxford Examiners, to enter with you into the Metaphysics of Aristotle, nor the extraordinary compliments with which they accompanied this refusal. (42)

Examiners at Oxford scared off by Hamilton's undergraduate grasp of Aristotle—a mere boy, and from Glasgow?[78] The important point is that proficiency in Aristotle—as ratified by Oxford dons—is here held up as a worthy credential for the chair of logic and metaphysics: "I believe you to be profoundly versed in the Aristotelian philosophy." (55).

What good is Aristotle? It is worth studying, and worth teaching, precisely because of its challenging abstractions and its ability to develop analytical prowess. One witness writes that Hamilton has the talent "to express himself in a way calculated to impart to his youthful hearers some relish for those *abstract* studies in which he is to initiate them." In this supporter's opinion, "possessed with uncommon acuteness, penetration and real philosophical genius, Sir William Hamilton, . . . is almost unparalleled in the profound knowledge of ancient and modern philosophy, and enjoys the advantage of great clearness in explaining the most *difficult and abstruse* objects of philosophical discussion" (38, 25; emphases mine). Abstract, difficult, and abstruse are positively charged adjectives when applied to the knowledge of Hamilton, and presumably to the subject matter of the courses he would teach.

Good scholarship is translated by another witness into good teaching: "I know something of the art of teaching, and I am convinced that the first requisite in a good teacher, is a complete knowledge of the subject he has to teach" (42). Perhaps reacting defensively to charges that Hamilton had difficulty putting his deep thoughts into a style understandable to undergraduates, another witness affirms the close connection between scholarly capabilities and pedagogical ones:

> It might be doubted whether so very learned and able a person was the fittest for an elementary instructor, and might not be in danger of passing too hastily over the necessary rudiments of science, hurrying his pupils, with too little preparation, into its higher mysteries. . . . The experience of all ages accordingly is in unison with the opposite doctrine. The most eminent scholars . . . have always been the most successful teachers. (31)

In Hamilton's testimonials, the proper education of youth is frequently mentioned as the primary task of the incumbent chair of logic and metaphysics, a claim that stands apart from efforts by Combe's witnesses to make phrenology into a philosophical machine capable of solving all the rest of the problems of the universe: "The sound instruction of our aca-

78. Hamilton's biographer puts this moment in a slightly different light: "Still, his examination, in the Oxford sense of the word, was not a brilliant one," although Veitch then adds that "the masters themselves seemed to feel his superiority" (*Sir William Hamilton,* 45).

demic youth, is no doubt the greatest object for which Professors are appointed. . . . But, next to that, is to be considered the duty of promoting the improvement of science itself" (32).

Not curing the insane or improving legislation, but teaching pupils and advancing knowledge, activities best pursued on a foundation of scholarly erudition and analytic prowess. The subtext in Hamilton's testimonials is the idea that a university education is, after all, not the same as the banausic training available at workingmen's institutes or trade schools. The professor of logic and metaphysics should cultivate students' intellect and give them the means of sound reasoning and judgment: "A professor of such vast and varied erudition, and of such high authority, could not fail to command the respect of his pupils, and to be looked up to as a safe guide in the direction of their studies, and in the formation of their opinions, in that important branch of their education" (26). Though no witness said it quite so neatly, the idea behind Hamilton's case for the chair was summarized by Veitch in his recollection of Hamilton's inaugural lecture: "It was refreshing, in an age of facts and practical applications, and narrow utilitarian aims, to find the cultivation of mind declared to be a higher end than the stocking of it with information. . . . It was shown that knowledge itself is principally valuable as a means of intellectual cultivation."[79] Significantly, Hamilton's claim to the chair is warranted not by the substance of *what* he knows, but *that* he knows a lot and knows it well.

But "utilitarian aims" are precisely what science should shoot for, if one accepts the testimonials written for George Combe: phrenology hits the mark consistently. The case for Combe rests not on his scholarly skills and erudition, but on the potential of his philosophical system to solve practical problems well beyond academe. In his introduction to the first set of testimonials, Combe defines the territory of logic to include its useful implications: "The Logic Chair . . . has embraced the study of the intellectual faculties of the mind and their applications" (*GC Testimonials*, vi). In the wide range of its practical utilities, phrenology betters all other philosophical systems: "The science of Phrenology is one of great practical value; and that no such system of Logic or Moral Philosophy, in which the principles of Phrenology are neglected, can henceforth be considered on a par with the science of the present age" (6). Combe himself introduces Bacon as his own scientific landmark, putting into the sharpest possible relief the inutility of Aristotelian science against the

79. Ibid., 204, 205.

practicality of Baconian science: "Lord Bacon inferred that the Philosophy of Aristotle was false because it was barren; and the same rule of judging would lead to a similar conclusion regarding the philosophy of mind as hitherto taught in the established universities" (vii). Combe here adds another Baconian virtue to his criteria for deciding whether a philosophy is true: it must be based on empirical observations of nature; it must receive widespread endorsement, not because of some expert's authority, but through the reasoning of ordinary people; and it must be usefully and successfully applied to practical problems. Indeed, its utility signifies its necessary truth, "like all other great truths, fraught with the most important consequences to human improvement" (26).

Few witnesses had any doubt that phrenology (alone) would make mental philosophy into something instrumental:

> The evidence upon which this science is now founded appears to me quite irresistible, and the means which it affords of simplifying and rendering really practical the philosophy of the human mind, so superior to all others, that I should consider no individual properly qualified for the Chair of Logic, who was either unacquainted with, or who disregarded, its principles. (3)

If utility is the gauge, the Scottish "common sense" tradition comes up short:

> Many years after I had neglected the study of mind, in consequence of having been disgusted with the utter uselessness and emptiness of what I had listened to in the University of Edinburgh, I became a zealous student of what I perceive to be the truth.
>
> Logic has hitherto remained an abstract and valueless subject, productive of few beneficial results beyond the mere mental exercise involved in its study.
>
> Logic and mental science, as at present taught, are inapplicable to any practical purpose, except serving as a species of gymnastics for exercising the mental faculties of the young. (7, 27, 129)

What had been touted in Hamilton's testimonials as the beneficial payoff of his erudite studies (improving pupils' "critical thinking") here becomes an embarrassing waste of time.

What is phrenology good *for?* It can explain things that other mental philosophies cannot touch: "From it alone we learn why two persons educated together, and subjected to the same moral and physical impressions, may be widely different from each other as to their dispositions, talents and acquirements. From it alone we learn why certain individuals should excel in one pursuit or branch of knowledge, and be dull in most others" (90). Rather than making students adept at handling abstract and abstruse ideas, now logic and metaphysics is a means for them to better understand their talents and limitations. This testimonial puts students at the university dangerously close to those who make up the very different audiences for Combe's lectures and examinations at workingmen's institutes: both get self-enlightenment from the skull. "As Education, properly considered, aims at the proper development and regulation of man's nature . . . , [so] Phrenology appears to me not only the plainest, but the most satisfactory guide yet discovered" (34).

But the more consequential benefits of a phrenologically constituted logic and metaphysics lay beyond university walls. The list of presumed beneficiaries of Combe's doctrine is, read differently, a list of his allies—for whom science was remapped, drawing them within its new boundaries. In order to get these many people so committed to phrenology that they would appeal to the town council on Combe's behalf, science had to leave the university: "The moralist, the physician, the legislator, and the teacher, are able to draw from [phrenology] lights to guide them in their practical duties" (vii). Officials from insane asylums lined up to sing the praises of phrenology:

> Until I became acquainted with Phrenology, I had no solid basis upon which I could ground any treatment for the cure of the disease of insanity.
>
> The influence of this light . . . extends equally to the functions of mind in a state of disease—giving new insight into the hitherto dark and unaccountable mysteries of insanity, and clearing up what was formerly in impenetrable darkness. As a medical man, I have derived the greatest benefit from the forcible manner in which the study of Phrenology has directed my attention to the functions of the brain in health and disease. (12, 16)

Prison wardens were just as appreciative—in this case a real "testimonial":

> A few days ago Sheriff Moir . . . told me of your intention to examine phrenologically some of the criminals in Glasgow jail. . . . I was very much pleased to observe, that while your examination of each did not average a minute, you instantly, and without hesitation, state the character, not generally, but with specialties of feeling and propensities, surprisingly justified by what I knew of them; and being aware that you had no access to them, nor means of knowing them previously, as they were taken at the moment promiscuously from numbers of other criminals, I was at once led to the conviction of the truth of the science, and to see eminent advantages of such knowledge to society, and more immediately in regard to criminal jurisprudence and practice. (42–43)

The truths of phrenology are useful even in the bailiwick of the town council itself, where the science "contributes to establish the surest foundation for legislation" (48). Those truths were presented as "directly applicable to . . . civil and criminal legislation" (26). In short, "the principles of Phrenology applied to the science of Political Economy [were] found strikingly useful" (52). Were the magistrates prepared for this anticipation of a technocratic society? Having just recently put control of the town's government in the hands of the people instead of established elites (via the newly instituted popular elections), were councilors now ready to turn over legislative decisions to the authority of phrenological experts? Maybe not—perhaps their concern that science was here overstepping its bounds worked against Combe.

Science with Combe became action, not thought. Perhaps not wishing to draw attention to his lack of formal, specialized training in the range of philosophical schools comprising logic and metaphysics, Combe's attestors only rarely mentioned his academic preparedness: "He has devoted himself for the last twenty years to the study of mind in all its bearings. His knowledge of the opinions of the metaphysicians is extensive, and would enable him to give a most comprehensive view of the history or progress of mental philosophy" (11). Such words would fit better in Hamilton's book of testimonials, for most of Combe's other witnesses find little reason to mention any philosophy but phrenology. And they are certain that the kind of progress afforded by this doctrine is not mere intellectual advancement or the growth of knowledge: "It is never possible for Mr. Combe's audience to doubt that the aim and object of his instruction is the benefit of mankind" (24). His work has paved the way for "a large increase of happiness of mankind" and the "alleviation of human suffering" (23, 26).

So why didn't this utilitarian message sway the practically minded Scots of the town council? Those who did not vote for Combe may have been impressed by the claimful attestations of those who put phrenology to practical ends, but other qualities of the candidate—such as his religious thoughts, to be considered next—put them off. Or maybe the magistrates were fearful that if phrenology were legitimated by inclusion in the university, its aggressive rationalist programs of reform would unsettle once again a Scottish society that had just begun to settle down—to the advantage of the new elite. Historians Wrobel and Cooter do not understate the matter when they describe phrenology's goals as "restructuring social life" or "the reformation of humankind."[80] Maybe humankind was just fine, or at least good enough, for the magistrates to prefer the candidate who consolidated recent improvements over one who would pressure for even more fundamental changes. And some of those changes, if phrenologists were in charge, might have been personally threatening. Did the Lord Provost really want to hear from Combe in the preface to his testimonials that "independently of rank, education or wealth, men differ from each other in the amount and kind of their intelligence" (xiv)? Would he or any other magistrate be eager to have Combe examine his head—perhaps to find his elevated rank, enabled in some cases by professional education, negatively correlated to "capacity to make sound political decision"? Perhaps Edinburgh's councilors were content to leave such a powerful science inside the university, as something just academic. Or perhaps the councilors judged these two candidates by the company they had been keeping. Hamilton succeeded swimmingly at the premier university in England; Combe drew swarms to lectures at workingmen's institutes and other venues of the middle and working classes. What kind of university did the status-conscious councilors want for their children? Perhaps they preferred one with an erudite Oxford-trained scholar in the chair—over a mercantile center for self-improvement.

Science/Religion

Combe stirred things up with his version of natural theology, a view that the workings of God were visible in the laws of nature, that the structure of the brain, for example, was evidence for God's design in the universe, and that Christianity could be given a more rationalist cast without loss of ecclesiastical authority. Since the publication of *The Constitution of Man,* however, Combe had been attacked by representatives of established religion for promoting blasphemous ideas. He was accused of being an atheist, a materialist, an usher of a soulless world (it's all in the brain) where individuals no longer had moral responsibility but could attain pre-Fall "perfection" by properly exercising their mental faculties. It is enough to know that the established church had taken a dim view of phrenology—whatever the Christian pretensions of its leading advocate—and Kirk leaders carried their weight to the town council in hopes

80. Wrobel, "Phrenology as Political Science," 125; Cooter, *Cultural Meaning of Popular Science,* 117.

of persuading them to appoint someone to the chair of logic and metaphysics whose faith was not in doubt. Even though by this time the University of Edinburgh was not under formal ecclesiastical control, appointments of faculty to chairs with pertinence to the church (say, metaphysics) would be scrutinized "in a culture in which religion was still far more important than science." As Grant remarks, "to have flown in the face of established religious opinion, or to have [been] deemed to have done so, would most certainly have been fatal to any public enterprise, and Combe was at pains to avoid this."[81] The task for Combe's witnesses was to reassure the magistrates that phrenology worked with Christianity rather than against it. Combe wrote as much when he solicited a testimonial from Thomas Wyse, M.P.: "The Church party here are now founding charges of heresy against my work on *The Constitution of Man,* with a view to defeat my pretensions to the Logic Chair; your testimonial to the Christian character of that work, is precisely the kind of document calculated to do me the most favor at present."[82] The strategy was not altogether effective.

Combe had succeeded in winning allies precisely because his rationalist, scientific approach to questions of morality, human behavior, and social progress was perceived to be an improvement on religious dogma and antiquated mental philosophy. Several witnesses were perhaps more aggressive than Combe would have wished in their assertions that the science of phrenology could—and should—move into provinces once thought to be exclusively religious. But there was little doubt in supporters' minds that phrenology offered a scientific basis for morality: "As a science of Mind, its doctrines inculcate morality, rationality, and religion. . . . In Phrenology, we find united the best exposition of the moral sentiments, and the most approved metaphysical doctrines heretofore taught" (20). Witnesses argued that the science articulated the relation between free will and determinism, thus inserting itself into the middle of questions about moral responsibility:

> Phrenology has alone afforded a satisfactory explanation of the long disputed doctrines of free will and necessity,—it teaches us to what degree we are necessitated to obey the impulses arising from organization, and how far and by what means we are free agents, to act as the superior faculties direct. (90)

81. Cooter, *Cultural Meaning of Popular Science,* 80; Grant, "Combe and the 1836 Election," 178. Cf. Saunders, *Scottish Democracy,* 429 n. 3 on the university's formal autonomy from the church.

82. George Combe to Thomas Wyse, May 20, 1836.

One testimonial even hinted that phrenology rendered the soul more intelligible: "Mr. Combe is eminently qualified to teach the manifestations of the *immortal spark* through the medium of its perishable instrument on earth" (67). Several witnesses simply went too far in their discussions of how rational phrenologists approach the matter of revelation, so much so that Combe appended footnotes to several testimonials to clarify "what he really thought." Even a mild move toward rationalism demanded Combe's intrusion:

> Science may do much to restore reason to its proper place, and even to render it susceptible to the power of revelation, and thus of that supernatural influence which accompanies divine truth: but it is to revelation, accompanied with this influence, we must now look for the true regeneration of man. (106)

Combe added: "Phrenologists assume the existence and authority of revelation" (104).

Three arguments may have been designed to mollify the Church Party (and they echo what other scientists-cum-blasphemists have said since the seventeenth century to defend the autonomy of their science). First, the church and science are not at loggerheads, neither rivals nor enemies but fellow travelers.[83] In the words of one witness, "Its leading principles appear to me to involve nothing subversive of morality, or incompatible with the doctrines of Christianity" (99). Another testimonial takes up the presumed threat of Combe's putative materialism and denial of free will: "The phenomena of which it treats are in my view of great importance in mental training, and no more inconsistent with human responsibility, or favourable to materialism, than any other phenomena of our physical condition, long known, and universally admitted" (110). A poignant statement by yet another of Combe's witnesses suggests that science and religion are not a zero-sum game:

> One of my greatest objections was removed by Sir William and Lady Ellis (now of Hanwell Lunatic Asylum [presumably as superintendent]), viz. that Phrenology interfered with the religion of the individual imbibing its principles. This was falsified most completely in their case; for while they advocated the science, they remained the same pious persons they had hitherto been. (54)

Second, the testimonials imply that truths found by phrenology or any other science are evidence for God's genius, that the Bible and the *Phre-*

83. George Combe to Alex Hood, May 27, 1836: "It is extremely injudicious, as well as morally wrong, to dash religion against natural science, as if it were a thing dangerous and hurtful."

nological Journal are two records of the same divinely shaped universe—merely different discursive styles. Science aids Christian religion by providing rational grounds and empirical evidence for its timeless beliefs.

> If we take the science as a scheme of mind founded upon observation of actual facts, and, in comparing it with Christianity, find it in exact harmony with both the doctrinal and perceptive parts of that faith, can we resist the conclusion that Christianity has here obtained the aid of demonstrative, in addition to testimonial evidence? Phrenology might be described as Christianity thrown into the character of a science; each is calculated to have great force in urging the other upon the convictions of mankind. (56)

It is doubtful whether Edinburgh ministers would have seen any desirability in having their religion "thrown into the character of a science." God needed no additional evidence from facts. But other witnesses pursued the same line:

> The impression which the novelty and interest of [Gall's] discoveries excited, alarmed the priests, who, from a false and ignorant view of the subject, were led to imagine that such a theory might lead to results inconsistent with their religious tenets; . . . I was, however, so much struck and impressed with the truth of his discoveries, which I considered so consistent with the wise simplicity and unity which mark so forcibly the laws of our Omnipotent Creator. (18–19)

In effect, the testimonials tried to establish the following syllogism: if religious beliefs are true, phrenology (and other sciences) will find them in the laws of nature; if such empirical confirmation is not found, those particular beliefs are not true—and thus cannot be part of God's infallible design and should not be part of Christian doctrine. Combe's attestors sought to distinguish good religion (beliefs harmonious with nature scientifically wrought) from bad religion unworthy of faith (dogmatic superstitions shown by science to be false). Thus one witness testifies: "In directing my attention . . . to Nature, I never once imagined that if I discriminated truth I could be deviating from scripture" (111). He stoutly maintains: "I have on principle confined myself to the investigation of nature, never doubting that, in so far as I may have discovered truth, Scripture will be found to harmonize with my doctrines" (113). In some testimonials, that logic is specifically applied to the contentious issue of revelation. Could science find evidence in nature for *that?*

> I regard Revelation as a sacred subject which ought not lightly to be brought into collision with philosophy. . . . It appears more advantageous to investigate nature *by herself first,* and to proceed to compare her phenomena with Scrip-

> ture only after being certain that we have rightly observed and interpreted them. By this method we shall preserve our minds calm and unbiased for the investigation of truth . . . and we shall avoid bringing discredit on Revelation by involving it in unseemly conflicts with natural phenomena. (111–12)

This witness may have made more space for science than churchmen would want—the sacred texts will be checked only *after* science has made its way into nature (and he carefully avoids taking a stand on what will be done in the event of a discrepancy). But the next testimony neatly summarizes the justification for Combe's hope that the councilors would see his phrenology as an ally to their Christianity: "If I advance only doctrines founded in nature and in accordance with Christian morality, I am entitled to the benefit of the presumption that they are also in harmony with all sound doctrinal interpretations of Scripture. If any of my views are at variance with nature or Christian morality, I am ready to give them up" (115).

Third, attestors offer confirmation that phrenology—by exercising their faculties for veneration and by guiding them more effectively to moral and proper conduct—makes people into better Christians:

> The moral results of his system may be said to be, that we best promote our own wellbeing when we venerate God, and obey the voice of conscience,—when we are temperate, industrious, and orderly, and exercise justice and charity toward our neighbors. These principles are not only enforced in the *Constitution of Man*, but they may be said to pervade every page of it. (58–59)

No churchman could argue with this list of virtuous behaviors. Phrenology is said to guide people to, not away from, Christian principles: "Phrenology contains the first principles of morals and of natural rights; and it is, in short, the safest guide by which to accomplish an arrangement of life that will conduct to happiness, in conformity to the wise and benevolent laws of the Creator" (86). Two testimonials describe how phrenological therapy can drive subjects to charity. The first witness speaks from his own experience:

> I lament the distance between us prevents me from personally offering you my grateful thanks for the great pleasure and instruction I have derived from your works on Phrenology, . . . assuring you . . . that they have increased and given energy to my charity, teaching me by my own infirmities to [be] compassionate and make every allowance for those of my neighbour. (14)

The second is a teacher who evidently uses phrenological principles in training his pupils:

> I have not merely employed such physical objects as tended to develop the *knowing faculties*, but have also habitually exercised the pupils in the use of their reflecting powers. In teaching morals, I consider *mere instruction* as very inferior *to training*. For instance, instead of *telling* a boy to be charitable, I direct his Benevolence [both the faculty and the organ] to a suitable *object;* instead of commanding him to be just, I exercise his Conscientiousness by making him act as juryman in the petty cases of the school. (35)

Still, all this talk of a science of morality—of finding evidence in nature for God's revealed design, of training moral people through the exercise of selected phrenological faculties—might have backfired as a case for Combe's conciliatory attitude toward religion. It could have been read as defensiveness, an obvious and disingenuous tactic for deflecting attacks on the unchristian character of *The Constitution of Man*. With Isaac Taylor in the running, whose religiosity was apparently never questioned, the Church Party may have had little reason even to consider whether Combe (as presented by his witnesses) had recanted his previous heresies.

But what of Sir William Hamilton on this score? I have mentioned that one magistrate—at the election meeting—sought to raise doubts about Hamilton's faith, but was castigated for doing so and the matter was dropped. There is quite literally no mention in his testimonials of the implications of the substance of Hamilton's updated "common sense" philosophy for religious beliefs or practices: logic and metaphysics as here conceived had nothing to do with religion (they occupied well-bounded, amply distanced cultural spaces).[84] The only mention of religion or morality in Hamilton's testimonials considered not his philosophy and science, but his character:

> Respecting his moral and religious conduct, so far as I know, it has uniformly been such, even from his earliest years, as would do honour to the purest heart, and such as the most scrupulous could not fail to approve. It was in

84. "Reid and his followers developed a grammar setting forth the meaning of key scientific concepts and delimiting the relationship between the spheres of religion and science. . . . Reid's concept of 'common sense' was built upon the foundation of a strict mind-matter dualism and a seemingly tough-minded realism. The dualism protected traditional religion from challenge by science, by dividing the legitimate spheres to be covered by mentalist and materialist accounts of phenomena and by insisting on a purely descriptive or proto-positivist view of science. . . . Reid explained how true science was concerned solely with the description of phenomena—science, he held, could never penetrate beneath phenomena to the ultimate reality. Religion thus was safe, for science could only butt its inquisitive head against the wall of common sense consciousness but never penetrate to any reality more fundamental" (Campbell, "Scientific Revolution," 355).

> consequence of the high opinion entertained of his character, that I solicited him some few years ago to become an elder of the church and a member of the kirk-session of this parish. (*WH Testimonials*, 46–47)

Such an endorsement might not have made Hamilton more attractive than Isaac Taylor in the eyes of those with Kirk allegiances, but it made him infinitely more appealing than Combe.

AND THE WINNER IS . . .

. . . Sir William Hamilton, but only sort of. He certainly got the chair of logic and metaphysics, but he failed to restore Scottish philosophy to its previous grandeur and enters history as a minor figure in the development of that discipline. Once elected, Hamilton immediately embroiled himself in a drawn-out rumble with the very town council that had chosen him—centered on the question of patrons' involvement with internal curricular decisions at the university. In 1838, Hamilton added a senior class in speculative philosophy—in effect, splitting off metaphysics from logic—and collected additional fees from students for this separate course. (His predecessor Ritchie had in fact dropped metaphysics altogether, because the church required its divinity students to take only the logic portion of the course.) But the town council asserted its authority and negated Hamilton's move, and he was eventually forced to teach logic and metaphysics in alternate years.

Worse, Hamilton suffered an attack of paralysis in 1844, "no doubt precipitated by his habit of sitting up writing or reading all night."[85] His mind remained sound, but his eyesight and speech were bothered, and he never regained full use of his limbs. He became an invalid, only eight years into his professorship. The affliction understandably limited the output of Hamilton's writings, and he died in 1856 at the age of sixty-nine.

George Combe lost the chair, but seemed to do better than Hamilton with the rest of his life. He emerged from defeat undaunted, more convinced than ever of the truth of his message. The day after the election, Combe wrote to Convener Dick, who had given him a vote: "I would have called personally to return thanks, but I go early to the country this morning, to attend the opening of the head of Mr. Robert Liston, who died yesterday." Science marches on. Combe did not leave so early that

85. *Dictionary of National Biography*, 8 (1917): 1114. Details on Hamilton's life after 1836 come from Grant, *Story of the University of Edinburgh*, and Saunders, *Scottish Democracy*.

he could not find time to write a long, rationalizing letter to Andrew Carmichael:

> I am not in the least disappointed at the result, as I knew it three weeks ago. . . . The election of Sir William Hamilton was carried entirely by the Whigs bringing their political influence to bear in his favor. The Church supported Mr. Taylor and Mr. MacDougall, so that I was left to the one of reason and of truth alone. Every effort was used to prevent my success. The professors . . . canvassed and wrote letters in favor of Sir William Hamilton. The partisans of the Church publish scurrilous and contemptible pamphlets announcing the unchristian tendency of Mr Combe's work. . . . Do not imagine that I am cast down or disappointed at this result. I have made a prodigious step in advance. . . . The men of the old school have gained the day. . . . They have canvassed Sir William Hamilton for retarding to the utmost extent of his abilities the progress of truth and the welfare of mankind, and they have excluded me because I have done exactly the reverse. Let them alone! The natural laws will give them and their University their reward in time. I am in good spirits, full of gratitude to all my excellent friends for their support.[86]

"Let them alone" he did, for after 1836, Combe spent less time in Edinburgh and generally withdrew from town society and its literary culture. He soon retired from his legal business, his wife's fortune and the continuing sales of his books being more than enough to support him. Combe lectured widely, making two trips to America (where some bad business ventures cut into his financial security, though he made public presentations only if guaranteed a take of $750), and then left for Germany just as "his" phrenological movement in Britain was splintering on the question "Is phrenology materialist?" He died in 1858, two years after his rival Hamilton.[87] Phrenology had a much longer life, both inside science and also, of course, in sideshows at carnivals and world's fairs.

Combe's own assessment of his defeat is reasonable: the Whigs chose "common sense" philosophy over phrenology, even though Combe had previously shown his loyalty to their politics of reform; magistrates in the Church Party were possibly more threatened than appeased by witnesses who told them that Combe stood at the helm of a movement to put Christianity in scientific dress. The deeper questions are why Whigs went for Hamilton's dusty philosophy over Combe's shiny and useful one, why the church saw in phrenology heresy instead of truth and reason, and

86. George Combe to Convener Dick, July 16, 1836; George Combe to Andrew Carmichael, July 16, 1836.

87. Details of Combe's later life are found in the *Dictionary of National Biography;* Davies, *Phrenology;* Grant, "Combe and the 1836 Election."

why Edinburgh's liberal professionals once committed to social reform now regarded Combe's programs as foolish if not dangerous. What was the interpretative milieu in which the votes were cast?

If I am right, Combe and his phrenological witnesses failed to heed Amen-em-apt's wise dictum. Their testimonials tread down the boundaries of so many *cultural* fields that calamitous defeat in the election was assured. To attract badly needed allies and to counter long-standing charges from Edinburgh's cultural elite that phrenology had no basis in fact or nature, Combe remapped science. His daring culturescape pushed out the frontiers of science far enough to include Gall's new inquiries, but also far enough to encroach upon realms of politics and religion. Moreover, Combe's map required that boundaries between anatomy and mental philosophy, and between professional expertise and the knowledge of ordinary people, be erased. The final vote in the 1836 election may measure the extent to which Combe's representation of science challenged powerful interests and established cultural institutions—not by its empirical claims about mind or brain or skull, but by its threatening (or just mystifying) rendering of culture. Its populist epistemology denied the monopolistic authority of professional expertise, which had only recently become a source of legitimation for the town's emerging political elite. Its intellectual imperialism—melding anatomy and mental philosophy—threatened professors at the university, whose livelihood depended upon the preservation of established disciplinary jurisdictions. Its intimations of technocratic politics and a thoroughly secularized morality may have pushed science too far, too fast, into territories where authority and advantage were not (at that time) routinely grounded on empirical observations of nature or logical reasoning.

As in other episodes of cultural cartography, the enduring winner in the competition for the chair of logic and metaphysics at Edinburgh in 1836 was science itself. Both Hamilton and Combe defined logic and metaphysics as "science," and each sought to attach "scientific" to their distinctive philosophies (while denying it to the rival). Neither could avoid this rhetorical move: to locate one's philosophy inside science blessed it with a certain cultural authority and gave it the appearance of reliably objective truth. Combe and Hamilton could agree that science had these core qualities, but they disagreed on just about everything else to be found in that cultural space. Hamilton's map kept science (or at least logic and metaphysics) well apart from religion, from politics, from the "knowledge" of Everyman, and he respected internal borders between various disciplines. Combe's map showed science oozing over into politics, religion, and lay knowledge, and it obliterated the line between anatomy and mental processes.

But when all is said and done, both cartographies were accurate. Science has certainly become, since the early nineteenth century, a mélange of what testimonials for Hamilton and Combe said that it was. Indeed, continued reproduction of the epistemic authority of science has probably depended on a flexible oscillation between these two mappings. Scientific authority is sometimes preserved by asserting its autonomy from religion, politics, and common sense; the resistance of science to commodification, and thus its autonomy, is secured via justifications that emphasize something other than its utilitarian practicality. But scientific authority is extended to new ontological domains and over different personal or practical decisions precisely as its usefulness and public accessibility is emphasized. Hamilton won not because his account is what science really is. At that place and time, his rendering of the culturescape was quite possibly better aligned with the interests of those who picked David Ritchie's successor. And there is nothing inaccurate about Combe's science. Indeed, a science that is empirical, practically useful, easily grasped by laymen, applicable to individual and societal decision making—irrespective of established disciplinary boundaries—sounds very normal and au courant these days. But not in Edinburgh that July in the chambers of the town council: Combe had the wrong science for the time and place, though many after him would win with it.

Chapter Four

THE (COLD) FUSION OF SCIENCE, MASS MEDIA, AND POLITICS

Is there anything fresh to say about cold fusion? Nuclear physicists and most electrochemists want nothing to do with it anymore, and—after a half-dozen postmortems, three master's theses, and a raft of scholarly papers—even historians and sociologists of science must surely be bored. How different was the scene on March 23, 1989, when the whole world watched and listened as Professors Stanley Pons and Martin Fleischmann told a press conference at the University of Utah that they had "successfully created a sustained nuclear fusion reaction at room temperature" by passing an electric current between a palladium electrode and a platinum anode submerged in an insulated tube filled with heavy water. Deuterium nuclei were said to have been driven into the palladium rod so densely that some of them fused together—with a release of energy in the form of heat, along with other signatures of nuclear fusion (tritium, neutrons). The claim interested physicists and chemists because nuclear fusion had heretofore required horribly high temperatures (like those found in the sun, where fusion happens au naturel) or powerful energy-consuming lasers. It interested everybody else in the world because, as the Utah press release put it, cold fusion has the potential of "providing humanity with a nearly unlimited supply of energy," as cheap as seawater and with no radioactive by-products.

For most natural scientists now, the case for cold fusion is closed and the verdict unequivocal: "pathological science."[1] Irving Langmuir's space for scientific fantasies was retrieved and put to cartographic use in de-

1. Langmuir, "Pathological Science."

nouncing the claim and in castigating its pitchmen, Pons and Fleischmann, as the episode quickly entered the annals of science next to other illusions: N-rays, polywater, and Martian canals. Exactly six months after cold fusion first hit the front pages, a historian and the director of the Brookhaven National Laboratory took to the *New York Times Magazine* to accuse Pons and Fleischmann of self-deception, as evidenced by these symptoms of pathological science: the discoverers had given too little effort to disconfirming their baby, and they had responded to others' failed replications with endless ad hoc complaints that some procedure wasn't done quite right.[2] The title of Gary Taubes's tell-all book hides little: *Bad Science.* He cites Langmuir's definition of "pathological science" as "the science of things that aren't so," a case "where there is no dishonesty involved but where people are tricked into false results by a lack of understanding about what human beings can do to themselves in the way of being led astray by subjective efforts [or] wishful thinking." Taubes adds more symptoms to the cold fusion disease: claims of unbelievable accuracy are used to justify the tiniest causes for supposedly gargantuan effects. And he gives Langmuir the last word on cold fusion: "There isn't anything there. There never was."[3] It was hardly coincidence, but rather good boundary-work, that *Physics Today* chose to reprint Langmuir's 1953 talk in its October 1989 issue.

Patient readers will by now recognize these accounts not as sociologi-

2. Crease and Samios, "Cold Fusion Confusion."

3. Taubes, *Bad Science,* 342–43. There are at least four other book-length contributions to the cold fusion controversy. Chapter 14 of Close, *Too Hot to Handle,* is titled "Test-Tube Fusion: Science or Non-Science," and begins by tracing the consignment of Pons and Fleischmann to the realm of pathological science back to CERN physicist Doug Morrison—who we shall meet at the Baltimore meeting of the American Physical Society. John R. Huizinga, the University of Rochester scientist who cochaired the Energy Research Advisory Board for the Department of Energy (thumbs down to special federal funding for cold fusion research), titles his book *Cold Fusion: The Scientific Fiasco of the Century* and devotes an entire chapter to pathological science and to showing how Pons and Fleischmann fit so snugly inside it. F. David Peat's instant book, *Cold Fusion: The Making of a Scientific Controversy,* does not mention Langmuir or pathological science, perhaps because it was written from a jury-still-out vantage. Eugene F. Mallove's *Fire from Ice* holds out hope that something big might yet come from Pons and Fleischmann's daring announcement, and perhaps predictably his discussion of Langmuir comes in a section titled "The Pathology of 'Pathological Science.'" Two shorter pieces also locate cold fusion in pathological science: Rousseau, "Case Studies in Pathological Science," and Rothman, "Cold Fusion." The ubiquitous retrieval of Langmuir's pathological science as a space in which to bury Pons and Fleischmann's claim illustrates perfectly how extant and available maps of culturescapes are given immediacy and interpretative force through their rhetorical reproduction.

cal analyses of the cold fusion controversy, but as the controversy itself—juicy cartographic data demanding interpretative work by the analyst. *How* did cold fusion end up in pathological science? So juicy indeed is the cold fusion case that it could easily become the mother of all scientific controversies, a hunch invited by the number of scholars in "science and technology studies" who have already sunk their teeth into it. Harry Collins and Trevor Pinch suggest that the arguments and rhetoric used to deny validity to the claim and credibility to Pons and Fleischmann could, in principle, "cut both ways": unwavering commitments to professional or disciplinary loyalties, difficulties in getting experiments to work consistently, and unabashed publicity-seeking might just as easily describe the naysayers as Pons and Fleischmann.[4] Bart Simon draws on Collins's concept of "experimenters' regress" to explore the contingent nature of replication in controversial science, focusing on the rhetorically constructed and interest-laden character of what was decided to be a competent and faithful reproduction of what Pons and Fleischmann had done.[5] James W. McAllister fastens onto the palpable disciplinary competition between physicists and chemists, each with their own competence, knowledge-base, and prestige at stake in the controversy.[6] Bruce V. Lewenstein uses cold fusion as a case to demonstrate the difficulties of doing history of science-as-it-happens, and looks at the role played by the mass media in carrying along the controversy.[7] Jo Gottschalk treats the media broadcast of scientific claims as a circumvention of ordinary processes of peer review.[8] Three communications scholars take a rhetorical perspective: Charles Taylor sees cold fusion as a site for practical demarcations of science from non-science; Alan Gross tells it as a story of "consensus threatened and renewed"; and, in an episode of the Public Broadcasting System's science show *Nova*, Dale L. Sullivan finds an almost biblical "epideictic rhetoric" in the ritualized excommunication of Pons and Fleischmann.[9] David Goodstein creates a new cultural space for the scat-

4. Pinch, "Opening Black Boxes," 505; Collins and Pinch, *Golem*, 57–78.

5. Simon, "Voices of Cold (Con)Fusion"; Collins, "Seven Sexes."

6. McAllister, "Competition among Scientific Disciplines."

7. Lewenstein, "Cold Fusion and Hot History"; cf. his "Public Electronic Bulletin Boards" and "Fax to Facts." Leah A. Lievrouw also picks up the mass media angle, but she sees the press as a vehicle for scientists to broadcast favorable images of themselves and their practices ("Communication and the Social Representation of Scientific Knowledge").

8. Gottschalk, "Cold Fusion."

9. Taylor, "Defining the Scientific Community"; Gross, "Renewing Aristotelian Theory"; Sullivan, "Exclusionary Epideictic." See also Garrit Curfs, "Experiment as Rhetoric."

tered few who continue to pursue Pons and Fleischmann's lead: "pariah science" because "nobody out there [in real science] is listening."[10]

Even I could not resist the easy sociological pickings served up by cold fusion. For a 1989 conference at Notre Dame, "The Social Dimensions of Science," I analyzed the first six weeks of newspaper accounts and found that neither experimental evidence nor theoretical logic alone were sufficient to keep Pons and Fleischmann's claim alive—but they weren't by themselves powerful enough to kill it either. Only when factual claims, instrument readings, and established models were ensconced in narratives that assigned credibility and drew out lessons—good stories of heroes and villains, dreams and nightmares, passions and foibles—could cold fusion live or die.[11]

Happily, not quite everything has been said about cold fusion. Theoretically, what interests me now is this: *Where* does adjudication of the truth or falsity of claims about nature take place? To extend Andrew Abbott's conceptual framework for studying the professions in general to the specific case of science: scientists seek to protect their "jurisdiction" over the "task" of deciding which natural claims are true and which are not.[12] During the first six weeks of the cold fusion controversy, that process of adjudication moved into venues—at once physical places and cultural spaces—not customarily thought of as part of "science." This raised vexing questions for the profession (and its audiences, consumers and patrons) about who really was in charge of deciding the fate of this natural claim, and how its fate would be decided. This episode of cultural cartography features demarcations designed to expand and diversify those with the authority to make scientific facts—which, predictably, gave rise to counterdemarcations designed to restrict that authority to science-the-real-thing.

In this chapter, I follow Pons and Fleischmann's claim from the initial press conference in Salt Lake City, to Washington and the halls of Congress one month later, and then one week after that to the annual meeting of the American Physical Society in Baltimore. I suggest that proponents of cold fusion sought to keep their instantly imperiled fact alive by transporting claims-making and claims-adjudicating processes into

10. Goodstein, "Pariah Science." In a later paper, "Undead Science," Bart Simon describes cold fusion—still pursued by some scientists eight years after—as a "ghost."

11. Gieryn, "Ballad of Pons and Fleischmann." Toumey also sees cold fusion as—for the public who watched it—the story of an "emotional whipsaw of hope and the dashing of hope" (*Conjuring Science*, 97).

12. Abbott, *System of Professions*.

places and spaces that consequentially expanded the domain of "doing science"—first into the mass media, then into the belly of national politics. Journalists and congressmen were absorbed into the processes of making a scientific discovery and of ascertaining its implications: getting the word out, and projecting technological and economic payoffs, became as vital to the survival of cold fusion as did evidence of heat, neutrons, and tritium from a bubbling cell in a Utah basement laboratory.

But then came Baltimore, where physicists restored the orderly ordinary boundaries of science, restricted those who could legitimately decide the truth and falsity of natural claims, and (by the way) killed the claim. The formal sessions and press conferences at the American Physical Society meeting are interesting in this context not for the specific arguments and evidence used to deny facticity to cold fusion, but for the arguments and evidence used to restore the customary places and spaces of science. Plainly, most folks on the street will tell you that TV news and congressional meeting rooms are not exactly science. But that commonsense, tacit, and stabilized divorce of science from mass media and from politics is precisely what needs to be explained sociologically. Physicists at Baltimore reproduced the familiar boundaries between science and the media and between science in politics, boundaries that had blurred as cold fusion was announced under bright lights in Salt Lake City and as its potentials were debated on Capitol Hill. After the physicists finished with cold fusion, the fact was false, and both press and politicians were rhetorically removed from the cultural space for real science.

The state motto of Utah quotes Brigham Young, who said as he led his tired Mormon flock into the abundant Salt Lake Valley: "This is the right place." It was the wrong place to ascertain the scientific truth or falsity of cold fusion, and so was Washington. Baltimore was the right place, at a gathering of specialists, speaking the language of nuclear physics, to an audience of competent credible peers and trusted assessors.

NORMAL SCIENCE IN UTAH: DISCOVERY THROUGH PRESS CONFERENCE

Press conferences are called when individuals or organizations seek to bring a message to the masses via the media of newspapers, magazines, radio, and television.[13] A successful press conference depends upon a

13. The media has also played a crucial role in drawing boundaries around science in controversies over mass extinction: see, for example, Clemens, "Of Asteroids and Dinosaurs."

press release sufficiently grabbing to compel the attendance of many media representatives, followed by the elaborated delivery of a message that simply begs for coverage in the next edition or on nightly news. If the message happens to be a knowledge claim about nature, the press release and press conference must accomplish three things. First, the claim must appear credible, so that the media have prima facie reason to consider it fact rather than fiction. Second, the claim must appear to have potentially huge consequences for everybody everywhere. Third, the claim must appear as a discovery here and now, a novel and original result of work done by nobody else. By all measures, the press conference in Salt Lake City, Utah, on March 23, 1989, to announce the discovery of cold fusion was a smashing success. But *how?*

The cold fusion press conference faced the following challenge: how to make the discovery look like normal science amid abnormal circumstances. The key to the claim's credibility was to locate it squarely inside the cultural space that has for centuries conferred authority on spokespersons for nature: science. To restate a point that has run through every episode of this book, the authority of science is instantiated and reproduced as knowledge claims, practices, instruments, and investigators are attributed credibility because they are located inside this space. The concerted effort to surround Pons and Fleischmann with credibility as "scientists" reestablishes the authority of science itself as a bounded territory of culture. However, that task was made difficult by several oddities—not least of which is the very idea of a press conference as a venue for making a knowledge claim, and especially one discovered at the University of Utah (rarely the fount of earth-shattering scientific discoveries). Those who called the press conference sought to create an atmosphere of normal science in order to give credibility to the claim, while at the same time accounting for the abnormal circumstances that might have raised doubts about the believability, significance, or originality of the claim. I shall argue, based on analysis of the five-page press release and of a transcription of the press conference, that the normalization of this event as science involved, in part, the translation of media representatives into codiscoverers of cold fusion and the translation of their videocams and tape recorders into instruments of research.

The crowded press conference was called to order by James J. Brophy, vice president for research at the University of Utah, who immediately removed any doubt about where, culturally, these goings-on should be located, welcoming the media "to share with me this exciting scientific announcement."[14] The media had already been put on alert that this was to be a science story, as the press release was titled "'Simple Experiment' Results in Sustained N-Fusion at Room Temperatures for the First Time," and started off: "Two scientists have successfully created a sustained nuclear fusion." Brophy handed off the microphone first to Chase Peterson, then president of the university, who has since resigned for reasons not unrelated to events set in motion by the press conference he addressed that day: "What is an experiment? An experiment is an informed probing of the unknown under controlled circumstances." Peterson's question-and-answer sequence was hardly "rhetorical." Rather, it attached cold fusion to controlled experiments (possibly the sine qua non of science since the seventeenth century),[15] as it ascribed expertise to two probers into nature's unknown. These experiments were triggered by just what you might expect if you were only mildly familiar with the ways of science: "Pons looked at isotopic separation in electrodes and was puzzled at certain results." Even those who had never heard of Thomas Kuhn would agree that normal science solves puzzles![16]

This mood of normal science was enhanced by descriptions of the discovery in a strange language that sometimes gets labeled "technical." The microphone was handed to Stanley Pons, purveyor of words not heard every day: "Deuterium, which is a component of heavy water, is driven into a metal rod, similar to, exactly like, the one I have in my hand here . . . , to such an extent that fusion between these components, these deuterons in heavy water, are fused to form a single new atom." Martin Fleischmann followed up with the kind of mathematical discourse that has become emblematic of science:[17] "If you want some quantitative information, we have run cells now generating in excess of 20 watts per cubic centimeter of electrode structure. . . . We have run cells of this

14. These extracts come either from the official press release or from my transcription of a videotape of the press conference.

15. Shapin and Schaffer, *Leviathan and the Air-Pump.*

16. Kuhn, *Structure of Scientific Revolutions.*

17. Porter, *Trust in Numbers,* argues that the authority of science in public settings may be traced back to the impersonality afforded by its reliance upon rule-governed and precise quantifications. I prefer to see the connection between science and "numbers" as contingent, variably made or severed as players—"scientific" or not—assign or deny credibility to claims. George Combe presented his phrenology as "estimative" rather than "exact" (chapter 3), and Sir Albert Howard came to prefer informal testimony over statistical results as evidence for the efficacy of organic fertilizers (chapter 5). But at an earlier stage in his life, Howard—and his wife Gabrielle—made every use of the latest measures of correlation to show the superiority of the new strains of wheat they had hybridized.

kind up to 100 percent of breakeven under very low-yield conditions, and got 60 percent of breakeven . . . when we generate over 20 watts per cubic centimeter—configurations actually suitable for that—and if you extrapolate that, we would be at several hundred percent of breakeven in this experiment." Genuine scientific experiments require sophisticated instruments and carefully calibrated measurements, in order to acquire needed precision and to control for spurious effects:

> Well, we insert into the device thermistors—devices for measuring temperature rise—we then measure carefully the amount of heat we put in by monitoring the voltage and the current . . . applied to the cell, so we know the joule heat which is being put into the cell. We then monitor the amount of heat built up in the cell with the thermistors I just mentioned, with the temperature measuring devices. This is put into a water bath, so that the outside of it is maintained at a very constant temperature. Then we calculate what is called the heavy water equivalent and the other thermodynamic constants which we have to know in order to be able to calculate the loss of heat from the cell, and then from those numbers we can come up with the figures which we have.

Like good scientists anywhere, Pons and Fleischmann show that the observed results (excess heat, neutrons, gamma rays) could only have been produced by a process worthy of a press conference: nuclear fusion. They describe the care taken to rule out alternative, less interesting explanations for the findings, like an ordinary electrolytic chemical reaction—if that's the cause, no dynamo. The press release states: "The scientists know their experimental result is fusion in an electrode because the generation of excess heat is proportional to the volume of the electrode. 'This generation of heat continues over long periods, and is so large that it can only be attributed to a nuclear process,' Fleischmann says." Pons drove home the it-must-be-nuclear-fusion point during the press conference, listing the telltale by-products that could not possibly result from anything chemical alone:

> The heat that we then measured can only be accounted for by nuclear reactions. The heat is so intense that it cannot be explained by any chemical process that is known. We have direct measurements of neutrons by measuring the gamma radiation, which builds up in a tank where one of these cells is under operation. We . . . have a gamma ray spectrum of the neutrons as they interact with the water to form a gamma ray and another deuterium atom in the water. In addition, there is a buildup of tritium in the cell, which we measure with a scintillation counter.

The revelation that this breakthrough science took place in the basement of Utah's chemistry building prompted several jokes and nervous embarrassment among university administrators—relieved by Fleischmann's justification in terms of good scientific practice: "There is a reason why we are in the basement, 'cause we want all that concrete above us, to help cut the cosmic ray neutrons out as far as possible, and we also see the gamma rays generated in the vicinity of the electrochemical cell. So they can only originate because of the neutrons coming through the glass wall into the water, reacting with water to generate [the] gamma rays we observed vertically above. So, basically, that's the story." The journalists and reporters must surely have agreed: What a story!

The credibility of the cold fusion claim was enhanced even more by assurances of Pons and Fleischmann's respect for the normative conventions of the scientific community. The press release stated flatly that "their findings [would] appear in the scientific literature in May," and Brophy responded laconically to a journalist's question about whether the work would be submitted to a professional journal: "The scientific journal paper will have much more detail than you are hearing this afternoon." Chase Peterson balanced the exactness and precision of the observed results with a properly scientific humility, expressing gratitude for all scientists whose hard work had enabled Pons and Fleischmann's discovery, while reminding everybody that absolute truth is not the stock-in-trade of science: "Does [scientific experiment] always give clear and full answers? No. Science grows like rings on a tree, each larger, but shaped by the ring it grew from. The full story of the research [that] Professor Pons and Professor Fleischmann will announce today will not be known for months or years, as others confirm, challenge, and enlarge their ideas in their data." Peterson was nothing if not prescient in anticipating the scrutiny cold fusion would receive immediately following the press conference, but his opening remarks would clash with later criticisms that Pons and Fleischmann obstructed the process of replication and certification by withholding important experimental details. That day in Salt Lake City, Fleischmann welcomed the involvement of his scientific peers: "The subject has to be fully researched; the science base has to be established. I would emphasize that it is absolutely essential to establish a science base as widely as possible, as correct[ly] as possible, to challenge our findings, to extend our findings." In the press release, Fleischmann separated the purely scientific implications of cold fusion from the engineering applications that really lured the press: "What we have done is to open the door of a new research area." On two occasions during the press conference, Fleischmann playfully chastised himself for departing from the sober role expected of those who speak with the weight of science: "Now 'indefinitely' is an emotive word [and later] Vice President Brophy won't like me using this emotive language." No matter that cold fusion could thoroughly transform human existence as we know it—the scientist remains calm, unemotional.

These extracts are likely to induce the tedium that comes from reading exactly what one expects to read. How boring to be told that the discovery was made by two scientists, piqued by anomalies in nature, struggling through five and a half years of controlled experiments with sophisticated instruments, their results expressed in technical argot, their findings submitted to peer-reviewed journals, their rejection of alternative explanations and cautious uncertainty about how the claim would withstand peer scrutiny, their concern with getting "the science" straight before engineering applications are undertaken, and on and on. But "boring" is precisely the sociological point: to give cold fusion the credibility worthy of a worldwide broadcast, the Utah gang shrouded the surprising events in the utterly unsurprising, droning on about the obvious features of normal science as exemplified by Pons and Fleischmann. This rehearsal of normal science is sociologically interesting not just because it lent cold fusion credibility, but also because it would differ from accounts of science every bit as "normal" offered up to Congress in Washington and by physicists in Baltimore. We shall see (a) that the sequential relationship between science and technology described here matter-of-factly by Fleischmann would become the centerpiece of debate about whether $25 million tax dollars should be invested in cold fusion, and (b) that the supposedly routine and obvious good laboratory practices of Pons and Fleischmann would be used by others to move the pair perilously close to the fringe of science, if not beyond the pale. There is no single set of attributes defining normal science, only variably selected subsets of attributes that work, more or less effectively, to advance identifiable interests in contingent circumstances. Trotting out tried-and-true images of science, plopping cold fusion inside the familiar space, worked well enough to give the claim enough credibility for subsequent reporting by media more respectable than the *Star* or *National Enquirer.*

The press conference still had much to do. If the claim to cold fusion was science all right—then why *these* scientists, and why Utah? The media had to be curious: how had Pons and Fleischmann managed to scoop the rest of the scientific community, especially geniuses based at centers of research excellence like Harvard or Stanford, MIT or CalTech? An effective answer required strategically careful presentations of self: Pons and Fleischmann must be ascribed the kind of credentials that would make them capable of such a discovery (in terms of training and experience), but also unique personal qualities that would account for the failure of all other scientists to reach cold fusion before them. Chase Peterson tackled both problems in his introduction of Pons and Fleischmann: "The breakthrough they will report today comes from the work of trained minds working at an old problem from a new perspective. This particular study examines a traditional problem in physical science from the chemists' point of view, specifically, that of electrochemists. This university prides itself, whether it be in creative writing, or dance, or chemistry, or genetics, or artificial organs, in a long tradition of intellectual freedom, intellectual excitement, and a willingness to try new ways to solve old problems."

The two scientists were given track records that made it not just plausible that they would discover cold fusion, but that they were uniquely suited to getting there first. The press release gave one-half page to their accomplishments: "Fleischmann has written more than 240 articles in the electrochemical, physics, chemistry and electrochemical engineering fields, and is regarded as one of the leading electrochemists in the world. He is a Fellow of the Royal Society of England. . . . Pons has authored more than 140 articles and has lectured throughout the United States, Canada and Europe." Their combined multidisciplinary experiences were described during the press conference as particularly relevant to the kind of research required for successful cold fusion: "The researchers' expertise in electrochemistry, physics and chemistry led them to make the discovery. 'Without our particular backgrounds, you wouldn't think of the combination of circumstances required to get this to work,' says Pons. Some may call the discovery serendipity, but Fleischmann says it was more accident built on foreknowledge. 'We realize we are singularly fortunate in having the combination of knowledge that allowed us to accomplish a fusion reaction in this new way.'"

Reciting Pons and Fleischmann's accomplishments and expertise could not in itself account for why other scientists—presumably with training at least as extensive—had not yet done a viable cold fusion experiment. The added ingredient was mentioned by Chase Peterson in his allusion to the distinctive environment of intellectual freedom at the

University of Utah—conducive to seeing old problems in new ways.[18] Pons and Fleischmann rode into the spotlight as maverick cowboys from the Wild West, risk takers willing to tackle the daring research that nobody else would waste time or money on. If the experiment itself exemplified normal science (done in a lab full of technical instruments, under carefully controlled circumstances), the decision to start such work was so crazy that no normal scientists would risk it. Some press accounts even hinted that strong liquor was present, the classic rationalization of dumb decisions. Improbable research begins in improbable places: "The research strategy was concocted in the Pons' family kitchen. The nature of the experiment was so simple, says Pons, that at first it was done for the fun of it and to satisfy scientific curiosity. 'It had a billion to one chance of working although it made perfectly good scientific sense.'" The press release mentions other improbable places now reconstructed as plausible excuses for why two rational and nonemotive scientists would bother to do an experiment with such low odds: a long drive through Texas and a hike up Mill Creek Canyon. Fleischmann is quoted in the press release: "Stan and I talk often of doing impossible experiments. We each have a good track record of getting them to work. The stakes were so high with this one, we decided to give it a try." So daring were these mavericks that they put up their own funding for needed equipment and supplies, a measure of sincerity but also perhaps of the unwillingness of more sober scientific authorities to provide patronage: "Stan and I thought this experiment was so stupid that we financed it ourselves, and I think it would be safe to say that we have burnt up about $100,000 in the process." The press release version differs little, suggesting perhaps how tightly the whole affair was choreographed: "They decided to self-fund the early research rather than try to raise funds out of the University because, says Pons, 'We thought we wouldn't be able to raise any money since the experiment is so far fetched.'" Other press accounts mentioned that the intrepid scientists had relied upon ordinary Rubbermaid kitchenware. Scientists at MIT or Stanford might have greater expertise or fancier equipment than Pons and Fleischmann, but none evidently had their moxie.

The press conference itself must still be accounted for. The media were told that the first cold fusion paper had been accepted for May 1989

18. At the start of his remarks to the press conference, Chase Peterson welcomed "free minds, which is what we're about here today."

publication in a peer-reviewed scientific journal—so why spill the beans now, weeks before the entire scientific community would have equal opportunity to ponder the fully detailed findings? Involvement of the media at this premature moment was a departure from normal science, and possibly a risk to the credibility so far secured. Who—back then—*didn't* have dark thoughts that this jumping-the-gun press conference portended trouble: Did Pons and Fleischmann fear the response of their scientific peers. If so, why? Was there a priority dispute lurking behind the microphones?

Not to worry: the Utah team offered three compelling reasons why the show had to start immediately, erasing for the moment any doubts that this might not be a science story worth broadcasting. First, the "implications" of this scientific experiment, as Chase Peterson described them, were so wonderfully heavy that the news had to get out to everybody as soon as possible. Having established that the successful experiment would advance pure science by "open[ing] the door of a new research area," the Utah administrators dropped the other shoe by conjuring the universal panacea that cold fusion could become. The press release was hardly bashful in spelling out how cold fusion would solve our energy problems, and much more. "Nuclear fusion offers the promise of providing humanity with a nearly unlimited supply of energy. It is more desirable than the nuclear fission process used today in nuclear power plants. Fusion creates a minimum of radioactive waste, gives off much more energy and has a virtually unlimited fuel source in the earth's oceans." The problems solved by cold fusion touch everyone, as Brophy announced: "It has been clear for three or four decades that the promise of virtually unlimited, radiation-free energy is something that is worth spending perhaps billions of dollars on. And if indeed, this scientific discovery proves to be [as] practical as appears to be, not only do we the world's population get a promise of virtually unlimited energy, [but we'll see that cold fusion leads to] the elimination of acid rain, reduces the greenhouse effect, and allows us to use fossil fuels in a way which is much more important than simply lighting a match to them. Those are very valuable chemicals which are the source for production of plastics and a whole host of other things." As if that were not enough to get media-consumers interested, Chase Peterson later added dying "trees in the Black Forest" to the list of things that could be helped by cold fusion. Pons and Fleischmann, emotions restrained by their scientific role, still managed to throw considerable caution to the wind. Pons saw no reason to equivocate: "We've been concerned primarily with the effect, the observation of the fusion event. I would think that it would be reasonable within a short number of years to build a fully operational device that could produce electric power or drive a steam generator or steam turbine, for instance." Fleischmann was more hesitant, but no less confident about the panacea at hand: "We don't know what the implications are . . . but it does seem that there is here a possibility of realizing sustained fusion in a relatively inexpensive device, which could be brought to some successful conclusion fairly early on." Such news could not possibly be hidden away on the pages of the *Journal of Electroanalytical Chemistry*.

The second justification for calling in the media is evidenced best by what was not said in the press release, and only obliquely mentioned in the press conference itself. Brophy explained why the release of news about the discovery could not wait until publication of the results in a scientific journal: "We have chosen to have a press conference this afternoon frankly because the results are so exciting—[and to] set the record straight, so to speak." What record needed straightening? Several early news accounts suggested that the unusual press conference was called to save scientists' lives. One cold fusion cell in Pons and Fleischmann's lab had exploded. With hints that enough leaks about the discovery could enable and encourage scientists elsewhere to attempt replication, the two decided to issue some experimental details as quickly as possible so that their peers would not blow themselves up. Interestingly, nothing was said about this meaning of "setting the record straight" in the press release or during the press conference—perhaps because Brophy had already admitted that many details of the set-up would not be revealed until publication of the first journal paper, implying that scientists hoping to try cold fusion for themselves would still be in the dark about exactly what to do even after the press conference and media blitz.

Another meaning of "setting the record straight" was inescapable: "The fusion technology is owned by the University of Utah, which has filed patent applications covering the technology." The press conference sought to make perfectly clear that cold fusion was a Utah breakthrough, a Pons-and-Fleischmann discovery, and Peterson said as much:

> They'll be asking those questions about where all the credit lies and where the ownership lies, and with the ownership, if it turns out to lie with the University of Utah, as we think it will, then we would do all in our power to have this exploited by ourselves and others, for the benefit of cheap energy with little cost to the world's ecology. We would also like it to come to the benefit of the economy of Utah. That's not always easy to guarantee because ideas

> aren't contained by borders—but perhaps ownership and patents are, if indeed there is ownership and if there are patents that come out of this.

This was quite a rhetorical dance! "Credit" referred to the scientific discovery (measurement of excess heat caused by nuclear fusion in a room temperature electrolytic cell) and belonged to those who accomplished cold fusion first. Ever respectful of the norms of the scientific community, these "ideas" about nature were treated as communal property, owned by no one, restricted by no borders, available to all for certification and extension. None of that applied to technologies spun off from the science: "ownership" of patents belonged to Utah, who would develop fusion energy for its own benefit and for the benefit of the rest of the world too.

Geoff Bowker has shown that patent claims sometimes elicit contentious "same/different" debates—not unrelated sociologically to the same/different debate over the placement of social science in the National Science Foundation (chapter 2).[19] The distinctive novelty of a technology proposed for patenting can always be contested by others arguing that it is "just like" an extant technology of their own. Surely aware that no newspaper or television news program would report a discovery made for the second (or *n*th) time, the Utah administration used the press release to set the record straight about the originality of the findings: "Scientists worldwide have searched for more than three decades for the ability to create and sustain nuclear fusion reactions, which are thought to be the ideal energy source. . . . Prior to the breakthrough research at the University of Utah, imitating nature's fusion reactions in a laboratory has been extremely difficult and expensive. In the Utah research, the electrochemists have created a surprisingly simple experiment." Pointing out that scientists around the world had long sought the discovery that only Pons and Fleischmann achieved made the media circus into even more normal science. Perhaps no one at Harvard or CalTech had tried cold fusion because it was just too simple to bother with—still another plausible interpretation of why Pons and Fleischmann improbably got there first. "Conventional nuclear fusion research requires temperatures of millions of degrees, like those found in the sun's interior, to create a reaction. The Utah research, however, creates the reaction at room temperature." No one could possibly miss the difference between the heat inside the sun and room temperature, seemingly making patent attorneys

19. Bowker, "What's in a Patent?" On the boundary between science and technology in patent writing, see Myers, "From Discovery to Invention."

superfluous. But still Pons elaborated those differences during the press conference, referring to but not naming others who might have thought the cold fusion discovery was their own:

> *Cold* nuclear fusion, for instance, is well known. Muon-catalyzed fusion has been known since Sakarov in the forties. . . . It's a totally different process, and that can be run at very cold temperatures. . . . We're twenty orders of magnitude higher up in that one parameter than they are with a conventional fusion device. . . . In a tokamak or one of those devices, they may be considerable smaller than we have here—I mean higher than we have here. We have very low values of the chemical potential compared to what they have, but we have a very large confinement time. So all the basic physical parameters are still met.

Pons was distinguishing Utah's room-temperature process from the even colder older nuclear fusion, muon-catalyzed fusion, and from the hot fusion pursued in huge tokamak reactors constructed at Princeton, for example.

Nothing more was said that day about muon-catalyzed fusion, but at least some of the journalists at the press conference knew what had been left out. It might well have breached scientific decorum to dredge up a priority squabble while announcing a startling breakthrough, especially when the rival was based just a few miles down the road and had previously exchanged ideas with Pons and Fleischmann. Sleuthful journalists would be left to find out about this compelling sidebar for themselves. The day after the press conference, Salt Lake City television station KUED broadcast its regular program *Off the Record* with a feature on the cold fusion announcement. The rival scientist absent from the press conference was revealed by Ed Yeates, science reporter at the local KSL-TV:

> I think you have to understand the background of what's been happening in this fusion research, between BYU [Brigham Young University] and the University of Utah. . . . Dr. Steve Jones of BYU has been working on the fusion project for years and years. But his specific experiment is not the same as what was revealed here at the university. We're talking about two completely different kinds of experiments. Dr. Jones has been working on something called [the] muon[-catalyzed] concept of fusion, which is not an electrical-chemical process. The University of Utah chemists have been working on this particular project completely separate from BYU—the two of them are not even related. . . . BYU thinks that there has been some kind of betrayal because, in March, Dr. Pons and other University of Utah officials went down to BYU and . . . talked with them, with Dr. Jones. [There] was apparently a

> verbal agreement, according to BYU, that when the results of these studies were to be released, that they would be released in a back-to-back fashion. That is, BYU's project and the University of Utah's project would be published back-to-back in a scientific journal, the peers could review it, and that's how it would be handled. It would not be handled in a news conference kind of setting. So when this happened, it surprised BYU—they felt somewhat betrayed because this verbal agreement had been made, and the University of Utah announced it to the press. . . .
>
> . . . Now, to understand even some more cloak-and-dagger . . . : in September of last year, the University of Utah applied to the Department of Energy for some research money for this project. As is done in the scientific community, according to protocol, when you send in a research project to the Department of Energy, they send it out for review among other scientific groups, and then it comes back to the Department of Energy, and they decide whether to grant the money. It so happens that this particular report went to Dr. Steven Jones of BYU, who also had a Department of Energy grant. So Dr. Steven Jones read the University of Utah report, which was somewhat secretive. . . . The University of Utah may have had a little bit of paranoia in this thing, believing that because Dr. Steven Jones had had a preview look at this very exclusive research: what is to stop the researcher from taking some of that information and incorporating it as part of his own research. . . . That kind of paranoia exists today in the scientific community, because of the highly competitive nature of scientific research now, and [the need] to get research money. Right or wrong, the University of Utah announced it to the press as a whole, and BYU in the process felt a little chagrined.

The media only later let the Steve Jones priority squabble out of the bag, and in assuming this responsibility saved the University of Utah team from being asked to delve into possibly sordid and unprofessional details of Pons and Fleischmann's relationship to the research at Brigham Young. At the same time, in broadcasting the priority issue, the media offered an unassailable justification for why the press conference really had to happen right then: after all, Jones and Brigham Young, pressed by the same competitive pressures for fame, research funds, and patentable energy panaceas, could have become trigger-happy themselves. Despite the intrigue created by a backyard dispute, Yeates and other reporters left the press conference with the University of Utah's intended message: Pons and Fleischmann's research was essentially unlike the muon-catalysis fusion experiments of Steve Jones, just as it was unlike the tokamak hot fusion programs, or anything else—it was *original.* Ironically, as doubts about room-temperature fusion began to grow, it would be Steve Jones who would tell all (in Washington, in Baltimore) just how dramatically different his experiments were from those of Pons and Fleischmann.

The third justification of what would later be derided as "discovery-by-press-conference" was in effect an invitation to the media to join Pons, Fleischmann, Brophy, and Peterson in *making* the scientific breakthrough—to become an instrumental and necessary part of what had begun almost six years ago with Jack Daniels in Pons's kitchen and a hike up Mill Creek Canyon. The media were asked to add their own instruments (videocameras, tape recorders, images, the printed word) to the technical equipment (calorimeters, scintillation counters, palladium rods) that had carried Pons and Fleischmann up to this exciting moment but could carry them no further. The press conference became part of science, as journalists were enlisted as vital allies in making a scientific discovery. Ethnomethodologists Garfinkel, Lynch, and Livingston have taken the elusive moment of scientific discovery upstream, displaying the kind of interpretative and discursive work needed one night at a British astronomical facility to convert an array of dots on an oscilloscope into a breakthrough (was there a mouse gnawing at some wires or is this—could it possibly be—the first empirical evidence of a pulsar?).[20] The cold fusion press conference took the moment of discovery downstream from the lab bench—suggesting at the very least that there is no moment of discovery, but rather a flow of practices, representations, and readings with nebulous starts and finishes.

What could reporters do that neither Pons or Fleischmann, nor Brophy or Peterson, could do for themselves? The media here became the late-twentieth-century equivalent of what Steven Shapin has described as the "gentlemen witnesses" so vital for attesting knowledge claims three centuries earlier.[21] Originally, a gentleman scientist might visit a colleague's lab to examine firsthand his experimental apparatus. This tradition soon gave way to a more removed peer review, as scientists learned about others' experiments through their representations in what would become scientific journals. At first, print accounts contained every minute detail of each instrument or procedure, though with time the scientific paper became increasingly elliptical as more and more parts of the experiment became routine, not requiring elaboration. In a sense, involvement of the mass media in scientific discovery extends that trend toward enlarged ellipses in reporting experimental details. Journalists at

20. Garfinkel, Lynch, and Livingston, "Discovering Science."

21. Shapin, *Social History of Truth*.

Salt Lake City became privileged witnesses to the physicalness of the experiment, privileged not because of their expertise or trustworthiness but because they were needed as conduits for a "second-order virtuality." Their broadcasts and stories carried images and words about what would be presented eventually in a scientific paper, another account of what actually happened in Pons and Fleischmann's lab. Details may have been lacking in these journalists' accounts, but details are *always* lacking—even in the most detailed scientific papers. The question is whether these representations of representations were sufficient to do one last thing the press conference was designed to do: to make it impossible for the best scientific minds at the leading research centers to ignore the claim to cold fusion.

How were the media brought inside science? A videograb from the media tour of the Utah lab adds yet another interpretative layer to the palimpsest of (a) Pons and Fleischmann's actual experiment; (b) their account of it to be published in a scientific journal; (c) the mass media's gloss of the scientific paper. It shows the introduction of the mechanical witness (the videocamera) into a laboratory full of research equipment and discoverers. At the press conference, journalists were put on alert by Brophy that "following the question-and-answer period, there will be an opportunity for the press to view experiments underway in the basement of the chemistry department." It was clear that all this hoopla was staged for them: "In order to conserve time, we confine questions [to those] from the press," Brophy told an auditorium filled in part with scientists, who might have been expected to ask more informed questions about the experiment than journalists. There was an unmistakable dumbing-down at the press conference, as Pons and Fleischmann tried to make their discovery accessible to the media, hoping that the unfamiliar technical language, mathematizations, and black-boxed instruments that reassured everybody that something scientific was going on would not also exclude journalists from active involvement. Now the breakthrough experiment became, in Fleischmann's words "a scaled-up test-tube with which you might be familiar from your high school background." The press release described it as only slightly more challenging: "a surprisingly simple experiment that is equivalent to one in a freshmen-level college chemistry course." The scientists sought to give reporters the confidence they would need to write their science stories: anyone could plainly see what was going on.

And see they did. Watching a videotape of the parade of journalists into the place of scientific discovery, we hear from off-camera a voice:

"Videograb." A collage of instruments vital for the discovery of cold fusion, assembled in Pons' and Fleischmann's laboratory in Salt Lake City. The two discoverers stand to the right, watching their graduate student, Marvin Hawkins, as he explains the fuel cells and measurement devices. In the foreground is another essential device: a television news reporter sets up his video camera to capture the goings on.

"Just carry on with your discussion. Put the camera here, great." Journalists are escorted through the lab by graduate student Marvin Hawkins, whose crowd control gives further privilege to the press: "We need to clear people out that do not [belong] up here," and later, "if we can keep the people in the hallway" (a uniformed guard polices the doorway to this site of intellectual freedom). Hawkins patiently explains the experimental apparatus: "What we have here are the electronics that are driving it. . . . Over here, we have a neutron detector, and the batteries that drive it. . . . Here we have a liquid scintillation system. This we are using to detect the tritium. . . . That's our calorimeter." He answers questions: "We have not detected $Helium^3$. . . . This is something we haven't had, OK?" Every effort is made to involve the witness in experiencing the raw stuff of experimental electrochemistry and nuclear physics. Hawkins asks repeatedly: "Anything else?" "Any specific questions?" And at the end: "All right gentlemen, thank you very much." It is impossible to know how thoroughly the media understood what they saw in the lab, but to judge from their subsequent storytelling, they understood it well enough to be convinced that Pons and Fleischmann had tinkered with nature and that cold fusion was no hoax. And they understood it sufficiently well that they could—with materials gleaned from the press release and from statements made at the press conference—confidently put together an account of cold fusion that did not lie. With that confidence, they could undertake the role in this discovery designed uniquely for them: to report the claim to other scientists faster and with greater saturation than any professional journal possibly could.

Counterfactuals are always dangerous, but consider for a moment what might have happened if the cold fusion press conference had been a complete flop. Suppose Pons, Fleischmann, Brophy, Peterson, and whoever wrote the press release had failed to establish the claim as credible, original, and significant. No articles about cold fusion in the *New York Times* or the *Wall Street Journal,* no images of Pons, Fleischmann, and their cells on television news—and without these, no scientists staring at blurry CNN representations of the experimental apparatus, guessing its exact makeup as they try to put together one of their own. Without a successful press conference, no one would have had a clue that something interesting was happening in a University of Utah basement, and even if an article were to be published in the *Journal of Electroanalytic Chemistry,* it would either go unread, be read and ignored, or be summarily dismissed as yet another anomaly in a long tradition of false hopes. Without a successful press conference, without vidoecams in the lab, no scientific discovery could possibly have emerged from Pons and Fleischmann's flasks.

But everything worked to perfection. The public now insisted that scientists elsewhere take a long look at Pons and Fleischmann's cold fusion—could it possibly be true? Why didn't anyone find this out before? Will this really offer ecologically friendly energy too cheap to meter? Before the public could formulate those questions, and before scientists outside Utah would bother to answer them for themselves, the media had to decide—as they crafted their stories—the credibility, originality, and significance of this knowledge claim about nature. What could possibly be more "scientific" than that?

SCALING UP: MR. PONS AND MR. FLEISCHMANN GO TO WASHINGTON

The four weeks following the Salt Lake City press conference were a rocky road for Pons and Fleischmann's cold fusion. "Fusion scorecards" were published in the mass media, one labeled a "credibility index": a confirmation here, disconfirmation there, confirmation retracted, then something weird. By the time Pons, Fleischmann, Peterson, business maven Ira Magaziner, and others appeared before the House Committee on Science, Space, and Technology at 9:45 a. m. on Wednesday, April 26, 1989, in Room 2318 of the Rayburn Office Building, response by scientists to the possibility of cold nuclear fusion was equivocal but leaning to thumbs-down.[22] Even the most successful replications could only partly reproduce Pons and Fleischmann's results (those who found tritium did not find neutrons, for example). The supposed successes were mainly from institutions not located on the cutting edge of natural science, and skeptics explained them away as the result of sloppy procedure or erroneous interpretation. Still, cold fusion was alive enough in the last week of April for freshman congressman Wayne Owens (of Utah) to float the idea for a bill that would provide $25 million of federal support for a research center dedicated to following up Pons and Fleischmann's lead. The hearings called to solicit testimonials from diverse experts polarized into two camps: those supporting an immediate congressional appropriation specifically for cold fusion research versus those urging a wait-and-see response.

At the Salt Lake press conference, the mass media had been invited to decide on the credibility, significance, and originality of Pons and Fleischmann's claim. A month later, a congressional committee was asked to decide quite literally on its *viability*. With financial support and encouragement, would cold fusion grow into the credible fact and energy panacea it had promised to be at the press conference? "Scaling up" now took on a double meaning: in order for Pons and Fleischmann to scale up the tiny quantities of excess energy (heat) produced by a kitchen experiment into amounts that could power the world, the financial and instrumentational and manpower investments in cold fusion research and engineering would themselves require scaling up. Deep-pocketed Congress may have been a friendlier place to seek institutional support than other customary sources of research funding—the Department of Energy or the National Science Foundation—that would have required review by peers who were rapidly lining up against the truth of cold fusion. Even so, the Utah team had to convince the Science Committee that their hearing room was indeed a legitimate venue for making decisions about the viability of a claim to natural knowledge. The pro–cold fusion lobby had to make the Honorable Robert A. Roe and his committee

22. Taubes provides the most thorough account of this four-week roller coaster, in his *Bad Science*.

members feel competent enough to decide whether $25 million invested in Pons and Fleischmann's claim was taxpayers' money justifiably spent. Just as the initial press conference sought to enroll the mass media in making cold fusion into a claim that could not be ignored by the public or by scientists, now the goal was to enlist Congress by giving these politicians an equally pivotal role in scientific discovery: deciding on the viability and implications of a claim, and thus steering the direction of future scientific inquiry.

How did Congress get made into a proper place for doing science? Based on analysis of a complete transcript of the April 26 proceedings, "Hearing on Recent Developments in Fusion Energy Research," I suggest that the lobby from Utah moved questions about the viability of cold fusion onto terrain that fell well within the métier of elected government officials. Whether Utah got $25 million for cold fusion research was not to be decided by techniques of calorimetry, evidence for signatures of nuclear fusion, or interpretations that would require discarding extant laws of physics. Boundary-work was deployed in the service of boundary-work: the border between science and politics was erased as members of Congress took it upon themselves to map out the relationship of science to technology. Congressman Roe's committee was asked to decide between two spatial models of science/technology: (a) opponents of immediate funding for cold fusion put forth a "sequential" model in which a technological application can be developed effectively only after the scientific knowledge on which it depends has been thoroughly scrutinized and established; (b) promoters of cold fusion put forth a "simultaneous" model in which scientific research and technological applications develop in tandem, interacting with each other as they move along parallel tracks. The choice between these two models did not depend upon expertise in either electrochemistry or nuclear physics, but upon expertise that most politicians routinely claim for themselves: putting the taxpayer's dollars on a winning horse, and being able to justify the bet even when the nag comes in last.

These two representations of science/technology will surely be familiar to those involved in the sociology of science. In their influential collection of readings on science and technology studies, Barry Barnes and David Edge contrast two analytical models that resemble closely the two lines of rhetoric heard on Capitol Hill that day.[23] A "hierarchical model"

23. Barnes and Edge, "Interaction of Science and Technology." The authors draw heavily on classic treatments of the subject by Layton: "Technology as Knowledge" and "Conditions of Technological Development." On the boundary between politics and technology, see MacKenzie, *Inventing Accuracy,* 295, 413; Vaughan, *Challenger Launch Decision.*

suggests that science precedes technological applications, that science is creative while technological applications are routine, and that science is driven by the search for truthful knowledge transcending the time and place of its discovery while technology is driven by the search for practical solutions to problems irreducibly contextual and contingent—in short, the sequential model adopted by those urging Congress to wait and see before legislating funds for cold fusion research. A "symmetrical model" points to the simultaneity of discovery (science) and invention (technology), an interactive exchange of information and skill among those working on both sides, and the idea that both discovery and invention are creative, knowledge-making pursuits driven by hoped-for solutions to real-world problems—the simultaneous model favored by the Utah gang. As befits my theoretical constructivism, the sociologist's task is not to provide evidence and reasoning for the superiority of one or the other of these abstract models. Instead, the analytic models synthesized by Barnes and Edge are considered as working rhetorical representations used contingently and concretely to advance pragmatically an interest-laden cause—in this case, to get money for fusion research, or to deny it. The viability of cold fusion hinged on the preferred model of the association between science and technology—a choice that politicians were asked to make, with the immediate course of scientific research hanging in the balance.

The Case for Funding

The case for cold fusion funding was heard first. The Utah dog-and-pony show was scripted by Cassidy and Associates, a lobbying firm with demonstrated success in obtaining congressional pork for universities and colleges. Each of the four witnesses had a distinct role to play: Pons would review "the science" of cold fusion, just in case any committee members wavered on questions of credibility and originality of the claim. Fleischmann would turn from science to technology—indeed, collapse them as simultaneous pursuits. Magaziner would justify this simultaneous model of science and technology in terms of policies that would maximize American global economic competitiveness. Peterson would bring cold fusion back from the marketplace to the laboratory, and discuss how the $25 million would be spent. As the discussion shifted from science (narrowly drawn) to public policy, Pons's testimony receded in importance: in the end, the Utah witnesses suggested that the truth or falsity of the claim to room-temperature nuclear fusion was immaterial to the decision about whether to invest tax dollars in this line of research and development.

Pons brought the accoutrements of experimental science to Capitol Hill: technical talk, schematic slides, quantifications, even a hand-held cell (synecdoche for the basement lab back in Utah). After wading through "high-pressure electrochemical phenomena," "metal lattices," "hydrostatic pressures," "thermistor," "palladium electrode," "platinum anode," "B20," "deuterium absorption," "hydrogen evolution reaction," "0.8 volts potential," "atoms and ions," "helium-3 and a neutron," "tritium," and "gamma ray spectrum," Pons gets to the punch line that the congressmen have been waiting for. "If we try to explain the magnitude of the heat in the conventional deuterium deuterium reaction . . . we find that we have 10 to the ninth times more energy from these thermal measurements than that represented by this neutron and tritium that we observe" (613–18).[24] In other words, Pons is saying that he and Fleischmann have produced a whole lot more heat (excess energy) than would be expected if this were interpreted as an ordinary chemical reaction, given the quantities of observed tritium and neutrons. Pons explicitly avoids discussion of the technological or economic consequences of their discovery, and at one point implies that not all the ramifications have been thought through: "We have not considered any weapons applications whatsoever. I imagine that there could be social problems as a new technology begins, but we have not considered any" (cross-exam., 1672–74). His is the pure science role, the disinterested explorer of nature, the expert in electrochemistry. Pons remains uncommitted on the key decision the committee is asked to make, distancing his clean unstoppable science from the taint of politics: "I think we should hear the rest of the testimony and then make a decision on whether the establishment of a center or the establishment of ongoing research should be made. As far as personally, I think we are going to continue with our research, irregardless" (cross-exam., 1726–30).

Pons's testimony was designed to surround the discovery with the presumptive credibility routinely assigned to claims dressed up in science. In a sense, it was an optimistic and hopeful rejoinder to doubts expressed by committee members at the opening of the hearings: "if this discovery is fully proven" (90); "the two men who may have discovered cold fusion" (71–72). But if Pons had been the only cold fusion supporter to testify,

24. House Committee on Science, Space, and Technology, "Hearing on Recent Developments in Fusion Energy Research," April 26, 1989, mimeograph. The hearings are cited in the text by line number.

the congressmen would not necessarily have felt invited to join in this scientific discovery—because "the science" (Pons's technical talk) had not yet been merged with the domains of technology and public policy. The remainder of the Utah delegation did just that (as they made Pons's claims to fact less and less pertinent). But Pons had been asked only to play the straight man—drawing on and reproducing the authority of science, wrapping himself in the proverbial white lab coat, seeking truth not money, looking at nature not markets—giving the claim just enough credibility that subsequent discussions of technological marvels and economic bonanzas would not seem completely groundless.

Fleischmann begins his prepared statement by reassuring committee members that they will not hear another long and probably incomprehensible scientific soliloquy: "Professor Pons has just given you the essential experimental details" (630–31). In no way, he continues, is the science of cold fusion complete: "It is quite clear to us that a vast amount of new research is required. We want to extend the science base of the investigation and, in that extension of the science base, look for the appropriate theoretical description" (814–18). This statement does two things: it implies, first, that Pons and Fleischmann are aware (as the congressmen surely are) that attempted replications of their finding have not been uniformly positive, and second, that at least some of the money for a cold fusion research center will go toward the kind of basic science that the two researchers and the University of Utah are equipped to pursue. But Fleischmann's job is to bring science back *together* with technology, which he accomplishes by hypothetically scaling up the experimentally observed excess heat into far larger and more impressive numbers: "In fusion research, you have to project to a viable technology. This is part and parcel of fusion research so far, so I think we are justified in making such a projection ourselves" (711–16). He admits that his basic research with Pons was driven not by idle curiosity but by the practical potential of an energy panacea: "We have been more interested in the conditions for the heat release, which is really the social side of our research, and I will tell the members of the committee that the social considerations have, of course, been very much in our mind" (642–48). Fleischmann restates the dream, giving new meaning to the Baconian maxim that truth is power: "Our motivation was social. If this is correct, then we have a source of energy which is clean, which avoids the pitfalls of generating carbon dioxide and sulphur dioxide" (cross-exam., 1491–99). Pure science begets pure technology—no dirty by-products. One plank in the "simultaneous" platform is nailed down: both science and technology are driven by a desire to solve the critical problems of human society—like finding a cheap, clean and abundant source of energy.

The rest of the model is hammered on, as Fleischmann suggests that technological development of a fusion energy source should not be delayed until scientific understanding of the physical phenomena is complete and secure. He emphasizes that science and technology move along parallel tracks toward a common destination: "Part of our objective here today is to illustrate to you that this would be an opportunity where science and technology—technological applications—could be investigated at an early stage in parallel rather than sequentially, as has been the practice so far" (833–40). During cross-examination, Fleischmann urges the committee to foster a new association between science and technology—to abandon the sequential model for a simultaneous one: "What we are looking to is the resources to widen the science base and the theory base, and to try to short-circuit the consecutive development of this project and to attempt, for once, to initiate a parallel technological development at an early stage. That is really the substance of what we are looking to the committee for" (cross-exam., 1737–43). Fleischmann boldly asks the congressmen to liberate technology from its debilitating dependence on pure scientific knowledge, cartographically replacing S→T with S||T. The short-circuit metaphor is especially apt for a discussion of how to move from iffy scientific findings to a permanent solution for the energy crisis.

But why should the committee have bought into such a reconfigured culturescape? And was a congressional hearing room really the place to decide between two models that surely were much debated among experts in academe? The politicians were empowered by the next turn in Utah's testimony—from science and technology to public policy. Although this move fell within Magaziner's purview, it was Fleischmann who started off in that direction: "In a society with high interest rates and inflationary pressure, it is essential to shorten the commercialization of any new idea. That is a high risk strategy and we must be willing to say, if it doesn't work, oops, curtail it. Cut it off" (1826–31). Neutrons, excess heat, thermistors, and palladium rods had now been linked not just to energy breakthroughs but to the lowering of interest rates and the slowing of inflation—social problems that belonged as much to Congress as to experts from economics. Increasingly, the choice to invest $25 million in cold fusion appeared not to be about nature correctly or incorrectly described, but about the nation's economic future. Magaziner put the committee members on familiar terrain: "What I am concerned about is what is American public policy going to be with respect to this invention?" (1960–61).

The man who would become several years later one of President Clinton's closest economic advisers advanced a "bad old"/"good new" trope, tying the promise of cold fusion to the abandonment of traditional economic development policies that did not work and stretching the bad old sequence from science to technology to absurdity: "In the former days, we had a kind of chain of events that took many decades from basic research to getting products. Basic research was done in the universities, then company or Government laboratory scientists read the papers, produced and began to think about new technologies. Then company product divisions began to engineer specific product prototypes to take to their customers. Then the customers looked them over and suggested modifications. Then these products were introduced to the market. It was all done sequentially, it all took decades" (2014–25). Magaziner warns that cold fusion should not be allowed to go down the failed path of another recent scientific breakthrough, ending with this trump card: "Well, if we do what we did with high temperature superconductivity, we will work for awhile to verify and test the science; then the Defense Department will sponsor some work on how it could be useful to them. . . . Then what will happen is OTA will undertake an 18-month study . . . and will report that the Japanese have blown past us again in commercialization of another new science" (2064–76). The loss of American economic competitiveness in the global market was a hot button in 1989, and Magaziner pressed it again, casting the two science and technology models in high relief: "Today, these steps don't move sequentially. They move in parallel. Even before basic science is proven, applied research begins, product developments are undertaken, market research is done, and manufacturing processes are working. That's the way the Japanese and Europeans are playing the game. We in America are not playing it that way" (2026–31). Editorial cartoonists had a field day: one showed a "Fusion Honda" to represent what the American economy would lose if it continued to pursue the outmoded public policies on science and technology that Magaziner wanted to replace.

Magaziner was not shy about admitting the downside of his "simultaneous" model—after all, cold fusion was a race, if not among scientists to discover it, certainly among nations to develop and profit from it. "If we dawdle and wait until the science is proven . . . or if we move cautiously and invest only in basic research, or only in defense applications, and wait for the spinoff, we're going to be much slower off the blocks than our Japanese and European competitors. Whether we approve of the way they do it or not, that's what they do, and they move quickly to commercialize" (2139–48). The scientific truth of cold fusion is germane to the decision to invest $25 million only in the sense that a racing form is germane to wagers at the track: despite the collation of tons of reliable information about horses/physical reality, no one knows for sure when or if cold fusion will cross the finish line. Posttime is nigh! Don't get shut out at the window! "But wait a second, you say. The science isn't even proven yet. . . . We don't even know whether it's really fusion for sure, although I would say fairly convincing arguments have been made to say that it is. . . . I do know that if it is [real], the implications are dramatic for the world and in particular for the nation that pioneers the products based on it" (2105–21). The private corporation becomes a model for the kind of risk taking that Magaziner urges as a new government economic policy: "I would suggest that you get the process going . . . as if it's going to succeed. Just like the most successful companies in the world, over half the projects they'll try to invest in don't work" (cross-exam., 2386–90). Two things are certain: you can't win if you don't bet, but not every bet will win.

The final part of the Utah pitch was to create a vision of a new kind of research institute that would embody the parallel development of science and technology. Utah hoped that the $25 million would be added to state and corporate investments in what they were calling the Cold Fusion Research Institute. In the committee hearing room, Magaziner gets ready to hand the baton off to Chase Peterson, whose job it will be to describe how the federal money will be spent: "The alternative is to form a research institute around this new science, which will engage both in basic research and, very importantly, in commercial development work" (2080–84). Peterson's prose takes an architectural turn that is both metaphoric and literal: "We may be obliged to build the first floor of commercial development, as well as the second floor of engineering development, while we are still building the basement of scientific confirmation and enlargement" (2217–20). The private sector appears willing to invest in a new-style research and development center, even though they are usually hesitant to assume the risk of unproven scientific hunches: "We are working with companies and have talked about numbers of dollars that might be put into this sort of thing, some after confirmation, some even during confirmation of the scientific significance of this" (2353–57). Once again, the committee is told that an immediate appropriation is vital: "We should move very quickly. We propose to have ideas for your consideration literally within the next week or so" (2371–73).[25] As cold fusion began to stink in the week following these hearings, some might have questioned Peterson's use of this metaphor: "Twenty-five million dollars would allow us to start the 'onion' growing" (cross-exam., 2361–62).

The Case against Funding

The case against immediate funding for cold fusion research was made on the same discursive terrain that had been mapped out by the cartographers from Utah—the matter belongs to public policy, not ontology. Seven witnesses urged patience, recommending that the decision to earmark research dollars for cold fusion should wait until the results of further replications were known. The linchpin of the opponents' case was not that Pons and Fleischmann were wrong in their electrochemistry and nuclear physics, but that the sequential model of "getting the science right before technological applications begin" was a wiser public policy than the simultaneous model just presented. At first glance, it seems odd for seven scientists to recommend that Congress not spend $25 million on scientific research. There are several plausible explanations: Perhaps these witnesses were convinced that the odds of cold fusion being true were so low that to provide funds for a likely failure would jeopardize funding for future, and possibly more promising, lines of scientific inquiry (once burned. . .). Some may have believed that the money for cold fusion would come out of extant budgets for other research projects, maybe even those they depended upon. Perhaps those who came out against an immediate decision placed a higher value on the professional autonomy and authority of science than on a $25 million plum: after all, the Utah appeal sidestepped peer review, certainly not unprecedented in the Reagan years, but still an obvious attempt to shift decision making on the direction of scientific research from scientists to politicians. Or perhaps the simultaneous model used to justify cold fusion funding compromised the salutary image that science was the fount of technological wonders, and that those wonders were fully dependent upon getting nature right first. Whatever the reason for their opposition to immediate and targeted cold fusion funding, the next seven witnesses each found it useful to defend a "sequential" model of science → technology.

25. On the necessity for "speed" in science, see Fortun, "Human Genome Project."

The first witness for the opposition was the scientist so visibly absent from the Salt Lake press conference one month before. Steven Jones, the physicist from Brigham Young University, started off by putting some distance between science and technology—an essential element of the sequential model. With a metaphor easily understood by politicians, Jones in effect put cold fusion back into the lab: "There is nothing to get excited about from an energy production point of view at the moment. But the gap between the bona fide fusion yield and energy production by fusion is roughly equivalent to that which separates the dollar bill from the Federal national debt, a factor of about a trillion to one" (3663–69). Jones speaks from the experience of his "own" muon-catalyzed fusion—which he carefully distinguishes scientifically from what Pons and Fleischmann are up to. Still, in both cases, he says, the gap between experimental finding and commercialization is huge: "We have made tremendous strides in the last decade in research on muon catalyzed fusion. But I hasten to add that commercial power production would require a ten-fold improvement. . . . It is not at all clear that we can bridge that gap" (3513–18). Muon-catalyzed fusion had a seven-year lead on Pons and Fleischmann's cold fusion, but it was far from being able to power anything: "It's accepted that this is a fact, that we did achieve it in 1982 by muon catalyzed fusion. If we had announced it to the world, the public would have expected commercial power around the corner. As we see now from the perspective of seven years later, this was certainly not the case" (3574–79). The sequential model is crisply stated: "I think there is a need for support for basic research which could potentially lead to applications but doesn't tie into those possibilities. Because ultimately, progress comes from those starts, the new information. Then applications come later. I think that's obvious" (cross-exam., 4594–4600). But it certainly wasn't obvious in the earlier testimony of those other guys from Utah.

Both on Capitol Hill that day, and the next week in Baltimore, Jones fit himself into a humble, sincere, modest, and understated role that contrasted favorably to Pons and Fleischmann's hard sell. Plainly, it is the role typically expected of a scientist—think back to Tyndall's characterization of Faraday in chapter 1—and though Pons tried to exemplify it at the congressional hearings, his downplayed science was lost in overheated talk of energy panaceas and economic boom. Jones opened up what would be the leitmotif of opponents' testimony: the sequence "first science, then technology" is the normal, time-tested route to the successful economic development of new knowledge. For example, how embarrassing it would have been if the first inklings of muon-catalyzed fusion energy had led to an immediate appeal to Congress for research funding: "There was contemplation of starting a cold fusion nuclear research center, requesting large amounts of money from the Federal Government. . . . Those plans were put on the shelf until scientific confirmation, wisely so, and then were deferred indefinitely as it became clear that, while the process of muon catalyzed fusion is interesting, the possibility of energy applications are distant" (3585–91). The sequential model may have cost selfless Steve Jones a flashy federally funded laboratory, but that was a small price to pay for not wasting taxpayers' money on a line of inquiry so tenuously linked to real-world applications.

The sequential model dismisses Utah's sense of urgency: "At this stage, we need just a few months, perhaps two months, to evaluate the significance, and in this case also the facts, of the scientific discovery" (3581–84). Time for what? To do science, not engineering, not development, not marketing: "When a new idea like this one hits the scientific filter, it's very carefully checked. Experiments are done, theories are created. . . . We are at that filter now. We have not passed that filter, and it will take a couple of months to pass or not the filter" (cross-exam., 4698–4704). Here Jones anticipates a line that will be heard often at the American Physical Society meetings in Baltimore: adjudication of knowledge claims, including assessments of which lines of inquiry are promising or dead-end, belongs to the jurisdiction of "science." Peer review is not a just a means for deciding which papers get published or which grants get funded: it is an exclusion of non-scientists from judgments of scientific merit or promise. And this is the normal scientific way: "I believe that funding for cold nuclear fusion should come by peer reviews from such organizations as the Department of Energy and NSF, in the established peer review way" (3737–40).

Clearly, Jones does not believe Congress to be the place to scrutinize scientific findings: "I hasten to add that peer-reviewed and published papers . . . must be first presented before we can accept these and understand these results in detail. The findings have not passed the scrutiny of other scientists. So there is still a great deal of work that needs to be done to confirm and certainly to understand this process" (3651–59). As a parting shot, Jones implies that providing immediate federal funding might even impede the normal scientific evaluation of the claim: "I would like to compare cold nuclear fusion to this little plant, which is starting to wither—that may have some significance as well. . . . Let's give it a chance to grow. I think adding too much fertilizer at this stage will be detrimental. I think we need to give it time, at least a couple of months, please, to see whether this is something that's a rose or a tree. If it should turn out to be a rose, we can then admire it for its beauty, even if we are a bit disappointed it was not a tree" (3744–62).

The remaining six opponents added nuance and examples to Jones's clear articulation of the normal sequence science → technology. Daniel Decker, chair of Jones's department at Brigham Young, suggested that a scientifically coherent explanation for the Utah experimental findings—even if the data were replicated—would need to be found before any technological or commercial development could take place (and theorizing about nature is surely the scientists' domain): "If the Pons-Fleischmann effect is duplicated, we are still left with the puzzling question—what is it? Several hand-waving theories have been reported through the media. . . . Either [theory] will require some very subtle reasoning but until that question is resolved, we shall not know if the Pons-Fleischmann effect represents a new source of energy for an energy-hungry world or just a fantastic battery." Decker puts the sequential model into a down-home metaphor: "I think there's enough evidence that we should put some money into trying to confirm the science and finding out what is truly the source of the heat that is observed. I'm not sure we're ready to do the technology yet. At least I think, if my grandfather were here, he would say to me, don't say 'gee haw' to the oxen before you attach them to the covered wagon" (cross-exam., 4535–41). George Miley, director of the Fusion Studies Center at the University of Illinois, points to a historical precedent where departures from the sequential model proved costly: "Now, I like the thought: let's take some risks and try to develop something quickly. But if we jump into something and it crashes—I'm reminded of the sad experience in the early days of fusion in Britain, where the ZETA experiment reported neutrons, hit the headlines of all the papers, and it was a very exciting time. Then it was discovered these neutrons weren't thermal nuclear; they came from instabilities. The fusion program in Britain was set back years, years. The same thing would happen here if we launch an endeavor and view it as a 'wildcat' oil well, I'm afraid" (3973–83).

Mike Saltmarsh, from the Fusion Energy Program at Oak Ridge National Laboratory, reminds the committee that none of Pons and Fleischmann's claims have yet been replicated in a Department of Energy lab. Still, the reason to withhold $25 million is not because the claim is false, but because more time is needed to verify it—a position that preserves the open-ended nature of scientific research, and its need for continued funding (it also avoids the implication that fusion energy per se is not worth the research funding that it now enjoys). "The scientific process of verification is far from complete [though] careful scientific scrutiny will eventually provide solid conclusions on the reproducibility of the original results and on their interpretations" (4106–10). Harold Furth, of Princeton's Plasma Physics Laboratory, also states that the jury of scientific peers is still out: "But the immediate need is verification of the reality of the thing. . . . Don't skip the verification stage. . . . If [verification] is done, then the truth will come out in fairly short order" (4884–97). Furth adds that the committee could end up with egg on their faces, if they assumed a reality that wasn't: "But I think, before you launch on anything ambitious, you really should know whether it's for real or not. And in this case, you don't know it. I don't know it; maybe nobody knows it. This could lead to a very severe embarrassment, and seen from a scientific point of view, the scientific community is not thrilled at contemplating that, and I would think the Congress would not [be], either" (cross-exam., 5191–97).

Had scholars from the sociology of science been brought in as experts to advise the committee on the wisdom of appropriating a quick $25 million for cold fusion research, outcomes might have been different. As it turned out, Congress provided no funding explicitly for a cold fusion research institute in Utah, not in April 1989 nor anytime thereafter. Evidently, the tried-and-true sequential model offered by those skeptical of cold fusion became the preferred map of the culturescape: scientists (not politicians or journalists) must get facts straight before technological scale-ups and capitalist dreams can begin. But Barnes and Edge have announced the *death* of that sequential model, at least as a heuristic: "At the present time advocacy of the hierarchical [sequential] model is in steep decline, and among those involved in the serious study of science and technology it may already be defunct. It is no longer plausible to conceive of technology as applied science."[26] The point is not that Barnes and Edge are wrong, but that sociological attempts to adjudicate once and for all between sequential and simultaneous models of science and technology are themselves misguided. Policy outcomes (to say nothing here of professional credibility, authority, or autonomy) often hinge on debates over just this relationship of science to technology. Because of that, *both* sequential and simultaneous models will—selectively and contingently—be exploited rhetorically by social actors seeking to advance

26. Barnes and Edge, "Interaction of Science and Technology," 149.

diverse interests. "Some of the crucial debates in today's politics . . . have at their core conflicting views of the science-technology relationship. . . . All our quests for a definitive analytical solution to the problem of the actual science-technology relationship are in principle doomed to futility."[27]

The simultaneous model of science and technology fit the immediate needs of the Utah team. It created an interpretative context where decisions about the viability of Pons and Fleischmann's claim were moved out of the territory inhabited by those most suspicious of its credibility—scientists, with their doubts about neutrons, light-water controls, and de novo or ad hoc theories of nuclear physics. The question of viability was moved into territory inhabited by those who had the money to enable scaling up—congressmen who felt competent when making informed judgments about policies for economic development (even if they knew nothing about calorimetry or gamma rays). The simultaneous model brought these politicians within science by giving them an essential responsibility in the making of a scientific discovery: to provide the wherewithal for turning a kitchen experiment into an energy panacea by materially steering research toward hydrogenated metals and such. Like the journalists at the end of the Salt Lake press conference, what could be more scientific?

But as with George Combe's representations of phrenology in old Edinburgh, boundary-work doesn't always work. Things would get worse for Utah a week later, as "the science that must precede the technology" got put through the "filter"—or rather, the wringer.

RESTORING ORDER IN NATURE AND CULTURE: PHYSICISTS AT BALTIMORE

The Utah claim to cold fusion probably died around 9:00 P.M. on Monday, May 1, 1989, at the Baltimore Convention Center, during the spring meetings of the American Physical Society, in the midst of an invited talk by chemist Nathan Lewis of CalTech entitled "Calorimetry, Neutron Flux, Gamma Flux, and Tritium Yields from Electrochemically Charged Palladium in D_2O." Reporting on a series of experiments conducted since the March 23 press conference, Lewis presented results sequentially for each of the supposed signatures of nuclear fusion listed in his title: null, null, null, and null. His talk came immediately after one by

27. Mayr, "Science-Technology Relationship," 162.

Steve Koonin, a physicist colleague from CalTech, who laid bare the theoretical impossibility of the experimental results announced by Pons and Fleischmann. It was, according to eyewitness accounts, a fatal one-two punch.[28]

But that was only half of it. The scientists in Baltimore certainly tried to restore order to *nature* by preserving their established theories of it—an order that would have been jeopardized had Pons and Fleischmann's claim been true. However, the mugging of cold fusion by this gang also depended upon the simultaneous and inseparable restoration of order to *culture*. Professional scientific meetings (like this one for physicists) had to be (re)established as the rightful place for adjudicating claims about nature—not in press conferences, not in newspapers or media broadcasts, not in congressional committee rooms. This restoration of scientists' authority over the adjudication of claims began with explicit rhetorical exclusion of the media and Congress from scientific space, and ended with a four-part rationale for why scientists alone must be the ones to settle the credibility, originality, significance, viability, and truth of cold fusion. Scientists who went public in Baltimore presented their own cultural map of normal science, with landmarks such as collaborative specialized expertise; accumulation of established understanding in the published literature; technical skill in experimenting with nature; and an absence of distractions by fame or profit. These attributes in effect justified science as the unique cultural space for judging claims about nature, but—most important—they also pushed Pons and Fleischmann (who were said to lack them) to the edge of that space. As they were marginalized into bad science, Pons and Fleischmann were made into object lessons showing everybody the dangers of doing science where it should not be done, with outsiders who have neither the expertise nor experience nor technical skill nor disinterestedness of real scientists.

The organization of the program for the Baltimore APS meeting was itself boundary-work. Scientists were scheduled to speak to journalists at designated press conferences, which came on either side of two special "scientific sessions" on cold fusion. Only representatives of the media were allowed to attend the press conferences, and only scientists were allowed to ask questions during the scientific sessions. Joe Redish presided over the first special session: "The rules for the questions: please,

28. Phil Shewe, undated letter to Bruce Lewenstein, Cold Fusion Archives, Cornell University, Ithaca, New York; cf. Taubes, *Bad Science*.

this is a scientific session, let us take questions from scientists only."[29] Forty papers filled the two scientific sessions, which lasted more than four hours; about a dozen scientists who presented papers chose to speak separately and directly to the press. "Science" and "the media" became at Baltimore discrete, if still mutually dependent, cultural spaces, and—as we shall see—Congress was kept a safe thirty miles away in Washington.

At the beginning of the first scientific session, Redish, a physicist from the University of Maryland, put the media in their place: "I would like to address some remarks to the ladies and gentlemen of the press. You are serving for us as interpreters to the general public, and it's important to understand how certain aspects of the current debate illustrate the *normal* processes of science." The media here are no longer co-constructors of fact but just news givers who must wait until the scientists themselves decide nature (sometimes by fits and starts). The microphone goes first to APS president James Krumhansl: "The American Physical Society has as its purpose the advancement and diffusion of knowledge of physics. . . . We do our part through open meetings and refereed scientific publications. It is in that spirit that we want to provide this forum for the discussion of the subject tonight, that is, cold fusion, its possibility and implications." Thus begins the representation of "normal science," different from the science created at the Salt Lake press conference, which by contrast will become increasingly "abnormal" precisely for its preemption of peer review and its avoidance of "open meetings." Robert Perry of Ohio State tells the media that scientists must finish with cold fusion before it goes out to the public: "I'm hoping it'll stay in the scientific arena until it's understood what's going on, and there's consensus about a physical explanation of what is happening." He is echoed by J. K. Dickens of Oak Ridge National Laboratory: "We shall continue [to do research on fusion] because we should stay in the scientific arena as long as it's not understood." Redish is even more explicit: "Usually, preliminary results are challenged internally by the working group, then by peer review, then often again in the open scientific literature by other groups, before science goes public." The Baltimore meeting becomes a self-exemplification of how normal science challenges and then challenges again putative claims to truth. The author of the thirtieth scientific paper

29. The press conferences were audiotaped, and the special scientific sessions were videotaped. My transcriptions of these tapes made up the raw data for this analysis. The tapes were provided courtesy of the Cold Fusion Archive at Cornell University.

to question the existence of cold fusion—"It's just part of the Greek chorus. Nobody seems to find anything"—does not evidently believe that such redundancy is wasteful or unnecessary.

Normal scientists are said to resist the temptations of fame by declining to take premature conclusions into the limelight. Moshe Gai, physicist from Yale:

> Some of you members of the press know that our experiment was under extreme pressure from the press to reveal what we have. We are very proud that we have never let anything go out to the press. There was a conscious decision. We decided that we would not go out to the press and use any of our results even though, you know, being only sixty miles away from Manhattan, it's very easy for somebody from any magazine to hop to our lab and try to get results out of us. Basically, the results were discussed in what I would call scientific channels—that's why I gave a seminar at Yale, and yesterday for the first time we discussed it in public.

Doug Morrison from the European Center for Nuclear Research (CERN), whose Baltimore presentation first put cold fusion into a space labeled "pathological science," was the center node in the lively electronic network that grew up in the wake of Pons and Fleischmann's announcement. Morrison suggested that many scientists reported to him their inability to reproduce one or another bit of the cold fusion discovery, but that they were reluctant to go public for fear that they had missed something. Such electronically mediated peer review was kept by Morrison *entre nous:* "They are an informal network, they are meant to be academically confidential. . . . It's only academic people. I don't give it to the press. I turn down banks and oil companies if they ask me for it." Only as a response to what Pons and Fleischmann had done at Salt Lake City—brought reporters and their videocams into the lab—does this rhetorical exclusion of the media from science make sociological sense.

The demarcation of science from journalism—and from Congress too—is made reasonable through a cataloging of the risks and dangers inherent in their mingling. Briefing rooms and corridors of political power become places where it is easy to deceive habitués with lies passing as fact—in the absence of patient, critical, and informed scrutiny. Morrison recounts a public lecture given by Fleischmann at CERN on March 30, when cold fusion was riding high: "The excess heat was enormous, about 10 watts per cubic centimeter, but there . . . was this dream, the perfect thing that everyone is looking for, that you get enormous amounts of energy, and very little pollution. . . . The lecture hall was filled

. . . great lecture, great enthusiasm. *Most people weren't physicists,* you know. Great applause, everything was working, everything was right." Congress had also fallen for smoke and mirrors. In Baltimore, Lewis talks to the media about the supposed 4:1 and 10:1 ratios of energy in/energy out reported to the politicians by Pons and Fleischmann: "After we confronted them with this [it became clear] that the 4 to 1 and the 10 to 1 were hypothetical and calculated values. They were not experimentally measured." When asked by a journalist: "Would the congressmen tend to think 'If you're 4 to 1, you put one volt in [and] you get four volts out?'" Lewis replies: "That's what people would tend to think." He chides the press for their presumptuousness in believing that they could do what scientists do as a matter of routine:

> Finally, if we're going to have . . . publication through press conferences, we should have peer review through press conferences too. And so I think that if we ever have another press conference on this subject, I want the reporters to ask anyone all of these questions. [In reporting,] you should be prepared if you claim neutrons to tell me what the temperature dependence of the detector is . . . ; about what the background signal from the room was, and about what the signal was relative to water control; what was the gamma spectrum at a wide range of energies, not only one small peak, so that the world can see what the rest of the peaks look like, and know whether or not the signal you were measuring was real or the background. . . . I called up Florida, told them they should make sure about chemical interferences being eliminated, and if the reporters did that at the press conference, we might have saved oursel[ves] having to worry about that.

This is savage cultural cartography: journalists cannot do science, and they should not try. Leave fact making to us scientists.

The coup de grâce comes at the end of Koonin's formal scientific paper, where he reminds his audience that *this* is the right place—Baltimore, the APS meeting—for deciding the credibility of claims. In his wildly bowdlerized Aesop's fable, Rhodes becomes Salt Lake City or Washington, and the famous leap belongs to Pons and Fleischmann:

> A certain man who visited foreign lands and could talk of little [else] when he returned home . . . said what wonderful adventures he had met with, and the great deeds he had done for all. One of the feats he told about was a leap he had made in a city called Rhodes. That leap was so great, he said, that no other man could leap anywhere near the distance. A great many persons in Rhodes had seen him do it, and would prove that what he told was true. "No need of witnesses," said one of the hearers. "Suppose this city is Rhodes. Now show us how far you can jump."

Pons and Fleischmann were not present in Baltimore to leap again before a more skeptical audience, and they were absent for the worst reason—as Redish explains: "I spoke to Dr. Fleischmann in England about ten days ago. I had invited him to attend this session as soon as we decided to have it—by fax, they were in Erice [Italy] at the time. . . . I did not hear from them. When I called, they were in Congress, and eventually I got a phone message saying that they felt that they had too much work with the congressional committee . . . , so they would be unable to attend." The implication: Pons and Fleischmann preferred to advance their claim in the abnormal venue of politics rather than leap into the normal company of their peers.

The cultural cartographers at Baltimore sought to demonstrate that when journalists and congressmen get involved in fact making, they manifestly impede the efforts of legitimate scientists to get the facts straight. Stories flew around about how scientists had tried to replicate the Pons and Fleischmann experiment using images gleaned from CNN or diagrams in local newspapers. Dickens reported sarcastically: "We initiated our experiment almost immediately after the announcement . . . , taking of course our scientific and technical information from the *New York Times* and the *Wall Street Journal.*" To the great amusement of his audience, Lewis described the difficulties introduced when vital details come from the media rather than directly from scientists-who-know: "We built an exact cell as best we could from all the press photos we had of the Pons/Fleischmann cell. We measured the ratio of their wrist to their arm, and got a good scale." *Can* science proceed effectively with the media performing such an integral link between researchers? Johan Rafelski, Steve Jones's collaborator from Arizona, said no: "That is all . . . not so clear in the current literature—*Wall Street Journal,* or perhaps *Business Week* magazine." Perhaps the epitome of fusion confusion resulting from scientists' reliance on media accounts came from the Argentinean physicist A. O. Machiavelli: "We started following the news in the newspaper, so it was rather difficult at the beginning for us especially because on the way down [across the] equator, somehow something got screwed up. We were thinking of *magnetic* fields in the experiments!"

The Baltimore scientists saw a second liability in journalists' encroachments on scientists-at-work: pressure. Indeed, Rafelski blamed the emerging cold fusion fiasco on overzealous journalists: "The amount of attention that we are receiving right now prevents us from actually coming to a conclusion in our own usual fashion. We are humans, thus there is human nature, and with the attention we are receiving, some of us may be led to announce preliminary results. . . . I think it would be much better for the field if the attention would diminish." Moreover, the press raises false hopes among the public, which puts even more pressure on scientists to turn out miracle upon miracle. Steve Jones reflected with relief on his decision not to call a press conference to announce an earlier "achievement" of breakeven:

> I was at the Idaho National Engineering Laboratory, and the management there was very excited about this result. This was our first experiment . . . and they wanted to have a press conference right [then], to announce to the world that we had through muon-catalyzed fusion achieved scientific breakeven. We're very glad we didn't do it then. . . . The scientists insisted that first we had to publish our results, and we did so in *Phys[ical] Rev[iew] Letters* in 1983. By the time we published, which was several months after this experiment, the interest had dwindled. . . . It's a good thing I think, too, because if we had announced it as a breakthrough, people would have naturally expected there to be commercial power right away. . . . We took a little precautionary note, and we waited.

Many physicists at Baltimore chose to shoot the messenger in holding the press accountable for Pons and Fleischmann's perversion of normal science.

On the same note, we hear Jones tell his colleagues that the premature intrusion of politics into science has equally catastrophic consequences. He replays the song he sang to Congress: "I don't think this tree needs a lot of fertilizer right now, and I realize that has a little double entendre [laughter]. What I meant was—it doesn't need a whole lot of money right now, it just needs to grow, and we need to have the standard level of grants to check this out." Robert Perry also spells out the dangers resulting from politicians' efforts to decide which scientific claims are sufficiently viable to warrant unusual congressionally mandated research funds: "I'd hate to see funding right now taken away from another area of physics, or another area of chemistry."

More or less conventional boundaries of normal science were thus shored up in Baltimore: keep the media out of the lab in order to permit a patient, redundant sifting of truth from error, without pressure from unrealizable hopes; keep politicians away from decisions about the potentials of one or another line of scientific inquiry, and let the experts themselves decide how taxes-for-truth should be allocated. But why did these physicists belabor the obvious? Doesn't everybody know that science is not journalism or politics—that it sits aloof? Pons and Fleischmann transgressed these cultural boundaries to keep cold fusion alive.

Their discovery could not have happened without university administrators, lawyers, journalists, and a professional lobbyist; and its immediate viability in late April 1989 lay in the hands of politicians. An increasingly promiscuous bunch of truth makers were coming to inhabit scientific space—until the physicists at Baltimore began their work of purification, tossing out the congressmen and the reporters, restoring the walls of their jurisdiction over natural fact.

But something more was needed to complete their case for normal science: they had to make clear what exactly it is about science that makes its genuine insiders not just different, but *better* at getting nature right. The physicists' answers shifted boundary-work from science/media and science/politics to the line between good science and bad. That is, in order to put into high relief those qualities that make scientists uniquely authoritative and credible chroniclers of nature, the speakers at Baltimore pointed to the lack of such qualities in Pons and Fleischmann. Though few scientists at the APS meeting were willing to move the two outside science altogether, their secretive behavior, tolerance of inexplicables, technical incompetence, and weakness for publicity were presented as flaws that explained their apparent error. After first blaming the cold fusion mess on nosy reporters and congressional meddlers practicing science without a license, the physicists in Baltimore next blamed it on Pons and Fleischmann's failure to do good science. But as in any other episode of cultural cartography, "good science" becomes whatever is useful for protecting or extending the interests of mapmakers—in this case, restoring sanity both to established theories of nature and to stable classifications of institutional domains.

I've Got a Secret

Because they were unwilling to share details of their experiments with other scientists seeking to replicate their findings, Pons and Fleischmann's science had to be bad: good scientists tell all, and they tell it to their expert peers. By the time of the Baltimore meetings, the secretiveness of Pons and Fleischmann was old news and fodder for jokes. Doug Morrison was told that the overhead transparencies for Fleischmann's talk at CERN were "TOP SECRET." Gai from Yale deadpanned: "It was about the time we received the first paper [by Pons and Fleischmann], which came—like everybody else['s]—from a fax machine, where you couldn't possibly read anything on the paper aside from the big handwritten note CONFIDENTIAL—DO NOT COPY!" Once the blurry faxes were deciphered, Koonin found that they still lacked details critical not just for replication efforts but for understanding Pons and Fleischmann's theoretical interpretation of their results. "As in all [their] papers, one wants to see this in a little more detail." Attempts to contact either Pons or Fleischmann generally proved futile. Indeed, Lewis said, Stan Pons had "refuse[d] to answer any of our questions." On the important question of measuring heat accurately in a solution that had varying temperature levels: "He wouldn't answer us. We asked people at Los Alamos to directly ask him, and he said: 'We don't have to stir our solution because there are bubbles made, and the bubbles will be enough to eliminate thermal gradients.'" However, it turned out that, according to the CalTech experiments, the thermal gradients were not eliminated by bubbles but only by stirring. On most occasions, being mistaken about an experimental set-up is a forgivable sin, and it is also one that probably most scientists fall into at some point in their careers; on this occasion, the unforgivable sin of Pons was holding back details that would have allowed other scientists to "witness" the goings-on.

So foreign was such secretive behavior to what is considered good science that some Baltimore physicists decided that it could not have been by choice—it must have been forced on Pons and Fleischmann. News reports hinted that university attorneys had insisted that Pons and Fleischmann restrict details about their experiment until patents could protect rights to the profits anticipated from a clean, cheap energy machine. Whatever its source, the secrecy surrounding cold fusion explains why Pons and Fleischmann needed those uncredentialed allies to keep their claim viable: by not spilling the beans with credentialed peers, the Utah fusioneers cut themselves off from the vast networks of distributed expertise that might have spotted mistakes early on. Koonin contrasted the limited range of expertise of the two electrochemists with the diverse competences of those elsewhere who sought to validate their work, "many of whom bring to bear expertise and resources far greater than were available at the University of Utah." Those inspired to attempt the replication of cold fusion expressed their pleasure in having an opportunity to work with such a diverse mix of scientific talent, as Perry said: "One of the best things that's come out of this for me is that I've been able to work with condensed-matter physicists, theoretical physicists, theoretical chemists, experimental chemists, metallurgical engineers—and it's required that breadth of expertise to actually give this problem a fair hearing." Gai also suggested that specialized expertise from several disciplines was brought to bear on the Yale-Brookhaven collaboration:

"Another thing we learned from Pons and Fleischmann is that one should not do something that one doesn't understand. Right away, we decided that we needed chemistry on board. . . . We cannot do the experiment unless we have people who thoroughly understand the chemistry, and who can collaborate with people who can thoroughly understand the nuclear physics." In other words, this was *good* science.

Deferrals to specialists on technical questions were often heard in the APS presentations. Lewis points to "Steve Kellogg, who is sitting in the front row and who will help me with my nuclear physics." Koonin shows becoming modesty: "I don't know, because I'm not a solid state theorist." Robert Perry, too, acknowledges his limits: "That's something you might want to save for an experimental nuclear physicist [who] understands the neutron detectors fairly well." Ironically, this construction of good science as multidisciplinary, collaborative, and respectful of specialized talent challenges a reading of the cold fusion controversy that appeared in the (misguided) media, that it was all about competition *between* chemists and physicists for the last word on nature. One cartoon showed a physicist and a chemist squared off against each other, each putting enough hot air into their microphones to run a small fan with "fusion energy." The matter of disciplinary competition did surface in Baltimore, but—as in Nate Lewis's presentation—only as a joke: "You're pumping the cell up with heat, and this is hard thing for me to explain to physicists in my experience, but I'll try to do it [laughter]. No chemist jokes now!" Against the phalanx of nuclear physicists, chemists, metallurgists, condensed-matter physicists, and solid-state physicists assembled in Baltimore, Pons and Fleischmann's deep but narrow expertise in electrochemistry must have looked meager indeed. Because of their reluctance to share the essential details with scientists who knew different things than they did, Pons and Fleischmann got stuck with allies who knew nothing (at least, nothing that would have permitted cold fusion to survive the Baltimore onslaught).

Do You Believe in Magic?

With so much talk in Baltimore of faith, magic, blessings, and miracles, one could almost have mistaken the convention as a gathering of mystics or theologians. Perry confesses that the Utah findings require "two leaps of faith"; Rafelski says that "we just need a small materials science miracle" to bring about this "blessing for humanity." Someone asks Rafelski later if there was any "black magic associated with preparing these electrodes?" Doug Morrison implies something mysterious about cold fusion: "And also, when you say 'something happened,' . . . it just [doesn't] make sense."

To be sure, the cultural space for religion or metaphysics did not enter the Baltimore Convention Center as a rival authority over epistemic questions—that was John Tyndall's cartographic burden (chapter 1), or George Combe's (chapter 3). Instead, the boundary was charted out to show *by contrast* what has no place in good science—unsubstantiated belief—and to show that Pons and Fleischmann just had too much of it. Rafelski reminds everybody that "science is about knowledge, not about believing." Koonin adds that "there are no mysteries in nuclear physics." In response to Pons and Fleischmann's disclosure that one of their cells had exploded (taken by them as evidence that something nuclear, not chemical, was happening), Lewis explains the mystery not with amulets or incantations but with conventional chemistry and even a little conventional "sociology": "We can find no evidence for anything other than conventional chemistry in these cells. If you pass current through water, you get hydrogen and oxygen [laughter]. . . . We know that [Pons] wasn't there [at the time of the explosion]. His son supposedly came in and turned the current down . . . on a cube that had been running for months, otherwise uneventfully. We don't know actually if his son turned the current down or turned the current up." Count Pons's son as another discredited ally of cold fusion, joining newspeople and politicians.

As mapped out that day, good science moves from the known to the unknown, from agreed upon understandings to speculation. From Koonin's perspective as "a theoretical physicist, the logical thing to do is start from a system that you understand, and one that we think we understand is the [behavior of] hydrogen." Pons and Fleischmann need miracles because they have chosen to ignore what everybody else in science takes for granted, as Lewis tells the faithful: "Based on the first law of thermodynamics, that 0.5 volts would be impossible to access in this system, while maintaining this at the same potential relative to the reference." Pons and Fleischmann's science goes bad because they fail to consult the published scientific record in order to learn what has already been said about their experiment and what is already known about their speculative interpretations. "Belief" has substituted for the scientific literature. Morrison complains facetiously about his crowded desk at CERN: "There is an enormous literature about palladium, palladium hydrides, metallic hydrides—my desk is stacked with all these things. People just haven't read the literature." R. D. Petrasso from MIT suggests that the need for miracles vanishes if one bothers to go back and read the pertinent literature: "We've looked at the literature. It's been very well established for the last fifty years." Lewis's own science is, on this score, good: "We wanted to test that assumption. The first thing we did was to read the literature—that was a good thing to do." It was evidently easy for those at the APS convention to answer David Hume's question, posed by Doug Morrison: "Would you rather believe that all the laws of nature are wrong, or that one man [two maybe] has made a mistake?"

Elementary Egregious Errors

Good scientists are skilled in the manipulation of nature, or at least they are competent enough to make effective and reliable use of the experimental instruments of fact making. As the crowd in Baltimore became more convinced with each succeeding presentation that cold fusion was not fact but artifact, Pons and Fleischmann were increasingly seen as just not knowing what the hell they were doing in the lab—and this "status degradation ceremony" was pulled off without risking either a libel suit from eager Salt Lake City lawyers or the public's confidence in science and scientists.[30] The climax to Koonin's scientific paper brought down the house: "My conclusion, based on my experience, my knowledge of nuclear physics, and my intuition, is that the experiments are just wrong. And that we're suffering from the incompetence and perhaps delusion of Drs. Pons and Fleischmann." The possibility of fraud was only hinted at in Baltimore—did Pons and Fleischmann knowingly deceive?—perhaps because available evidence did not indicate sordid pasts, or because such an admission could shake the public's confidence in the honesty of scientists. I suspect that the epistemic authority of science was better served by admitting only that doing experiments like this is difficult and that not everybody in a white lab coat is equally skilled.

Cold fusion experiments required an array of more or less sophisticated equipment, from calorimeters to scintillation counters. Moshe Gai suspected that Pons and Fleischmann may have been led astray from nature by their use of inadequate technical instruments—to say nothing yet about how these were used: "When we read this paper, it really left us with a lot of worries. We're nuclear physicists, in our laboratories,

30. Garfinkel, "Conditions of Successful Status Degradation Ceremonies."

been doing nuclear physics for years and years. Myself, I'd been measuring neutrons for years, and it was very clear to me that inadequate equipment was used by Pons and Fleischmann." Almost all of the scientists who reported an inability to reproduce some feature of cold nuclear fusion suggested that the instruments used by Pons and Fleischmann lacked the resolution levels needed to reach the empirical conclusions reported in their paper, or alternatively, that the Utah team had failed to calibrate their instruments in a manner consistent with accepted experimental procedures.

Throughout the Baltimore sessions, every one of the indicators of nuclear fusion reported by Pons and Fleischmann is traced back to a goof, either in the experimental procedure, in the calculation of results, or in the presentation of findings. Supposed detection of helium and tritium is said to come from a failure to control for background noise. Nate Lewis addresses both artifacts: "They actually detected much more helium than they would have made, if all of the power were due to D + D fusion to produce $helium_4$. And that almost certainly suggests that that helium predominantly, if not all of it, came from contamination by laboratory air. $Helium_4$ is very common in chemistry laboratories because of the liquid helium superconducting magnets that are used for NMRs." Oops. Lewis also addresses the University of Florida's reported confirmation of tritium: "If you don't neutralize the sample or eliminate peroxide, which also causes flashes in most common chemical scintillation cocktails, you find large counts that might seem to be tritium. We know in this case they are not, simply (a) because there is no tritium in the water itself to begin with really, and (b) our machine has a coincidence method that can tell us the chemiluminescence of this from the actual tritium decay—and these flashes are over 90% confidence to be coincidences, and therefore chemiluminescence." Oops, oops.

Supposed excess heat is also traced back to sheer incompetence. Lewis reports: "If you don't stir the solution, you clearly have gradients, and these gradients are large. . . . It's not straightforward to convert temperature gradients into actual measured power errors. We're balancing the resister versus the cathode here versus here, and we're seeing just what those errors are—but they're obviously going to be very large. And could probably account for all the data." Moreover, it appears that the Utah experimenters' calculation of excess heat was not based on experimental observations results but on expectations derived from specious assumptions: "Instead of dividing it by the total power in, they divided it by an imaginary quantity: the current times the assumed voltage, *half a volt.* Not the voltage that they measured, just the voltage they assumed. . . . And we know that they assumed this, and never measured it." This arithmetic sleight of hand led to the 4:1 and 10:1 excess energies Pons and Fleischmann had presented to Congress.

The interpretation of gamma ray lines has been messed up in two ways. First, Pons and Fleischmann are accused by David Bailey of Toronto of having misread (or intentionally doctored?) the peak value on the energy scale, as reported here by Koonin: "Even funnier about that gamma ray line: it doesn't peak in the right place. . . . I've tried to indicate here as best I could on their energy scale the location where you would expect to see a gamma line at 2.224 MeV [the value claimed by Pons and Fleischmann]. Their line doesn't peak there; rather, it peaks at 2.200—or very close to that. What makes 2.200 gamma rays? It turns out that there is a decay of radon [that peaks here]. . . . Of course, it depends on how much radon there is in your laboratory. I don't know how much radon there is in their lab, but I do know that they mine uranium in Utah [laughter]." The second problem is identified by Petrasso of MIT: the overall shape of the (truncated) gamma ray spectrum reported by Pons and Fleischmann does not fit with the spectrum widely understood by physicists, raising serious doubts about the accuracy of their claim to a peak at 2.224 MeV. "Does the spectral shape make sense? And the answer to that question is also 'no,' because of the effect of the Compton edge, which even within the very restricted range of the data, should be quite evident." To translate Petrasso's graph: the Compton edge appears as a large shoulder just to the left of the peak claimed by Pons and Fleischmann; it is simply missing from the Utah graphs (another chunk of well-established scientific knowledge overlooked).

Accusations of technical incompetence were thus introduced to account for the discrepancy between Pons and Fleischmann's claims and what just about everybody at Baltimore thought they knew about nature. It was vital to show how the Utah scientists had ended up with findings that—if confirmed to everyone's satisfaction—would have forced wholesale rewriting of elementary textbooks in physics and chemistry. But those accusations also justify a cultural cartography that excludes journalists, politicians, and random family members from legitimate scientific research. Even scientists in more or less good standing—beneficiaries of technical training with years of experience at getting experiments to work—can screw up: imagine how hard it would be for those without advanced degrees or rafts of published peer reviewed papers. Still, isn't it obvious that it takes a skilled scientist to do science properly? The cold fusion controversy provided an occasion to restore obviousness to the boundaries of science—a response to their apparent permeability. The almost endless boring display of technical virtuosity and instrumentational prowess at the APS meetings—"we have a sensitive neutron detector; it's approximately 10^7 times more sensitive than we needed to measure the rates reported by Pons and Fleischmann"—instantiated scientists' monopoly over the production of reliable and valid knowledge about nature by giving reasons why nobody else should even try.

Downstream without a Paddle

Finally, Pons and Fleischmann were faulted for their obsession with the downstream implications of their knowledge claim (panaceas, pork, patents, profits, prizes). Good science takes claims back upstream, asking how they were made rather than where they are going. Without implying that Pons and Fleischmann had merely concocted cold fusion for fame and fortune, Morrison suggested that their better judgment might have been sacrificed to base interests that enter science at great peril to truth: "Utah is very special. As you know, it has serious economic problems. They are delighted by fusion. In the same way as California has a Silicon Valley, in Utah they want a Fusion Valley." Meyerhoff of Stanford poetically distanced upstream troubles with calorimetry from downstream desires for a federally funded cold fusion research institute: "Tens of millions of dollars are at stake / Dear sisters and brothers / Because scientists put a thermometer / at one place and not another."

Good science seeks truth without the distractions that Pons and Fleischmann brought onto themselves by their strategic cooptation of the media and Congress. The scientists who came to Baltimore to challenge Pons and Fleischmann did so with empty hands. As Perry explained, "I'm an innocent bystander. I don't have any axe to grind in all this, so that's why I'm here." And as for the pivotal Nate Lewis: "I'm an objective scientist, so hopefully that will please you." Interests and prejudices (financial, political) are located far downstream from the machinery of truth making, down where Pons and Fleischmann led gullible and untutored journalists and congressmen. Lewis went on to say that the possibility of an energy panacea was an unfortunate distraction, making it difficult to appreciate the genuinely scientific implications of Pons and Fleischmann's claim: "If the explanation were fusion, even if you could not make electrical power out of it more than got in, that would still be an important scientific advance." Dreams float downstream—not a place, according to Morrison, where hopeful scientists should seek allies: "And there is this wonderful dream of this energy which you can [use] without pollution, *but,* I mean: What are the facts?"

Cold fusion says something about what happens when the familiar becomes weirdly unfamiliar. Nature was, for a time, turned upside down (or so it looked from the vantage of the bedrock assumptions of most physicists and lots of chemists). Culture, too, took unrecognizable shapes, as science appeared for awhile as media circus and political football. Order was soon restored: Pons and Fleischmann disappeared from the nightly news; Congress declined to give Utah funding for a cold fusion research institute; only a few scientists continued to try replications or extensions of the Utah experiment; fact became artifact. But this particular order is neither secure nor inevitable: those who next seek credibility for their unfamiliar ideas about nature will redraw once again the cultural boundaries around science—in order to diminish the epistemic authority of naysayers, and to empower fellow travelers with the authority to decide truth. Cultural cartography does not happen on autopilot: the "familiar" demarcations of science from politics or the press become familiar only because they are explicitly and emphatically reproduced on occasions—like this one—where somebody had a stake in messing them all up.

Chapter Five

HYBRIDIZING CREDIBILITIES: ALBERT AND GABRIELLE HOWARD COMPOST ORGANIC WASTE, SCIENCE, AND THE REST OF SOCIETY

Besides maps, my other avocational obsession is flower gardening. I grow everything from achillea to zinnia, often starting from seed harvested each fall and potted up in a cold frame come spring. All of my flowers are grown organically: the soil in which they live is fed several times each year a rich mash of nutrients, particulates, and microorganisms from my compost pile. As piles go, mine is an informal one: I have never taken its temperature, nor do I worry much about achieving the proper ratio of carbon to nitrogen. Mostly I just dump stuff in a heap on the edge of a wood: grass clippings, leaves, plant trimmings, coffee grounds, spoiled fruit, potato peels, egg shells.

Everything I know about the manufacture of humus—the brown gold that my flowers love—comes from reading those green classics *Compost This Book!*, *Worms Eat My Garbage*, and *Let It Rot!* Thirty parts brown stuff for carbon to one part green stuff for nitrogen; layer and stir; keep it wet as a wrung out sponge; provide air throughout the pile; turn periodically; watch the temperature go up as microorganisms, worms, and bacteria gobble up the waste and convert it into humus; no smell or flies if it is all working according to plan (and if you don't put in any meat or dairy products). I feel good when the authors tell me that it really doesn't matter exactly what I do: rotting is rotting, and it all benefits my plants and the earth.

Sometimes, when I have a computer keyboard in my hands instead of a shovel, my thoughts still turn to the compost heap, and I wonder: Is

I thank Alyssa M. Kinker for substantial help in preparing this chapter.

there anything scientific going on in there—or maybe even sociological? Does composting have a place in the culturescape? This final episode will suggest that just as composting is itself a mix of heterogeneous ingredients, so too does it mix up and even decompose the cultural spaces and boundaries through which we interpret it.

During the gardening off-season, I read about flowers, horticulture, dirt. In Eleanor Perenyi's *Green Thoughts,* all things came together—composting, science, culturescape, sociology—in the person of Sir Albert Howard (1873–1947). Perenyi handed me the prolegomena for this chapter:

> Sir Albert had been an agronomist in British government service in India, and is now considered to have been the founder of modern organic farming—which could be described as a hybrid of western science and eastern agricultural wisdom based on poverty. For Sir Albert had what was rare in the early part of this century, a respect for "native" methods and needs. He noticed, for example, that plant and animal diseases were more prevalent on chemically farmed government lands than on those of Indian peasants too poor to buy artificial fertilizer. On the other hand, the peasants' yields were low in spite of their intelligent rotation of crops, which helped maintain a surprising degree of fertility. These observations came together in his decision to improve on local methods by devising a way to recycle the nutrients available in plant wastes on the spot. The result was the Indore system of composting, so called after the Indian state where he worked, and the original model for the scientifically layered heap we know.[1]

It was the student of cultural cartography, not the organic gardener, who salivated at this passage. Compost was placed in the liminal spaces between Western science and Eastern wisdom, between universal and local, between knowledge and practice, between traditional and modern. Surely there would be a credibility contest lurking around here, with stakes and interests won or lost depending on how lines between cultural territories were drawn.

A search of the secondary literature in agricultural and environmental history turned up almost nothing specifically on Albert Howard, nor has much been said by my peers in science and technology studies. Sadly, I have so far failed to locate an archive of personal letters and correspondence, which would surely provide richer insight into his sociologically tantalizing life. I did find a biography of Albert written by his second wife, Louise, who happened to be the sister of his first wife and long-

1. Perenyi, *Green Thoughts,* 42–43.

time scientific collaborator, Gabrielle. From there, I was led to Albert and Gabrielle's published papers and books, which now make for a three-and-a-half-foot stack on my office floor (and comprise the data analyzed below).

Albert Howard was born a Shropshire lad in 1873, son of a farmer. He took a first in the Natural Sciences Tripos at Cambridge, finishing in 1897. His first professional appointment was as mycologist in the Imperial Department of Agriculture for the West Indies from 1899 to 1902, after which he returned to England to teach at the South-Eastern Agricultural College at Wye (1903–5) and to marry Gabrielle in 1905. Gabrielle (b. 1876) was the daughter of a merchant, and like her husband, she had the benefit of a Cambridge education. The first year of their marriage, the two of them began a long tour in India, first at the Imperial Agricultural Research Institute at Pusa (1905–24), then at the Institute for Plant Industry, which Albert himself founded at Indore in 1924. Gabrielle died in 1930, and one year later Albert returned to England for good. His retirement only nominal, he became a tireless evangelist for composting and organic agriculture.

I follow Albert and Gabrielle Howard on their journey from the Barbados, to Wye, to India, and back to England—crossing the world as they also cross cultural spaces, hybridizing plants (both had specialized in botany as students) as they hybridized several new sciences. My own compost pile was a way for me to get into the lives of these two scientists.[2] I left them with a much clearer picture of the diverse tactics available for scientists to reproduce and expand their cultural authority.

AN ALMOST CONVENTIONAL SCIENTIST: CAMBRIDGE, BARBADOS, WYE, 1899–1905

Albert starts his career with studies of disease and ends it four decades later with studies of health—one of his many crossings of and through the culturescape. As a student at the Cambridge Botanical Laboratory, Albert begins with an investigation of fungal diseases in *Tradescantia*—

2. In a way, the compost pile becomes for Albert and Gabrielle Howard a kind of "boundary object," defined by Star and Griesemer as something "which inhabit[s] several intersecting social worlds and satisf[ies] the informational requirements of each of them." Boundary objects are "plastic enough to adapt to local needs and the constraints of several parties employing them, yet robust enough to maintain a common identity across sites" ("Institutional Ecology, 'Translations,' and Boundary Objects," 393). For a recent application of the concept to the human genome project, see Balmer, "Political Cartography."

which I grow as a pretty shade-loving flower, pink or purple or white, with reedlike foliage. The choice of research site is itself odd, when viewed through the lens of his subsequent achievements. *Tradescantia* (or any merely flowering plant) is so useless, and Louise Howard will later note that Albert has little time for flora with no economic consequence: "the true farmer does not care for flowers."[3] Inspiration for this research is simply the appearance of an epidemic disease among plants growing in the greenhouses of the University Botanical Garden, far removed from the hungry peasants, commodity markets, and imperial agriculture that will occupy him later. The "long white conidiophores" are a curiosity, made interesting only by science, a puzzle whose proper explanation will be its own end.[4]

Albert attacks the problem with conventional science, boring and predictable—except for its palpable contrast to how he will define and pursue science later on.[5] The cause of the observed trouble is external to the plant and autonomous from it: a fungus, maybe *Botryosporium* or *Cladosporium* or *Hormodendron.* And that is exactly the question at hand, a matter of nomenclature or classification: where does *this* fungus belong in the taxonomy that scientists have laid over nature? Albert evinces little interest in getting rid of the disease or in achieving healthy plots of *Tradescantia.* Indeed, the little flower all but disappears as the analytic report ensues: "If we follow the classification logically, the present fungus falls naturally into section I, *Amerosporeae,* because the elliptical or oblong conidia are continuous, i.e. non-septate; but it is evident on comparing it with Cladosporium (which Saccardo puts into section II, the *Didymosporeae* with typically uni-septate spores) that our Fungus comes into that or a closely allied genus."[6] There is much controversy surrounding the taxonomy of these fungi, and debate rages among botanists in England, France, and Germany (to judge from Albert's copious references to leading mycological authorities). It matters to science

3. Louise E. Howard, *Sir Albert Howard in India,* 137.

4. Albert Howard, "On a Disease of *Tradescantia fluminensis* and *T. sebrina,*" *Reports of Section K of the British Association for the Advancement of Science* (1899), 923; Albert Howard, "On a Disease of *Tradescantia,*" *Annals of Botany* 14 (1900): 27–38. Hereafter, all citations of Howard's work, as well as that of Gabrielle Howard, identify the authors by their initials.

5. On British agricultural science (and debates about genetics) in the first half of the twentieth century, see Brassley, "Agricultural Research"; Olby, "Social Imperialism"; Palladino, "Politics of Agricultural Research," "Between Craft and Science," "Wizards and Devotees"; Russell, *History of Agricultural Science.* On botany in Barbados at Howard's time there, see Galloway, "Botany in the Service of Empire."

6. AH, "On a Disease of *Tradescantia,*" 34.

whether *Cladosporium* is a parasite (getting nourishment from living plants, and at their expense) or a saprophyte (getting nourishment from dead organic matter): everything in its (conceptual) place.

How does Albert seek to clear up the fungal taxonomic confusion? Experiment is the key: observe, isolate, purify, intervene, control, vary, record, interpret—just as "the scientific method" would have it. The laboratory is the site of truth making, not fields, and certainly not wild nature. The pure science of Cambridge requires pure cultures of these whatever-fungi: "A large quantity of Tradescantia material [it doesn't sound like a pretty flower anymore] was collected, ground up with sand, and the extract filtered through a Pasteur filter. The filtered extract was then collected in minute capillary tubes, sterilized when drawn, the fine sealed ends of which were broken as they were pushed into the tissues of the plants." Nature is invaded, sliced and diced by science to yield truthful nomenclature. *Tradescantia* au naturel is something to be observed—"in the case of the naturally growing host plant [in the university greenhouses] infection started either on the upper side of the leaf as a small semitransparent dot, or from the margin"—but it cannot be properly seen there by naked eyes: only inside the lab, under the microscope, does scientific reality come into view.[7] Healthy plants are experimentally infected with pure fungus; technical procedures are repeated, then varied; and differences noted—but nothing decisive can be said. With all of the modesty, tentativeness, skepticism expected of the man of science, Albert ends: "Hence it is evident that a long research will be necessary to clear up the numerous difficulties accumulated around the Fungus *Cladosporium* and its allies."[8]

Albert's transit to the West Indies is a transit to scientific space still conventional but noticeably different from what has occupied him at Cambridge. It will be only the first instance of a new geographic place becoming a different cultural space for Albert (and, soon, for Gabrielle). He is still preoccupied with disease and its fungal origins, but now in an organism whose significance crosses botany and political economy. Barbados is "almost entirely dependent on the success of its sugar industry," and his research on the infections of sugarcane is inspired in part by "economic considerations."[9] Research conclusions are a hybrid of theoretical

7. Ibid., 31; AH, "On a Disease of *Tradescantia fluminensis*," 923.

8. AH, "On a Disease of *Tradescantia*," 36.

9. Daniel Morris, preface to *Lectures to Sugar Planters* (London: Dulau and Co., for the Imperial Department of Agriculture for the West Indies, 1906), vi; AH, "On Some Diseases of the Sugar-Cane in the West Indies," *Annals of Botany* 17 (March 1903): 407.

explanations and "remedies" to lessen the risk of disease (on these, more later). The scientific point of one paper is less modest and tentative now: "There can hardly be any doubt that this *Diplodia* is identical with the species discussed in this paper."[10] Lingering questions of taxonomy prevail: just after mentioning a finding that would most interest planters and British investors in sugar stocks ("It is clear that this fungus and not *Melanconium* is the cause of the 'rind' disease of the sugar-cane"), Howard speaks instead to his fellow botanists ("On referring this fungus to its systematic position it is evident that . . . it must be placed in the Fungi Imperfecti and that it falls into Corda's genus Colletotrichum").[11] In another paper, he titles a section "One Fungus or Two?" but a later section "Prophylaxis." Disease-free sugarcane fields are as much the point as subtle distinctions between mycelium.

The conventional scientific method goes with Albert from Cambridge to the West Indies. His research on sugarcane (and also cacao) is classically experimental: an econobotanical problem is observed in planters' fields; nature is gathered up; brought into the lab for purification, isolation, variation, control; findings become facts; scientific remedies change planters' unhealthy practices. Initial observations in the fields are not to be trusted. "An investigation *on the spot,* however, showed that it was highly probable that the [cacao] trees believed to be dying through the attack of the thrips were really being killed by Fungi." But "to the *naked eye* there is little difference between this fungus and the Melanconium stage of the 'rind fungus.'" So "experiments which are described below were undertaken to follow the development of this form under *artificial conditions,* and to prepare pure cultures."[12] Experimental controls exclude theoretically uninteresting bits of nature: in one series of experiments, flower pots of previously sterilized and moistened coral sand are placed in the middle of dishes filled with water "to keep away ants." QED, and field observations then merely ratify the facts that have already been manufactured experimentally in the lab: "It will be seen therefore that the development, under artificial conditions, of the Fungus of the cacao tree corresponds exactly in all its details with that found on the sugar-cane so that morphologically regarded the two forms are identical."[13] He seeks to move his facts from the local to the universal by

10. AH, "On *Diplodia cacaoicola,* P. Henn.: A Parasitic Fungus on Sugar-Cane and Cacao in the West Indies," *Annals of Botany* 15 (December 1901): 689.

11. AH, "On Some Diseases of the Sugar-Cane," 389.

12. AH, "On *Diplodia cacaoicola,*" 684, 683 (my emphasis).

13. AH, "On Some Diseases of the Sugar-Cane," 401; AH, "On *Diplodia cacaoicola,*" 692.

comparing and contrasting these findings in the New World to scientific reports coming from Java, Surinam, Antigua, and (with the highest authority) from the metropole at Kew.[14]

Albert does not undertake Mendelian crosses of sugarcane or cacao during his four years in the West Indies, but instead crosses the boundaries of conventional science and begins hybridization of a new cultural space. Artificial conditions (controlled purity in the lab) enable both the explanation of natural problems (in the fields) and their remedy, but the distinction between natural and artificial at times gets blurred. "Experiment" itself may not belong uniquely to science, but rather the procedures inside the lab replicate those in nature. Albert tells how the high trade winds in Barbados "bombard" the sugarcane crops with fungal spores just at the time the crops are ripening, a natural intervention that mirrors what was done indoors. "The result of the above culture experiments with this phase of the Fungus is in close accord with its behavior in a natural condition."[15] In a later paper, he describes the possibility of systematic field comparisons of a first and a second planting (in different seasons, with different amounts of rainfall) as "infection experiments on a large scale with the fungus *Marasmius.*"[16] First nature itself experiments, now the West Indies planters—in their routine agricultural practices—do the same: if "experiments" are the key to certain knowledge, must they happen only in the purity of the laboratory, performed only by the expert man of science?

Albert's observations of these sequential plantings of sugarcane also lead him to rethink the "causes" of disease: in research back at Cambridge, disease occurred because something nasty from outside (fungus spore) got inside *Tradescantia;* here in the Barbados, the environmental conditions in which a plant is grown affects its susceptibility to fatal attacks of fungus. Shouldn't the wise botanist study the plant in situ rather than remove it to the lab for decontextualization? Albert notes in the fields that sick canes are much rarer during the first planting of the cycle in December, when the soil is a friable tilth and rainfall abundant; they

14. The Royal Botanic Gardens outside London. As a preface to his attempt to universalize his findings about fungus in the West Indies, Albert writes: "It is a matter of universal experience to find that during damp and wet weather, articles such as boots and shoes become covered with 'mould,' and that this mould disappears when the articles on which it grows are moved into a dry place" (*The General Treatment of Fungoid Pests* (Imperial Department of Agriculture, West Indies, no. 17 [1902]: 1–2).

15. AH, "On *Trochosphaeria sacchari,* Massee: A Fungus Causing a Disease of the Sugar-Cane Known as 'Rind Fungus.'" *Annals of Botany* 14 (December 1900): 626.

16. AH, "On Some Diseases of the Sugar-Cane," 405.

are far more common during the second planting in February, when the soil is closely packed and rainfall scarce. "In the case of the first crop, conditions favor the cane. . . .In the second crop, everything helps the fungus." This finding would be almost inconceivable in the controlled and artificial conditions of the laboratory, where the seasonal timing of planting makes no difference: the plant-in-context yields the secrets of disease (and of health).

Thus, the faint beginnings of what would—with time and geographic mobility—become Albert's "resistance theory" of healthy plants, people, and society are heard from the West Indies: "Canes planted in December . . . were able to resist the fungus."[17] In a lecture to sugar planters, Albert explains that "cane roots can only develop healthily and work normally in properly drained soil in which the mechanical condition is suitable." Indeed, by heeding the conditions in which plants grow (soil, rainfall in fields), Albert is able to clarify the theoretical confusion of mycological taxonomy: "There is a very large group . . . the members of which are partly parasitic and partly saprophytic according to conditions."[18] Moreover, Albert's discoveries in the cane fields link botany to economics, and presage his aversion to chemical fertilizers: "Large sums of money are annually spent on artificial manures for these diseased canes which can obviously have no effect. . . . Successful economic experiments on the subject, on a sufficient scale to satisfy the planter, would doubtless greatly hasten these reforms."[19] Albert has crossed the boundaries between lab and field, natural and artificial, disease and health, plant and conditions, botany and political economy, observation and experiment. . .

. . . but he is not yet Albert of Indore, the holistic catholic scientist who learns from peasants that organic agricultural waste is the key to prosperity. Far from it: he is still something of the laboratory hermit, the fungus specialist he will come to despise, lamenting that "frequent absence from the laboratory and pressure of other duties prevented a complete study of the Fungus."[20] Nothing of scientific utility is gleaned from

17. Ibid., 405, 403. That the seed for these insights was planted in Cambridge is evidenced by Albert's expression of gratitude to a professor of botany there: "We have, therefore, a striking example of the influence of the environment on the result of the struggle between the host and the parasite, and a confirmation of the views brought forward on this subject by Marshall Ward" (405).

18. AH, "The Fungoid Diseases of the Sugar-Cane," in *Lectures to Sugar Planters* (London: Dulau and Co., for the Imperial Department of Agriculture for the West Indies, 1906), 161, 154.

19. AH, "On Some Diseases of the Sugar-Cane," 407.

20. AH, "On *Diplodia cacaoicola*," 684. Looking back on his West Indies experience from England in the year before his death, Albert writes: "In Barbados, I was a laboratory hermit, a

the West Indies planters, who are described as "superstitious" and who are chastised for picking the weakest strains of cane to plant. Even when the planters intuitively seem to have got it right—for example, as they discriminate between red and black soils variously suited to ratooning—they have not really: "The comparative failure of ratoon crops was found on examination to be due, in most cases, to a definite root disease." Reliable knowledge flows from scientist to planter, from Britain to colonies, from theory to practice, from lab to field—as the preface to *Lectures to Sugar Planters* makes abundantly clear: "The planter of to-day is in a far better position than his predecessors, for he has within his reach the services of men who are possessed of special knowledge in several branches of science, and this knowledge they are prepared to devote in a practical manner with the view of improving the sugar and other industries."[21]

The heaps of post-harvest sugarcane are discussed almost exclusively as a fertile home for fungus to lay in wait for the next crop. Planters are derided: "Little attempt is made in Barbados to keep the Fungus under control. During crop time large piles of 'rotten cane' covered with countless millions of spores are to be found in most of the estate yards of the Island." The same is true for old cacao pods, and Albert's remedy is identical: "Collect and burn all diseased pods." Tons and tons of potential humus—up in smoke? Only rarely in these papers from the West Indies does Albert even hint that organic waste is more than a fungus factory: "The greatest care should be taken in selecting the trash used for mulching only from healthy fields."[22] And he does not hesitate to recommend a spraying of fungus-infected leaves with an ammoniacal solution of copper carbonate, a synthetic remedy he will later abhor.[23]

A three-year layover in England follows, during which time Albert serves as botanist to the South-Eastern Agricultural College at Wye in

---

specialist of specialists, intent on learning more and more about less and less" (AH, *The Soil and Health: A Study of Organic Agriculture* [New York: Devin-Adair, 1952], 1; originally published in 1947 as *Farming and Gardening for Health or Disease*).

21. AH, "Fungoid Diseases," 155; AH, "On Some Diseases of the Sugar-Cane," 404, 392–93; Morris, preface to *Lectures*, vi.

22. AH, "On *Trichosphaeria*," 626; AH, "On *Diplodia cacaoicola*," 698; AH, "Fungoid Diseases," 164. Albert seems to lack confidence in sugar planters' ability or willingness to examine their cane trash for signs of fungus, even though he recognizes that the cuttings are valuable for "mulching young canes." He writes: "Before, however, planters can be advised to burn their trash generally . . . it will be necessary to prove that the harm done by the parasites on the cane trash is greater than the benefits resulting from its use." Evidently, Albert left the Barbados before such experiments could decide the issue ("The Field Treatment of Cane Cuttings in Reference to Fungoid Diseases," *West Indian Bulletin* [Bridgetown] 3, no. 1 [1902]: 85).

23. AH, *General Treatment of Fungoid Pests*, 31.

Kent and meets and marries Gabrielle, his future colleague as well as life's companion. Within the year, the two are ready for their most fateful passage—to India.

SCIENCE SEES EVERYWHERE FROM NOWHERE: PUSA AND QUETTA, 1905–1924

Albert and Gabrielle arrived in 1905 at Lord Curzon's new Imperial Agricultural Research Institute in the northern state of Bihar on the Gangetic plain, he with the title "Imperial Economic Botanist to the Government of India," she becoming "Second Imperial Economic Botanist to the Government of India" eight years later. The immediate task was to map out the contours of the hybrid space "imperial economic botany": whatever kind of science would answer such a call? To say that the job at hand was to improve crop production in India said nothing necessarily about which crops to work on, how they should be improved, or even what "improved" might mean. All that took Albert and Gabrielle almost twenty years to work out, and in the end, two investigators trained in fungal diseases and plant genetics had fashioned a promiscuously crossed science that was as well suited to the practical improvement of Indian crop production as it was unrecognizable to their Cambridge dons doing pure research.

Step one: observe *what* is being grown *where*—and *how*. "The first step in the improvement of Indian tobaccos [is] necessarily the study of the various races at present cultivated in the country," and "a botanical survey of the indigenous [sugar] canes has been started."[24] Such a systematic survey will provide both a baseline against which to measure subsequent progress as well as the resources (natural, cultural) needed to effect such improvements. What do Albert and Gabrielle see as they gaze upon India from the vantage of imperial economic botany? Their working ontology has three domains: *crops* (fruit, wheat, tobacco, indigo, oilseed, fodder, tomatoes, sugar, rice); *conditions* (natural ones such as soil, water, temperature, drainage, wind, humidity; political and economic ones such as labor supply and training, consumer demand, transportation infrastructure, global markets); *practices* (irrigation, manuring, green

24. AH and GH, "Studies in Indian Tobaccos," no. 1, *Memoirs, Department of Agriculture in India,* Botanical Series, 3 (1910): 2; AH, GH, and Abdur Rahman Khan, "The Economic Significance of Natural Cross-Fertilization in India," *Memoirs, Department of Agriculture in India,* Botanical Series, 3 (1910): 128.

manuring, cultivating, draining, planting out, harvesting, packaging, shipping, storing).

But they also see that they are not the first to ontologize Indian agriculture: the cultivators (as they are always described) have already chosen crops, confronted conditions, and developed practices. A cartographic relationship must be charted between the exotic knowledge and method that Albert and Gabrielle bring with them from the Metropole (rules of inheritance, experimental tools) and the indigenous understandings and skills of Indian cultivators. If the challenge of imperial economic botany is to improve crop production in India, does this imply that indigenous agriculture is deficient—that imported science is here to provide the remedy? Or can the knowledge and practices of Indian cultivators be harvested for scientific consumption? If cultivators are to be listened to and learned from, what improvements are left for science to make? "Imperial" *could* suggest the supplanting of obviously inferior native tradition by superior modern science: one space pasted over the other, obliterating it fully.[25] This is not Albert and Gabrielle's map: as they cross from England to India, they effect a cross of modern science and traditional wisdom—part of their hybridization of imperial economic botany.

Locating Indigenes Ambivalently

Albert and Gabrielle approach the knowledge and practices of Indian cultivators as Darwin approached the finches of Galápagos: they investigate the survival of features (for example, a hoe) that have allowed adaptation to conditions that might otherwise lead to extinction. For two botanists, a natural selection model so capable of explaining the diversity of species becomes a portable and transposable first-approximation for grasping sometimes shocking differences between Indian and Western agriculture. What the local cultivators know and do should be respected by interloping scientists as having survived the test of time—they must be doing something right: "The wheats at present in cultivation in this vast Empire, in which a civilized agriculture has been practiced from time immemorial, represent the survival of types most fitted for the con-

25. On the imperialist role of science in India, cf. Baber, *Science of Empire;* Gilmartin, "Scientific Empire"; Kumar, *Science and Empire,* and "'Culture' of Science"; Yang, *Limited Raj*—and more generally: Adas, *Machines as the Measure;* Petitjean et al., *Science and Empires.* A review of the historical literature on science and imperialism is provided in the exchange between Palladino and Warboys, "Science and Imperialism," and Pyenson, "Cultural Imperialism." The starting point for debate is Basalla, "The Spread of Western Science."

ditions of the various tracts. Nature has eliminated the unfit, and the experience of past centuries, handed down by tradition, has taught the cultivator what soils and what tracts are most suitable for this crop."[26] Cultivators read their environmental conditions and choose appropriate crops and tools: routine use of a "seed drill" is found only where the soil moisture reaches down to a sufficient depth, gram is planted only in an "open soil," and indigo is known to be inferior if grown on highly manured land.[27] Indigenes have apparently improved their own crop production by developing practices such as crop rotation, adding organic matter to improve water retention in the soil, and cultivating fields during the hot weather season—as "referred to in *Punjab Proverbs.*"[28] However backward other Indian agricultural practices might appear to a modern English eye, the natives' reluctance to embrace the new is in fact a conservatism healthy for their long-term well-being. "The annual crop was always at the mercy of a variety of circumstances quite beyond the control of the cultivator. It was not surprising, therefore, to find him conservative in his outlook, and to discover that his leading idea in crop production was to play for safety."[29]

With such a valorization of indigenous knowledge and skill, Albert and Gabrielle pave the way for a flow of ideas from native to scientist, from India to Britain, from colony to home—yet another raw material (like gold or cheap labor) in the spoils of imperialism. Cultivators become the teachers of scientists, their hard-won bits of wisdom valid before science arrives or its experts reach consensus. "Although it was not till 1888, after a protracted controversy lasting thirty years, that Western science finally accepted as proved the important part played by these crops in enriching the soil, nevertheless centuries of experience had already taught the cul-

26. AH and GH, *Wheat in India: Its Production, Varieties, and Improvement* (London: W. Thacker, 1909), 22.

27. Ibid., 8; AH, GH, and Abdur Rahman Khan, "Some Varieties of Indian Gram," *Memoirs, Department of Agriculture in India,* Botanical Series, 7 (1915): 219; AH and GH, "Second Report on the Improvement of Indigo in Bihar," *Agricultural Journal of India* 10 (1915): 170.

28. AH and GH, *Wheat in India,* 30; AH, "A Suggested Improvement in Sugarcane Cultivation in the Indo-Gangetic Plain," *Agricultural Journal of India* 7 (1912): 42; AH and GH, "The Improvement of Indian Wheat," *Agricultural Journal of India* 8 (1913): 29.

29. AH, "The Improvement of Crop Production in India," *Journal of the Royal Society of Arts* 68 (1920): 556. Frontier fruit dealers are said to have "absolutely no time for experiments or for anything else beyond a day's work. To reach such men, patience is essential and they must be given ample time for new ideas to sink into their consciousness. . . . They are certain to be frankly skeptical at first and to exhibit that conservatism which is so valuable in protecting the race from disaster" (AH and GH, "The Improvement of Fruit Packing in India," *Agricultural Journal of India* 15 [1920]: 54).

tivators of the Orient the same lesson. The leguminous crop in the rotation is everywhere one of their old fixed practices."[30] So well do indigenous farmers understand their surroundings that subsequent scientific analysis might only prove redundant. "The occurrence of these terms in the vernacular shows that the cultivators fully realize the existence of many different types of soil of varying suitability for the growth of crops. Possibly a modern soil survey would do little more than express in a scientific way what is already understood and applied by the people."[31] Albert suggests that for lately arriving scientists, patterns and variations in Indian agricultural practices present a "number of natural experiments repeated year after year" to be learned from.[32] Indeed, if science and modernity had listened to the cultivators and heeded their wisdom, certain problems of agriculture might have been avoided. "The alkali lands of to-day, in their intense form, are of modern origin, due to [perennial irrigation] practices which are evidently inadmissible, and which, in all probability, were known to be so by the people whom our modern civilization has supplanted."[33]

Such a figuration of native wisdom leaves little room for science to improve crop production, though that is what Albert and Gabrielle have been shipped to India to accomplish. Indigenous knowledge and practice is sufficiently valid that their task is not to replace it with science, but to adapt science to the different conditions of Indian culture and nature. "The present agricultural practices of India are worthy of respect, however strange or primitive they may appear to Western ideas. The attempt to improve Indian agriculture on Western lines appears to be a fundamental mistake. What is wanted is rather the application of Western Scientific methods to the local conditions so as to improve Indian agriculture on its own lines." Just as Indian agriculture is successfully tailored

30. AH, *Crop-Production in India: A Critical Survey of Its Problems* (Humphrey Milford: Oxford University Press, 1924), 36–37.

31. AH and GH, *Wheat in India,* 13.

32. "We have before us an old civilization, with a corresponding volume of traditional experience for the growth of crops. . . . In India, things agricultural have had time to settle themselves. The great growth factors have left their impression on the characters and distribution of cultivated plants. . . . These circumstances greatly assist the investigator in the study of the physiology of crop-production and in the deduction of some of the factors which are in operation" (AH, "Improvement of Crop Production in India," 560).

33. Albert quotes approvingly from F. H. King, *Irrigation and Drainage* (London, 1900), in *Crop-Production in India,* 49. It is largely owing to King's exposition of native Chinese and Japanese farming practices that Albert will begin to formulate his own ideas about composting and intensive agriculture. Alkali lands no longer appear suitable for growing crops because of the build-up of harmful salts. Albert will later consider this problem anew from Indore.

to its own soil and climate, so must imported agricultural science be tailored to the extant cultural understandings and practices of the cultivators. The improvements rendered by science will be incremental: though the importance of thorough cultivation to achieve "soil of proper texture . . . ha[s] been known and acted upon since the dawn of history," there is no question that "modern investigations have naturally added to the knowledge possessed by the ancients."[34] Just what does science arrive to add?

Cultivators have most assuredly not figured it all out for themselves: Albert and Gabrielle also suggest that native truths are winnowed from falsities via a trial-and-error empiricism that cannot yield the theoretical principles common to modern science: "In ancient systems of agriculture, such as that of India, it must never be forgotten that the cultivator is generally sound in his procedure; the difficulty is to perceive the scientific basis of his practice." Science can find the *principles*—generalizable, universalizable, transportable, applicable explanations—that undergird those instances where native cultivators finally hit on something that worked (presumably after many failures): "So far in the history of agriculture the methods of cultivation and the design of implements have been very largely the result of trial and error. Frequently no guiding principle beyond experience has been recognized."[35] The natives cannot explain why they do what they do in their farming because their knowledge is neither an induction from experimental observation nor a deduction from theory, but rather something more like unreflective instinct. Cultivators in a particular region are said to have "a strong instinctive preference for rice, although the profits obtainable from sugar are very large. It is probable that the ryots' preference for rice has a real scientific basis. . . . Sugar-cane needs expensive nitrogenous manure." In another district, where the soil is so packed that not even implement-assisted brute force can open it up, the cultivators have "unconsciously" devised a functional alternative: they grow a plant (*rahar*) with a very deep and extensive root system that "does the work for nothing." Indigenous knowledge is intuitive, not rational: in Bihar around Pusa, "people seem to be born with a special *papri*-breaking sense" (*papri* is a crust that forms on the surface of fields after flooding).[36] With its principles, science can speed

34. AH, "The Influence of Weather on Wheat," *Agricultural Journal of India* 11 (1916): 353.

35. AH, *Crop-Production in India,* 187, 24.

36. Ibid., 124, 57; AH and GH, "Soil Ventilation," *Bulletin, Agricultural Research Institute, Pusa,* no. 52 (1915): 17. "Although the Indian cultivator knows nothing of *Azotobacter*—one of

up or circumvent trial and error so that agricultural improvements can be achieved expeditiously: "Up to the present, rotations of crops, the applications of manures and of irrigation water, the preparation of the land for the seed, as well as methods of intertillage, have all been settled in the most empirical fashion. Such knowledge as exists has either arisen from experience or has been obtained indirectly as the result of innumerable costly experiments. A shorter and more scientific road."[37]

Worse, Indian cultivators do not always do the right thing, even when instinctively they know what the right thing would be. Science displays the errors of indigenous ways, offers correctives, and so improves crop production. The Howards' descriptions of cultivators' practices are replete with complaints about how they waste valuable water (in part because of poorly designed or sloppily implemented irrigation systems), how they do not select next year's seed from the best plants, how they fail to aerate the soil with sufficiently deep tillage, how they miss the proper time for planting, how they disfigure fruit on its way from orchard to market, how they kill young transplants, and on and on.[38] One might wonder how the natives managed to survive from "time immemorial"

---

the organisms concerned in . . . nitrogen fixation—he is aware of the fact that a well-managed monsoon fallow, in which the surface is kept stirred, means a good wheat crop, and that the land properly rested and tilled will manure itself" (AH, *Crop-Production in India,* 37).

37. AH, *Crop-Production in India,* 59

38. "The local practices are wasteful and unscientific in the extreme. Water is thrown away in all directions; there is no effort to conserve the soil moisture and to make the best use of what is, to the wheat crop, a most timely and well distributed rainfall" (AH and GH, "The Saving of Irrigation Water in Wheat Growing," *Agricultural Journal of India* 11 [1916]: 23). "At first, [seed] was obtained from Java, where the supply for Bihar was grown by natives, who naturally paid no attention either to the type or to methods of selection. . . . Java indigo is a mass of heterozygotes" (AH and GH, "Some Aspects of the Indigo Industry in Bihar," part 1, *Memoirs, Department of Agriculture in India,* Botanical Series, 11 [1920], 27). "It will be equally clear that the present practices in Bihar in growing Java indigo are about the worst that could be devised and that, in the past, the indigo plant has never had a proper chance. There has been no attempt at proper cultivation and nothing has been done to increase the air supply by means of surface drainage" (AH and GH, "Second Report on the Improvement of Indigo in Bihar," 173). "The ryot commonly leaves his preparation for wheat till the last fortnight before sowing and during the process a large amount of moisture is lost" (AH and GH, "The Improvement in the Yield and Quality of Indian Wheat," *Journal of the Bombay Natural History Society* 21 [1911]: 198). "The national custom, of doing everything on the floor, again asserts itself. . . . There is little wonder, therefore that the final product presents a battered and bruised appearance, and that good fruit is so rarely seen in India" (AH and GH, "Some Improvements in the Packing and Transport of Fruit in India," *Agricultural Journal of India* 8 [1913]: 248). "Much of the unevenness of the tobacco fields in Bihar is due to the want of care in taking out the plants from the nursery at planting time. The plants are pulled up by hand, and, in the process, practically all the young roots are destroyed" (AH and GH, "The Improvement of Tobacco Cultivation in Bihar," *Bulletin, Agricultural Research Institute, Pusa,* no. 50 [1915]: 6).

with such rotten agricultural practices! Science can do better, and Albert and Gabrielle do not hesitate to brag about improvements they have wrought. "During the last year when the failure of the autumn rains almost destroyed the tobacco crop of the cultivators in the district, no difficulty was experienced in growing good tobacco at Pusa"—because of the scientific system of furrow irrigation employed at this outpost of modernity.[39] When hot-weather cultivation is consistently and systematically employed, the "botanical area at Pusa" yielded forty bushels of wheat per acre compared to nineteen bushels on average in cultivators' fields.[40]

Such invidious comparisons are sometimes accompanied by a condescension and paternalism that might be expected of those working under the title "imperial." Albert suggests that "bright boys" be selected to pick fruit so that it will be less "difficult to train [them] to judge the proper ripeness of the peach by the color on the shady side."[41] Presumably with a great deal of effort, new procedures can be moved from scientists to cultivators so that "they do not involve European supervision once the Indian staff has been properly trained."[42] The Howards are less sanguine about the possibility of encouraging cultivators to plant trees as a future source of fuel in order to free up cow dung for use as a soil amendment: "It is idle to expect to change practices in India which have been in vogue from time immemorial."[43] At their most pessimistic, Albert and Gabrielle fall back on old assumptions: "Hence the well-marked fatalism of the people, the general stagnation of village life and the absence of any desire on the part of the cultivator to improve his condition. Anything ap-

39. AH, "Furrow Irrigation," *Agricultural Journal of India* 3 (1980): 260.

40. AH and GH, "The Milling and Baking Qualities of Indian Wheats," no. 3, *Bulletin, Agricultural Research Institute, Pusa,* no. 22 (1911): 18. "In spite of [a very short monsoon] and the failure of the winter rains, a crop of over 25 maunds of wheat to the acre was grown at Pusa without irrigation. The cultivators' crops failed almost entirely and a famine was declared in the District" (23). "The average yield per acre in the botanical area at Pusa is more than twice that obtained by the people" (AH and GH, "Improvement in Yield and Quality of Indian Wheat," 196).

41. AH, "Second Report on the Fruit Experiments at Pusa," *Bulletin, Agricultural Research Institute, Pusa,* no. 16 (1910): 23.

42. AH, "Some Aspects of the Agricultural Development of Bihar," *Bulletin, Agricultural Research Institute, Pusa,* no. 33 (1913): 2. For certain fibre crops, evidently, the transfer of knowledge can never be so complete that the imperial gaze is no longer needed: "It was found that great care was necessary in thinning to prevent the coolies and boys from removing all the seedlings which differed from the bulk of the culture. If left to themselves they invariably removed everything which differed from the bulk of the culture and so unconsciously rogued the plots. It was necessary to watch the thinning of the cultures personally and to point out which plants were to be removed."

43. AH and GH, *Wheat in India,* 64.

proaching high morale cannot therefore be expected under such conditions."[44]

But there is one more flaw in the accumulated wisdom of cultivators so huge that its solution becomes the cornerstone of Albert and Gabrielle's cartographic construction of imperial economic botany. Significantly, it is a flaw that indigenous farmers share with everybody else involved with crop production in India. They all lack the capacity for the *big picture.* "The [tobacco] growers are often relatively poor men with small capital and of limited outlook"; "the small cultivator is after all a person of a somewhat limited point of view"; "the cultivators and zamindars are so intent on their own small areas of land that they cannot be expected to evolve a scientific scheme of drainage for the country-side."[45] The Howards' hybrid science will be designed to eclipse the provincialism that prevents cultivators from getting the most out of their own agriculture. It will, at the same time, transcend the equally restricted visions of Indian railway operators, British consumers of Indian wheat, botanists, geneticists, mycologists, traders in commodity markets. Albert and Gabrielle's science relativizes (as it respects) these separate and narrow viewpoints, uniquely affording the pair a chance to look everywhere from nowhere in particular. Their own perspectives and interests are displaced by a trope littered throughout their scientific writings from Pusa and Quetta: "From the point of view of A . . . From the point of view of B . . ." (where A and B are the cultivator, the market, the botanist, the plant, the soil, India, or Europe). Albert and Gabrielle's science of imperial economic botany ascends to that Archimedean point from which it will be possible for them to translate and (re)align the self-interested tunnel visions of everybody else into harmonious projects of improving crop production in India.

But what does this science consist in? The very meaning of "improving crop production in India" depends on standpoints that Albert and Gabri-

44. This passage was actually written in 1927 from Indore, in a book that summarizes the Howards' accomplishments at Pusa: *The Development of Indian Agriculture* (Humphrey Milford: Oxford University Press, 1927), 11.

45. AH and GH, "Saving Irrigation Water," 2; AH, "Soil Erosion and Surface Drainage," *Bulletin, Agricultural Research Institute, Pusa,* no. 53 (1915): 16; AH and GH, "Drainage and Crop Production in India," *Agricultural Journal of India* 14 (1919): 387. Elsewhere, Albert emphasizes that what natives possess is *local* knowledge: "Obviously the first condition of progress, in dealing with an ancient agriculture like that of India, is to gather together the existing knowledge and experience and to set this out in concrete form in its relation to the main growth factors. This is always necessary in India, where there is little or no indigenous literature on such questions, and where knowledge is handed down *in the various localities* mainly by word of mouth" (AH, *Crop-Production in India,* 169; my emphasis).

elle must discern and construct: first cultivators, then botany, finally the market. The cultural space "imperial economic botany" will be found in and through their representation and alignment of the goals, interests, and improvement strategies of these three parties. In a talk addressed to budding imperial economic botanists, the Howards explain that "they must look at the questions from three points of view—that of the scientific investigator, that of the cultivator, and that of the trade. Science is the instrument by which the advance is made. A first-hand knowledge of practical agriculture and the cultivators' point of view suggest the problems to be attacked. The uses to which the final product can be put or in other words, the requirements of the trade, gives the direction in which the advance can be most profitably made."[46] Albert and Gabrielle define their new hybrid science so that it is the unique solution to problems of crop production as these would be seen from the perspectives of cultivators, botanists, and traders.

Cultivators' Standpoint

What do the cultivators want from the science of imperial economic botany? Simply put, they want immediate practical solutions to the problems of crop production—solutions that are both within their modest means and have payoffs in the marketplace. The Howards develop an imperial economic botany that is tailored to satisfy this account of cultivators' construction of "improving crop production in India." How so?

Their science must be applied rather than pure, directed toward the solution of immediate real-world problems of Indian cultivators rather than toward the pursuit of truth wherever it leads. Interesting dichotomies are used to create the border between pure and applied science. They chastise a fellow scientist because "the main portions of his work . . . w[ere] not of an *economic* character, but dealt with the *scientific* as-

46. AH, "Improvement of Crop Production in India," 573. Cf. AH, "The Application of Botanical Science to Agriculture," *Agricultural Journal of India*, Special Indian Science Congress Number, 11 (1916): 25. Only a new and enlarged science is up to the task: "These problems are not so simple, and often cannot be solved by the ordinary methods of academic research. Many of them can, however, be dealt with successfully if attacked simultaneously from several standpoints provided that the investigators themselves are fully qualified for the work" (AH, "Application of Botanical Science to Agriculture," 26). The problem of drainage, for example, "involves a multitude of other interests—those of the cultivator, the landowner, the revenue authorities, the engineer and the sanitarian. . . . The essence of drainage, for the plant's point of view, is the maintenance of the oxygen supply of the soil water" (AH, "Recent Investigations on Soil-Aeration," 378).

pect of the various rust fungi found growing on the wild plants in the Simla District." Useful improvements are valued more than theoretical advance: "[A useful] investigator would produce results which, instead of furnishing a scientific explanation of well-known agricultural methods, would lead practice and so place the whole subject on a higher plane." Albert challenges the suitability of a model where pure science discovers, then technology applies—a map also ripped apart by Pons and Fleischmann in their congressional defense of cold fusion (chapter 4). A concern for practical achievements that immediately benefit the cultivator must always inform the pursuit of all scientific knowledge: "Science and practice must be combined in the investigator who must himself strike a correct balance between the two. . . . There will be little or no progress if practical agriculturists are associated with pure scientists in economic investigations. . . . The reason why such co-operation fails is that without an appreciation of practice, the scientist himself never gets to the real heart of the problem." In the end, "first-hand practical experience will thus assist towards producing a proper relationship between the scientist on the one hand, and the practical man on the other."[47] The "proper relationship" is for scientists to bring their powers to bear on agricultural problems that cultivators confront now, to let practical utility become the paramount justification for imperial economic botany: the boundary pure/applied is transcended.

No improvements discovered and suggested by science are practical unless they can be put into effect by cultivators themselves—with sometimes meager material and cultural resources. Albert suggests that with respect to the intensive cultivation of fodder crops, "there is no point in discovering improved methods which cannot possibly be applied," and later adds that all suggested improvements in crop production must be "within the general limits imposed by labor and capital." Feasibility of cultivators' adoption of some innovation becomes an important criterion for deciding which lines of research to pursue at the experiment stations. So, Albert explains, "in view of the fact that artificial manures are not likely to be within reach of the Indian cultivator, no experiments have been started with these substances." "Artificial manures" are the chemically synthesized formulations of N-P-K (nitrogen, potassium, and phosphorus) that the committed holistic organicist Albert will rail against years later back home in England. Interestingly, he and Gabrielle are

47. AH and GH, *Wheat in India,* 90 (my emphasis); AH, "Green-Manuring with Sann," *Agricultural Journal of India* 7 (1912): 83; AH, "Application of Botanical Science to Agriculture," 20.

indifferent about the use of natural or artificial manures during their years at Pusa (as I shall show in a moment), and their strongest objections to chemical fertilizers come out of concern not for Mother Earth and human health but for the thin wallets of native cultivators, who simply cannot afford to participate in such a money-intensive agriculture. Ironically, some experiments into the effects of *natural* manures on crop performance become "mere counsels of perfection," owing to the scarcity of farmyard waste, which is often used instead for fuel.[48]

The same principle—scientific improvements of crop production must fit the means of the cultivators—applies as well to the development of innovative farm technology. The Howards poke fun at their peers who do not appreciate this lesson: "The experimental farms speedily became museums of all kinds of European implements which were for the most part quite useless in a country like India."[49] Improvements must also be within the cognitive, not just the economic, means of the cultivator (previously depicted as of limited or restricted vision). In selecting or hybridizing new varieties of wheat, care must be taken to make the preferred strains easily recognizable in the cultivators' fields, so that inferior plants can be "rogued" out and seed will be saved only from the best type.[50] Moreover, imperial economic botany should strive to capitalize upon the agricultural advantages unique to the Indian cultivator. Gabrielle suggests scientific means to develop a "truly indigenous industry" of dried vegetables—taking advantage of "the power of the Indian sun and the strong drying winds so characteristic of the western frontier regions."[51]

48. AH, "The Improvement of Fodder Production in India," *Agricultural Journal of India* 11 (1916): 393; AH, "Application of Botanical Science to Agriculture," 22; AH, "First Report on the Fruit Experiments at Pusa," *Bulletin, Agricultural Research Institute, Pusa,* no. 4 (1907): 11; AH and GH, *Wheat in India,* 52.

49. AH and GH, *Wheat in India,* 77.

50. "In the work of replacing the existing crop by an improved variety it would be an obvious advantage if the new wheats possessed some easily recognizable field character, such as color of chaff, which would readily differentiate them from the crop as ordinarily grown by the people. . . . Pusa 12 has another advantage in addition to its yielding power and quality, namely, its characteristic appearance in the field which distinguishes it at once from the country wheats" (AH and GH, "Pusa 12," *Agricultural Journal of India* 10 [1915]: 2, 6).

51. GH, "The Sun-Drying of Vegetables," *Agricultural Journal of India* 13 (1918): 616. "It is probable that a small industry in dried vegetables will arise" (GH, "The Sun-Drying of Vegetables," *Bulletin, Fruit Experiment Station, Quetta,* no. 6 [1918]: 1). In the Quetta paper, Gabrielle's scientific analysis even extends to suggestions about how to cook the dried veggies: spinach should soak for twenty minutes, French beans for thirty, and so on (19). Albert and Gabrielle often published papers by the same title in different journals, and sometimes made slight changes from one version to the next.

Part of the task of imperial economic botany is to develop techniques to persuade cultivators that scientific improvements are both within their means and consistent with their interests. Albert and Gabrielle are convinced that the ordinary inscription device of science—the journal article—will not do the trick: visual displays and demonstrations are required: "The publication of results and addresses to the planting community are not sufficient to ensure the successful adoption of new methods on the estates themselves. Demonstration of actual results on a field scale . . . are a step in advance and tend to stimulate interest." This may have something to do with the restricted cognitive abilities of cultivators and planters: "The agricultural public judges largely by eye, and is not trained in the rapid digestion and understanding of printed reports." Or perhaps the imperial economic botanist simply becomes more credible when he or she faces the "same" challenges that routinely face the cultivators: "The agricultural investigator must also pass through this ordeal with credit to himself before he can hope to establish his position and hold his own with the tillers of the soil." Whatever the reason, science must take its message to the people in the palpable language of *display,* a principle that will only be fully developed as Albert and Gabrielle later construct their own institute at Indore: "There is no doubt that a great increase in the yield of the present area of the alluvium under wheat and other crops is possible using only the means now possessed by the people. To bring this to the notice of cultivators there is one method beside which all others sink into insignificance and that method can be summed up in one word—*Example.*"[52]

Not even impressive demonstrations, however, will be sufficient to convince cultivators to adopt scientific ways if those ways cannot be translated by them into immediate personal gain. Enhanced earnings or reduced labor are their own persuasion. "It is difficult to make [the native farmer] take an interest in questions like improved quality unless an enhanced price can be obtained at once. Every cultivator, however, can understand the meaning of a good crop and of a variety which can be relied upon to produce a yield above the average." Indian cultivators are evidently no less rational actors than, say, farmers in Britain: they line up to procure seed of the most improved varieties of wheat, and at one estate "the factory servants [ask] to be paid in wheat instead of in money." Labor-saving improvements will also be adopted: "These are preferred

52. AH, "Some Aspects of the Agricultural Development of Bihar," 18–19; AH, "Application of Botanical Science to Agriculture," 20; AH and GH, "Milling and Baking Qualities of Indian Wheats," no. 3, 32.

by the cultivators to beardless wheats, as they do not as readily shed their grain and can, therefore, be left standing at harvest time for some days after they are dead ripe—a practical point of some consequence where labor is scarce and dear at harvest time, as in some districts of the Punjab."[53]

Imperial economic botany is shaped—in its choice of topics to investigate, in its selection of improvements to display to the cultivators—by an interest in enabling the Indian farmer to obtain "better prices in the long run," to get "a fair share of the profits of his labors," to find "the best possible market," and "to extract from the soil the largest possible profit."[54] The Howards attribute these motivations to cultivators and adopt them as their own goals for their hybrid science. Their personal interests and professional stake are displaced by the interests of the farmers who will be the immediate beneficiaries of their improvements in crop production. Indeed, "improvement" here translates into more money and less labor for Indian cultivators.[55] The successful imperial economic botanist will "experience a sympathy and understanding of the cultivator and of the grower's point of view." Albert and Gabrielle measure the distance they have already traversed from Cambridge, and even from the West Indies or Wye: "The raising of crops [a practical rather than theoretical endeavor] is a most useful discipline for a young investigator fresh from the university, and it also serves rapidly to remove any intellectual arrogance he may possess in his attitude towards the farmer or cultivator."[56] Needs and wants of cultivators become the mandate for their science.

53. AH, "Improvement of Crop Production in India," 556; AH and GH, "Pusa 12," *Agricultural Journal of India* 10 (1915): 3; AH and GH, "The Varietal Characters of Indian Wheats," *Memoirs, Department of Agriculture in India,* Botanical Series, 2 (1909): 9.

54. AH and GH, *Wheat in India,* 52; AH and GH, "Some Improvements in the Packing and Transport of Fruit," 260; B. C. Burt, AH, and GH, "Pusa 12 and Pusa 4 in the Central Circle of the United Provinces," *Bulletin, Agricultural Research Institute, Pusa,* no. 122 (1921): 13; AH and GH, "Improvement of Indian Wheat," 34.

55. Although I am jumping ahead in the story, the Howards are especially keen on improvements that do double duty, for example, making wind conditions better for young fruits trees by means of a windbreak that bears profit for the cultivator: "Where sufficient shelter cannot be obtained from mud walls, thick hedges of Persian roses form a very efficient substitute. . . .The flowers also represent a considerable source of income" (AH and GH, "The Improvement of Fruit Culture in Baluchistan," *Bulletin, Fruit Experiment Station, Quetta,* no. 9 [1918]: 13). I shall suggest that such double-duty improvements are illustrative of successful alignments of diverse interests achievable by imperial economic botany.

56. AH, "Application of Botanical Science to Agriculture," 19.

Botany's Standpoint

If all that were the sum and substance of imperial economic botany, there would be no need for hybridizing a new variety of science, no need for crosses of and through multiple cultural spaces. But the discipline of botany has its own call on the First and Second Imperial Economic Botanists, and they answer with scientific products very different from those they provided to cultivators: universalizable principles, valid truths, discovered facts, revised theories, quantifiable justifications. For much of the Pusa period, the demands of botany and those of Indian cultivators are easily aligned: learning more about the inheritance of characters in wheat translates easily into more profit for farmers in the market. But by the early 1920s, Albert and Gabrielle have begun to chafe under two sets of expectations that become, for them, increasingly inimical. By the dawn of the Indore experiment in 1924, the conventional science so enthusiastically prosecuted in the years before will have become more a problem for than solution to crop improvement in India.

Botany's standpoint encompasses the triple ontology mentioned above: plants, conditions, and practices. For each, scientific principles are extracted from experimental evidence and theory, findings that themselves constitute the improvement of crop production in India.

Plants

The Howards' botany begins with the systematic survey of what plants are grown where in India, and how. A research focus on the plant then moves through an identical sequence, in crop after crop after crop: observation of plant varieties grown by the cultivators; selection of the most promising country types; isolation of a consistently reproducible pure strain; hybridization of an improved new variety via crossings that may or may not conform to Mendelian principles; genetic stabilization of hybrid varieties that will come true from seed.[57] The cultivator and his interests are quickly left behind: the site of imperial economic botany shifts from fields to laboratory. Science begins to refine a promiscuous reality: "Little

57. "This tedious separation of the mixed wheats of a Province into their ultimate constituents is an example of that drudgery in scientific work which is so often necessary" (AH and GH, "Improvement in Yield and Quality of Indian Wheat," 190). A short summary of Albert and Gabrielle's work on the improvement of wheat at Pusa was written up after their move to Indore: "The Improvement of Indian Wheat," *Bulletin, Agricultural Research Institute, Pusa,* no. 171 (1927).

useful information on the quality of the wheat grown in the various tracts can be gained from an examination of the crop as grown by the cultivators as their fields are so mixed and often contain at least a dozen different botanical varieties often belonging to two subspecies."[58] The fields of India provide only the raw materials for scientific extraction and processing into knowledge and improvements: "The wheats of any region . . . supply material which may well turn out to be a veritable gold mine, for the exercise of systematic schemes of selection." The lab at Pusa afforded Albert and Gabrielle the opportunity to see things that no cultivator could see in his fields, using technical instruments to determine characters of wheat vital to their scientific comprehension. "It was also comparatively simple to separate classes 1, 2 and 3 from each other by the aid of the microscope. By examining a large number of plants in each culture it was not difficult to see whether the felting was uniform or not, and whether one or both kinds of hairs were present."[59]

What marks the botanical standpoint is not just observation of plants but intervention—manipulations of found crops, improved scientifically through selection and (more invasively) hybridization. This account of the rudiments of plant genetics attests to the novelty of Albert and Gabrielle's efforts in botany—they are, in this respect, on the cutting edge (even though they find themselves at a mere provincial outpost rather than at Cambridge):

> Hybridization [is] the application of the principles which have been founded on the work of Mendel. By means of this method, it is possible to create new wheats combining the desirable qualities of both [parents]. . . . In modern wheat breeding it is desirable to use for crossing only pure lines, i.e. progeny of single plants. . . . It is found in breeding that all these qualities [quality of grain, yielding power, etc.] behave as unit characters and pass over as a whole to the various hybrid generations. For example, when a rust liable and rust resistant wheat are crossed, the plants of the first hybrid generation are all rust-liable like one of the parents. In this generation the rust resistant character of the other parent is latent or recessive. Rustiness, on the other hand, is said to be dominant. In the second generation, however, splitting takes place in the proportion of three rusty plants to one rust-resistant plant. These latter in succeeding generations breed true as regards this character. It is possible,

58. AH, H. M. Leake, GH, "The Influence of the Environment on the Milling and Baking Qualities of Wheat in India," no. 1, *Memoirs, Department of Agriculture in India,* Botanical Series, 3 (1910): 199.

59. AH and GH, "On the Inheritance of Some Characters in Wheat," part 1, *Memoirs, Department of Agriculture in India,* Botanical Series, 5 (1912): 45, 11.

therefore, by crossing to introduce the character of rust-resistance into a wheat wanting in this quality.[60]

Here, the immediate practical benefits of such hybridization (along with the pecuniary interests of cultivators) recede in emphasis, as attention is drawn to the machinery for building into wheat the character of rust resistance. The Howards do not ask if rust is so significant a problem for cultivators that it merits such hard work to remove, nor do they ask if the improved hybrid seed would be within their meager economic means. The discourse is scientific botany, and the audience is Albert and Gabrielle's peers in the discipline: facts are made for them. "Color of skin is one unit, quality of endosperm another, and there is no necessary correlation between them. The wheat breeder can handle these two points as separate Mendelian units." Theories of inheritance patterns developed and tested by botanists the world over are confirmed (or revised) by Albert and Gabrielle's research at Pusa: "In all our crosses up to the present where one of the parents has been felted, hairy chaff [in wheat] has proved to be a dominant character thereby confirming the previous observations of Tschermak, Biffen and others."[61] The improvement of crop production suggests "the importance of systematic botany in the training of plant breeders of the future"—and so this science becomes part of the Howards' imperial economic botany.

For most of the first decade at Pusa, the Howards extend the botanical thread that now stretches from Cambridge to Barbados to Wye to India: experiments on selection and hybridization of improved varieties of wheat and other crops become interesting because of their taxonomic implications (just like fungus in the West Indies). Their papers from this period consist mainly of elaborate classifications of wheat varieties found and made, with detailed descriptions of the characters that make each different from the next—often accompanied by pictures or line drawings, as well as statistical charts showing inheritance patterns through successive generations. Certainly the search is on for a "better wheat," but until hassles of nomenclature are settled it will not be possible for Albert and Gabrielle to define or describe the distinctive qualities of the one line that proves best. And there *are* problems: "The demonstration of the complex nature of felting is of importance from the systematic point of view. Wheats with different sorts of felting can no longer be

60. AH and GH, "Improvement in Yield and Quality of Indian Wheat," 191–92.

61. AH and GH, "Milling and Baking Qualities of Indian Wheats," no. 3, 8; AH and GH, "Varietal Characters of Indian Wheats," 9.

placed in the same botanical variety and the present accepted classificatory schemes will need some enlargement." To say the least! The Howards anticipate social constructivism by at least a half-century when they realize that all botanical classifications are just made up rather than a mirror of nature: "All that could be done was to group the plants round arbitrary standards, but many cases arose where a plant could, with equal justice, be placed in either of two groups."[62]

A migraine taxonomic headache stems from trying to determine the suitability for Indian wheat of classification schemes developed elsewhere (Europe, North America, Australia). The science that the Howards bring from the West demands inclusivity and universality: systematic botany cannot tolerate different taxonomies for wheat grown in different parts of the world. But for wheat, the imported taxonomies do not travel well: "Eriksson has laid great stress on density [referring to spikelets on the rachis] in his classification of European wheats and in fact divides up his botanical varieties by means of this character. The conditions in India are such that his classification could not be usefully adopted."[63] The goal is not to devise a distinctive classification perfectly suited to the unique conditions of India, but to transcend local particulars with an inclusive and comprehensive taxonomy that sorts all wheat from Europe, Asia, and America into its proper pigeonhole. "The need for a revised systematic scheme for wheat is obvious, but such a scheme must be based on examination of wheats from all parts of the world and to devise a new scheme for the Indian wheats by themselves would only add to the confusion and would prove of no ultimate advantage."[64] No advantage to whom? Indian cultivators might prefer a taxonomy based on the characters most easily

62. AH and GH, "On the Inheritance of Some Characters in Wheat," part 1, 19, 8. "It is at this point that divergences are to be met in the literature on the classification of cultivated wheats. These differences of opinion arise from the varying degree[s] of importance assigned by investigators to the characters used in separating the types. None of the existing schemes of classification appear to us to be beyond criticism" (AH and GH, "Varietal Characters of Indian Wheats," 5).

63. AH and GH, "Varietal Characters of Indian Wheats," 21.

64. GH, "The Wheats of Baluchistan, Khorasan, and the Kurram Valley," *Memoirs, Department of Agriculture in India,* Botanical Series, 8 (1916): 11. The same issue comes up in Albert and Gabrielle's efforts to improve the tobacco crop: "Into the merits of this controversy it is impossible for us to enter as we have not at our disposal in India the material on which the rival [taxonomic] systems are founded, and we have therefore, been unable to make use of either. . . . It is certain both from theoretical considerations, and from the evidence already obtained from accidental cross-fertilization, that the number of forms obtained would be numerous. . . . Under these circumstances we considered that it would be best to arrange the Indian tobaccos in a provisional scheme of classification based on easily recognizable characters to facilitate identifi-

recognized by them in the field, without microscopic aid—even if such characters are less useful in the theoretical development of one universal classification.

The alignment of the distinctive needs of cultivators and the requirements of systematic botany for a generalizable scheme will engage the Howards throughout the Pusa period, and their tack is homologous to their strategy for finding the best variety of wheat or tobacco: cross exotic taxonomies from the West with the observed peculiarities of crops in India. Only some of the systematic botany from Cambridge will survive the transit to Pusa. For example, in research on wheat, Albert and Gabrielle struggle with the distinction between genotypic and phenotypic characters—a plant's genetic constitution (inherited qualities) as opposed to its outwardly visible physical appearance (which is also affected by environmental conditions such as sunlight or water). In their terminology, "botanic or morphological characters" must be distinguished from "field or agricultural characteristics," and attention to both is needed for the proper classification of varieties. The labels themselves suggest a hybridization of theory and practice, of science pure and applied. Botanical characters (for example, presence or absence of awns, felted or smooth chaff) "remain constant with the change of environment" and are "determined in the laboratory from properly developed specimens"; field characteristics (for example, erectness of the ear, straw length and strength) "are less conspicuous and cannot be distinguished in the laboratory" but only from field comparisons of pure cultures "grown side by side and under uniform condition." Extant classification schemes have not "given adequate emphasis" to field characteristics, focusing most of their attention on what can be seen and stabilized only in the lab. Imperial economic botany therefore moves the controlled conditions of the lab out to the experimental plots—preserving uniformity throughout distinctively Indian growing conditions but revealing a different set of variations among the types of wheat. Albert and Gabrielle are doing not just systematic botany, but also something economic: "The necessity for considering the agricultural characters in classifying and studying the wheats of any tract cannot be overestimated. It is the agricultural characters which render a wheat useful or otherwise, both for indigenous or export use."[65]

---

cation and reference [e.g., the habit and shape of leaf]" (AH and GH, "Studies in Indian Tobacco," no. 2, *Memoirs, Department of Agriculture in India,* Botanical Series, 3 [1910]: 77–78).

65. AH and GH, "Varietal Characters of Indian Wheats," 23; AH and GH, *Wheat in India,* 167, 171.

The Howards' expanded taxonomy for wheat is justified via practical utility rather than theoretical necessity or even tidiness.

Theoretical issues are nevertheless given their due. The Howards challenge an "acclimatization theory" which suggests in Lamarckian fashion that characters of wheat acquired when grown in radically different conditions can be passed on to later generations. Proponents have suggested, for example, that when a white wheat is introduced to some regions in India, the seed collected from it and then sown will (over subsequent generations) produce red wheat—evidence for "genetic" changes from acclimatization. Albert and Gabrielle argue instead that because "red wheats are hardier than whites," the color change is simply due to selection and relative survival rates.[66] Gabrielle finishes off the debate by asking: "Is there any acclimatization effect beyond this?"[67] A second theoretical debate centers on the incidence and implications of "natural crosses"—hybridization sans human interference. In this case, the Howards use India to theoretical advantage by exploiting its climatic contrast to England: "Although natural crosses are very rare in the damp climate of England, it by no means follows that such occurrences are equally rare in other wheat-growing countries." They go on to provide evidence of occasional natural crosses in the far dryer winds of Lyallpur: "It seems quite possible that this wheat, which is unique in India, may have arisen from a natural cross between a dwarf and macaroni wheat. This supposition is supported by the fact that we have found a dwarf wheat to be the female parent in some of the natural crosses found by us." The significance of natural crosses is "from the systematic standpoint"—note the trope—"that great care must be exercised in assigning a varietal position to any aberrant forms met with unless they have been shown to breed true from seed. In many cases where natural crossing occurs, these forms are likely to be mere Mendelian combinations of the characters of two existing species or varieties which in the next generations would give rise to a large series of forms."[68] An interest in such issues as acclimatization and natural crosses shows that Albert and Gabrielle sustain their engage-

66. That is, when seed is harvested from a field of supposedly pure white wheat, there is probably "contamination of the cultures, from stray seed left after a previous crop or from manure or irrigation." Because the red seed will over time reproduce better than the white, the observed color change in the aggregate will occur (AH and GH, "Varietal Characters of Indian Wheats," 13).

67. GH, "The Role of Plant Physiology in Agriculture," *Agricultural Journal of India* 18(1923), 218.

68. AH and GH, "Varietal Characters of Indian Wheats," 57, 35; AH, GH, and Khan, "Economic Significance of Natural Cross-Fertilization in India," 326.

ment with theoretical science back home even as they face very different contingencies and aberrations on the road.

Moreover, the displays and demonstration plots used by the Howards to convince the cultivators to adopt their scientific improvements will not, in themselves, persuade their peers in systematic botany. They resort to the ordinary inscriptions of that realm: printed texts in professional journals, full of quantitative analyses of experimental evidence. A different voice for a different audience:

> The distinction between the above two kinds of inheritance in [wheat] bearding is of great interest. In one case the simple Mendelian scheme is followed, but there is no question of dominance in the F_1 which is clearly intermediate. In all these cases however one of the parents is only beardless in the ordinary sense of the term, short tips to the glumes being present. In the other case the F_1 is almost beardless while in the F_2 the 15:1 ratio is obtained when all plants with beards, whatever may be their length, are separated from the absolutely beardless ones.
>
> In No. 83 (type XVIII) the [tobacco] flower is large, the calyx loose and baggy and the corola limb crumpled. In No. 60 (type I) the flower is medium in size, with no median constriction, the calyx is tubular with long, acute teeth and prominent midrib, while the corolla limb is deeply divided and not fully expanded.[69]

Sometimes the Howards' statistical tables seem fully unnecessary, and border on the quantophrenic:

The F_2 generation of a cross between parents with simple felting

Trail	Total Plants	Felted	Smooth
1	134	134	0
2	271	271	0
3	254	254	0
4	305	305	0
5	256	256	0
Total	1,220	1,220	0

SOURCE: Albert and Gabrielle Howard, "On the Inheritance of Some Characters in Wheat," part 2, *Memoirs, Department of Agriculture in India,* Botanical Series, 7 (1915): 283.

But not even such authoritative discourses as statistics and botanical argot will prove consistently useful for imperial economic botany. Some-

69. AH and GH, "On the Inheritance of Some Characters in Wheat," part 1, 33; AH and GH, "Studies in Indian Tobaccos," no. 1, 17.

times no words can adequately capture sublime differences between wheat varieties: "It is hoped that the photographs will make up for any deficiency in the description." And, near the end of their work in Pusa, Albert and Gabrielle find that certain elements of improved crop production resist statistical accounting: "No one has yet given a satisfactory quantitative expression to the various units which make up yielding power."[70]

What is the goal of imperial economic botany, when the plant is considered from the botanist's standpoint? In a word, *purity*. For two researchers whose science is fast becoming a promiscuous mix of discursive styles, methods, goals, and interests, it is ironic that Albert and Gabrielle make a fetish of purity in their work on breeding wheat, tobacco, and other crops. The improvement of crop production in India here becomes the accomplishment of a pure line—a variety that breeds true in generation after generation, holding fast the desired characters: "The first condition of wheat improvement in India was the isolation and growth in pure culture of the types already in the country. . . . Pure cultures are necessary for all wheat experiments both for breeding purposes and also for manurial, cultivation and variety trials. . . . Everything therefore depends on this preliminary work."[71] The same goal turns up repeatedly: "stability," "uniformity," and "maintaining the type." Improvement of crop production depends upon the reduction of diversity, the removal of "aberrant plants," avoidance of contaminating "vicinism."[72] The language used to describe this relentless pursuit of purity has an almost Mary

70. AH and GH, "Studies in Indian Tobacco," no. 2, 67; AH, "Improvement of Crop Production in India," 558. Albert's comment about the limits of statistical evaluations follows close upon the heels of his observation that "accurate mathematic investigations to determine which set of varieties is the highest yielder, which may prove of great use in a country like England, are often inapplicable to Indian conditions where the results are only of academic interest" (557). In an earlier paper on the effects of weather on wheat yields, Albert implies that not all aspects of imperial economic botany are ready for mathematization, and he again resorts to the familiar trope "standpoint": "Sometimes, however, the matter is gone into in greater detail and attempts are made to treat the subject from the statistical standpoint and to apply mathematics thereto. The results obtained can hardly be said to be convincing. Apart from the skepticism with which many people regard attempts to prove a case by means of numbers, a little consideration shows that the subject is one to which, in the present state of knowledge, a mathematical treatment can hardly with confidence be applied" (AH, "Influence of Weather on Wheat," 351).

71. AH and GH, "Varietal Characters of Indian Wheats," 1.

72. See, for example, AH and GH, "Studies in Indian Tobacco," no. 2, 60, 61, 64; and AH and GH, "The Production and Maintenance of Pure Seed of Improved Varieties of Crops in India," *Agricultural Journal of India* 7 (1912): 169, 173. "Vicinism" is when different varieties are planted so close together that the neighbors promiscuously interbreed (i.e., a natural cross), with the likely dilution of qualities in the next generations.

Douglas ring to it: "Crossing botanical varieties is a very dangerous and unscientific proceeding as the botanical variety in India especially is complex and often consists of a large number of wheats differing in field characteristics, rust-resistance and in the quality of the grain."[73] So risky are the products of vicinism that lesser varieties of wheat must be rooted out, and monopoly granted to the most improved lines: "It was decided to aim at the complete replacement of the country wheats by Pusa 12 or Pusa 4."[74] It is not enough to breed a better tobacco; the better tobacco must beget itself forever: "By combining these desirable qualities with those of an indigenous tobacco, which is robust and possesses a suitable habit of growth, a very great improvement might be effected. . . . In hybridization lies the greatest chance of producing a *permanent* improvement."[75] Imperial economic botany effaces contingencies by standardizing the crop, building desired characters into the plant itself.

Why is crop improvement translated into purity? The Howards pursue the pure line not for themselves but for others whose interests they have allowed to define imperial economic botany. Purity is vital for their botanical colleagues "at the various stations" around the world, who need "pure lines of known gametic constitution" in order to make their own additions and revisions of Mendelian laws of inheritance in plants: "There may very easily be useful exchanges of pure lines in Northern Europe and again in North America." Purity is vital for the imperial markets, which depend upon the raw agricultural products of colonies like

73. AH and GH, *Wheat in India,* 139. I am referring, of course, to Douglas's wonderful anthropology of symbolic classifications, *Purity and Danger.* Sometimes promiscuity in crops led to potentially fatal dangers, which purity could prevent. In later research conducted during the last years at Pusa, the Howards examine a form of paralysis known as lathyrism, once thought to afflict humans who ate a certain vetch (*Lathyrus sativus*). The illness was found in fact to be caused by another leguminous weed (*Vicia sativa*) whose seed resembled that of *Lathyrus;* chemical analysis of *Vicia* showed that it contained the poisonous alkaloid divicin (GH and K. S. Abdur Rahman Khan, "The Indian Types of *Lathyrus sativus* L.," *Memoirs, Department of Agriculture in India,* Botanical Series, 15 [1928]: 51–53).

74. Burt, AH, and GH, "Pusa 12 and Pusa 4," 7–8. "An attempt should be made to get the cultivators in the neighborhood to take up these two sorts and to replace the existing wheats of the country round the factory by the improved kinds" (AH, "Some Aspects of the Agricultural Development of Bihar," 15).

75. GH, "Studies in Indian Tobaccos," no. 3, *Memoirs, Department of Agriculture in India,* Botanical Series, 6 (1913): 48. The Howards rarely miss the chance to advertise the wonders that modern science hath wrought: "The behavior both in the field and in the mill of Pusa 106, which was obtained by crossing Muzaffarnagar white and Pusa 6, shows that it is possible to combine high quality and high yield by the application of modern methods of plant breeding. Such a combination had long been considered impossible but the work done at Cambridge and Pusa shows that this position is an unsound one" (AH and GH, "Milling and Baking Qualities of Indian Wheats," no. 3, 27).

India: "The preparation of finished indigo, in a standard form of high purity, suitable for Home dyers" is required by the British clothing trade, and the "modern industry" of Britain's millers and bakers also "demands . . . uniformity." Purification of Indian crop production brings cultivators into the orbit of industrial modernity, but they can realize their own goals (more profit, less work) only if the Howards' science can improve upon the status quo: "Indian produce lacks the uniformity and evenness which modern industries demand."[76] Thus, purity is vital for cultivators as well as those they trade with. As Albert and Gabrielle choose to represent the interests of cultivators, commodity traders, and professional colleagues, their alignment becomes child's play for imperial economic botany—the wheat heralded as Pusa 12.

Conditions

For the Howards, a botany that only looks *at* the plant is deficient: "There are limits to what can be accomplished by plant breeding alone." A fuller botany looks at things *from* the plant's leafy standpoint and asks what it wants or needs: "A much larger field of investigation lies in the determination of the physiological needs of such crops." Such an expansion of botany may, however, disturb the alignment of interests stabilized in the improved hybrids of wheat and other crops. While each may breed true under the controlled conditions of an experimental station, will their agricultural characters (grown-out physical features) remain the same when planted in farmers' fields with different soil or climatic conditions? "In crop-problems, it is not the plant alone that has to be studied but *the plant in relation to its environment*."[77]

Albert and Gabrielle describe the heredity versus environment matter as a "vexed question," but remain judicious: "Such fluctuating variability may be inherent in the plant, or may be directly due to the influence of the environment on the character under consideration, or indirectly to the effect of the environment on the general vigor of the plant."[78] They debate with themselves throughout the articles from Pusa and Quetta, searching for the best strategy to improve crop production in India. At

76. AH and GH, "On the Inheritance of Some Characters in Wheat," part 1, 43; AH, *Crop-Production in India,* 62.

77. GH, "The Improvement of Fodder and Forage in India," *Bulletin, Agricultural Research Institute, Pusa,* no. 150 (1923): iii; AH, *Crop-Production in India,* 186.

78. AH, H. M. Leake, and GH, "The Influence of the Environment on the Milling and Baking Qualities of Wheat in India," no. 2, *Memoirs, Department of Agriculture in India,* Botanical Series, 5 (1913): 49; GH, "Studies in Indian Tobaccos," no. 3, 29.

times, it seems expedient to adjust the plant to diverse environmental conditions—and back to the lab and experimental plot they go, selecting and hybridizing.[79] At other times, it seems expedient to adjust the environmental conditions to the plant—by changing agricultural practices such as soil amendment or irrigation.[80] As is their wont, the Howards cross genetics and environment to construct an imperial botany that is a hybrid of the hybridization of plants and the scientification of cultivators' practices.

The question of the variance in crop production explained by the plant and by conditions is put to the test of experiment, in proper botanical fashion. Albert and Gabrielle link up a network of agricultural experiment stations (and sometimes private farms) throughout the vastly different ecological regions of India, enabling them to compare how the same variety of wheat or tobacco works under different conditions.[81] In addition, experiments are moved out from research plots to "real" farms: "Several planters in Bihar have arranged an extensive trial of the new hybrids and selections under estate conditions." Given that even "the best of the varieties . . . perform erratically in their behavior and the variation in yield from year to year is considerable," Albert and Gabrielle must inquire into the sometimes subtle soil or climatic conditions that

79. "We cannot remove the soil of a locality and substitute one of better quality from elsewhere. One element in the system is therefore fixed. As regards the other unit—the plant—nature has provided us with a considerable range of scope. We can select among the various types of root-systems available and find the one which will connect the plant and the soil in the most efficient manner" (AH, *Crop-Production in India,* 56). Or hybridize one that performs consistently well no matter where grown: "It is clear that no matter how unfavorable the conditions were under which the wheats were grown, quality has not been lost. . . . Pusa 12 proved the strongest wheat and did best at all stations producing good loaves even under adverse conditions" (AH, Leake, and GH, "Influence of Environment on Milling and Baking Qualities of Wheat," no. 2, 88–89). Or at least select the most fit varieties for a certain environment, as in fruit trees: "The stock must be suited to the environment in which the fruit tree has to live, and the more extreme the climate, the more care is necessary in selecting the right stock" (AH and GH, "Improvement of Fruit Culture in Baluchistan," 6).

80. AH and GH, "Studies in Indian Tobacco," no. 2, 62–64: "That the physical condition of the soil, cultivation, manuring and climate have the most marked effect on the quality of the cured [tobacco] leaves and their suitability for various purposes cannot be denied. . . . Soil, climate, moisture and food materials, no doubt, influence the quality and suitability of the leaf for various purposes, but we have not found these causes lead to the breaking up of the type."

81. "The conditions, both as regards soil and climate under which wheat is grown in India, vary very widely" (AH, Leake, and GH, "Influence of Environment on Milling and Baking Qualities of Wheat," no. 1, 197). "Several pure lines, of widely different quality, [are being] grown at various stations. The stations have been selected so as to include as many as possible of the most important wheat tracts of the Indo-Gangetic plain. . . . It will be possible to determine how far milling and baking qualities are affected by environment" (AH, Leake, GH, "Influence of Environment on Milling and Baking Qualities of Wheat," no. 2, 56).

often produce unexpectedly large changes in performance. No contingent detail of the environment is too small to escape their attention: "The influence of the hedge had been under-estimated, and although the last line was twenty feet from the hedge, the plants developed very slowly, and were so stunted that they had to be rejected." The Howards are intrigued, as am I, by "edge effects," in which plants grown around the edge of a plot experience different conditions than those grown in the middle. In particular, plants at the edge sometimes benefit from tapping an increased supply of oxygen from the well-aerated soil of embankments that often surround the field.[82]

Slight environmental differences matter for the plant—so what? If the improvement of Indian crop production depends in part on the vicissitudes of soil, water, and wind, what kind of science is needed to do the job? An imperial economic botany defined from the plant's point of view requires of the scientist a "feeling for the organism."[83] During experimental trials of new varieties, there must be "continuous examination of the plant" in order to watch exactly how it is responding to climatic and soil conditions. One must acquire "first-hand experience of growing wheat under Indian conditions." Albert and Gabrielle write: "We have to deal with a plant which adapts itself in the most intimate manner to the method of cultivation, to the available moisture and to the food materials in the soil."[84] The imperial economic botanist recognizes that the plant must "be used as an interpreter of agricultural conditions." The plant is to be learned *from* (not just controlled, in the laboratory), much as scientists glean wisdom from cultivators.[85]

Albert and Gabrielle graft this touchy-feely science onto the very different rootstock that they brought from Cambridge. Distinctions are drawn between genetics and up-close observation, between laboratory

82. AH and GH, "Milling and Baking Qualities of Indian Wheats," no. 3, 29; AH and GH, "The Economic Significance of the Root-Development of Agricultural Crops," *Agricultural Journal of India* 12 (1917): 17; GH, "Studies in Indian Tobaccos," no. 3, 47. Their observations of "edge effects" lead Albert and Gabrielle to conclude that perimeter plants "grow strongly and well while those in the center show all the signs of oxygen starvation" (AH and GH, "Soil Ventilation," 32). The same effect is noticed for fruit trees (AH, "First Report on the Fruit Experiments," 4).

83. The title of Evelyn Fox Keller's biography of the Nobel prize–winning maize expert Barbara McClintock is *A Feeling for the Organism.*

84. GH, "Plant Physiology in Agriculture," 212; AH and GH, "Improvement in Yield and Quality of Indian Wheat," 189; AH and GH, "Studies in Indian Tobaccos," no. 1, 14. "The general agricultural conditions of the various wheat tracts have to be studied in detail and experience of the actual growing crop has to be obtained first hand" (AH and GH, *Wheat in India,* 7).

85. GH, "Plant Physiology in Agriculture," 213, 214.

and field as sites of truth making. For Albert, "to grow a crop to perfection in India [is] nothing more than lessons in physiology learnt by experience through a long period of time. . . . A first-hand knowledge of practice is necessary and nowhere is it so important as in plant breeding where practice is quite as valuable as an acquaintance with the methods and results of genetics." The artificiality and formalization of statistics-based genetics can at times hamper the improvement of crop production: "A constant observation of the growing crop by a fully qualified observer will lead to the deduction of the factors on which yield depends far more rapidly and accurate[ly] than can be done by such a mechanical method." Moreover, the skills and temperament required for successful laboratory botany are different from those needed for field experiments: "Anyone who has experience of field experiments knows how exasperating they are. . . . Climatic conditions cannot be controlled, some of the soil factors are uncertain and the results in succeeding years have a most annoying habit of being contradictory. The scientific man trained in the exact habits of laboratory investigation, where conditions can be regulated, is apt to be skeptical and to lose patience."[86]

Science, like the plant, must adjust its means to conditions of the local environment—the result: a different sort of knowledge is made. "The knowledge which comes from continuous contact with growing plants is of course not of the same order as that which may be called 'test-tube evidence,' but we venture to think that, rightly used, it is of the greatest value to the investigator and can be employed in advancing crop production." What does it mean to "advance crop production" now? The cultivator's standpoint is not forgotten by imperial economic botany: sustained, firsthand, continuous engagement with crops growing in real fields provides a kind of knowledge more immediately practical for increasing yield, expanding profits, or saving labor. Recall Albert and Gabrielle's distinction between botanical characters and agricultural characters in wheat: both are needed, they argue, for proper systematics of the species. But "these field characteristics are not recognized by systematic botany [though] they are of far greater importance to the cultivator than color of chaff, etc."[87] The Howards cultivate a feeling for the organism because this allows them once again to realign the interests of farmers in India and botanists in the West: imperial economic botany is both the sophisti-

86. AH, "Application of Botanical Science to Agriculture," 18, 19; GH, "Plant Physiology in Agriculture," 210.

87. AH and GH, "Soil Ventilation," 3; AH and GH, "Improvement in Yield and Quality of Indian Wheat," 189.

cated hybridization of a new strain and a deep familiarity with how the stabilized varieties will perform in a cultivator's fields, where F_2 inheritance statistics are less vital than annas and rupees.

The risk, of course, is an exaggeration of the uniqueness of conditions—not just in one or another cultivator's field, but in India *tout court.* Botany's standpoint is less interested in irreducible contingencies than in universalizable principles, less interested in what makes India unique than in how differences between India and the West can be melded into a more inclusive understanding of growing plants. Albert and Gabrielle must balance the desire of cultivators for local knowledge with immediate application and the desire of botany for abstract principles that transcend the local and immediate.[88] Imperial economic botany will succeed only if it can keep the pathways between cultivator and botany free and clear for transit in both directions: India—as a natural experiment in growing different plants under different conditions—is a resource for testing and expanding botanical knowledge; Mendel, Cambridge, and microscopes are resources for enlarging the yield and profits of cultivators' crops. The transit is sometimes laborious: Western classifications of wheat do not match comfortably to what Albert and Gabrielle have found on the ground in India. And even initially successful passages can end in failure: "Attempts have been made to improve the wheat and tobacco crops by the introduction of exotics, but in no case have these been taken up enthusiastically by the cultivators over large tracts of country. One reason of this general want of success appears to be due to the severe conditions imposed by the Indian climate"—specifically, the huge seasonal swing between hot-and-dry and the monsoon, which force plant and cultivator to conform to a natural calendar so different from that in Europe or North America.[89]

Is India "unique"? "The agricultural conditions of India and its problems are entirely different from anything to be found in the West. The investigators speedily realized that they were in a new world."[90] But not completely new: laboratory skills and principles of genetics are still helpful in selecting or hybridizing plants that can increase yield in Indian

88. "The principles established not only apply to local conditions [in Baluchistan] and to similar tracts in neighboring countries but it is considered that they may [be] of use to investigators who may have to deal with loess soils and with a set of climatic conditions somewhat outside the beaten track" (AH and GH, "The Agricultural Development of Baluchistan," *Bulletin, Fruit Experiment Station, Quetta,* no. 11 [1919]: 1).

89. AH, *Crop-Production in India,* 63.

90. AH, "Improvement of Crop Production in India," 555.

conditions. Or are they? What bubbled up during Albert and Gabrielle's years at Pusa was a sense that the "conditions" of India—here, natural and cultural—were sufficiently different from England that just maybe a *new science* would be required to improve the country's crop production adequately. Their ongoing hybridization of imperial economic botany will, after 1924 and the move to Indore, give way to efforts to construct a new science from the ground up—Indian ground.

Practices

But Albert and Gabrielle are not quite ready to leave Pusa, for there are more standpoints yet to be folded into imperial economic botany. Agricultural conditions (soil, water, wind) may be scientifically dealt with in two ways: by engineering a variety of plant that overcomes deficiencies of the environment, and by engineering scientific principles for how to improve the conditions directly—through such practices as irrigation or soil amendment.[91] These are presented as complementary approaches to improving crop production in India: "Very little attention has been paid to methods of cultivation. It has been tacitly assumed that the practices in vogue are more or less perfect and that any line of advance in improving crops must necessarily begin in the laboratory. Our experience at Pusa has been the direct opposite and we have found that a considerable degree of improvement is easily possible."[92] Interestingly, the Howards draw a boundary between "science" (as lab-based genetics work on selection and hybridization) and "agriculture" (the improvement in practices for growing plants in fields). Undeniably, both are important components of imperial economic botany. The gains made by the hybridizer are eviscerated if the better varieties are grown in conditions that compromise their built-in desired qualities.[93]

91. The typical language seeks to universalize *practices* by translating them into *methods* conforming to scientifically grounded principles, as for example: "Heavy waterings reduce the proportion of grain to total crop. This principle is well known and confirmed by numberless experiments" (AH and GH, "The Saving of Irrigation Water in Wheat Growing," *Bulletin, Fruit Experiment Station, Quetta,* no. 4 [1915]: 6).

92. AH and GH, "First Report on the Improvement of Indigo in Bihar," *Bulletin, Agricultural Research Institute, Pusa,* no. 51 (1915): 15.

93. "As far as crops are concerned, progress can best be made by botanists, well grounded in pure science, who, at the same time, possess sufficient aptitude to master agriculture as an art and who also have the type of mind to be found in the successful inventor" (AH, "Application of Botanical Science to Agriculture," 26). But Albert and Gabrielle suggest that the transit of high-yield varieties into cultivators' sometimes inadequate fields is not always simple: "With defective preliminary cultivation and insufficient soil moisture, these late potentially high-yielding wheats do not reach maturity before the onset of the hot weather has begun to diminish

The botanist's goal, then, with respect to agricultural practices, is to become "master of the plant": "By modifying the external conditions, he can make it yield the largest possible quantity of the particular product desired. The soil can be directly influenced by the agency of man by means of irrigation, aeration, cultivation and manuring."[94] Such a goal easily aligns cultivator and botanist: cultivators' indigenous practices are replaced by scientific methods that allow unfavorable conditions to be improved (enlarging yields, perhaps); botany learns more about the mechanics of plant growth through experimental interventions to alter the conditions in which they grow. As it happens, it is in this botanical study of the effects of agricultural practice on plant growth that Albert and Gabrielle will make their great discovery: if Pusa 12 has been their most famous achievement (noted even on the floor of Parliament), the dependence of plant growth on soil fertility and aeration becomes the fact that they most hope to universalize.

The range of agricultural practices examined by Albert and Gabrielle is enormous. They develop a novel system of "furrow irrigation" in which fruit trees are encircled by depressed rings connected separately to a main trench linked in turn to the source of water (parallel circuit as opposed to series). Among the benefits of this "furrow irrigation" system (which provides a kind of lateral percolation) over the customary routine flooding of an entire field is that the fruit can be manured with much less risk of subsequent damage from termites. They also devise a widely adopted system of surface drainage, resulting—in one case—in high rents for land that has previously "not [been] let to tenants at all."[95]

But the Howards give even more attention to "hot-weather cultivation" of the fields, an exemplary cross of pure science and applied, botany and agriculture, theory and practice, local and universal:

> It has become apparent that the first condition of success in wheat production in Pusa consists in hot-weather cultivation. . . . Increased fertility . . . is to be found in the increased supply of available nitrogen due to alteration in the soil flora. . . . The work of Russell and Hutchinson at Rothamsted will be found to

the moisture in the soil. The result is a low yield, often of rather poorly filled grain. The experiment station results are thus reversed. . . . Experiment station results, therefore, must be used with caution" (AH and GH, "Pusa 12," 4).

94. GH, "Plant Physiology in Agriculture," 205.

95. AH, "Second Report on the Fruit Experiments at Pusa," 10; AH, "Recent Investigations on Soil-Aeration," 382. "Flooding" also "destroys the porosity and the surface runs together easily. Under the dry, hot winds which are frequent at Quetta, irrigated land sets on the surface into a cement-like mass, which cracks in all directions and rapidly loses its moisture"—none of this a good thing for the crops (AH and GH, "Saving of Irrigation Water," 2).

> apply to the alluvium of the Indo-Gangetic plain. Whether this theory will be verified or not it is too early to venture an opinion. As regards the practical value of hot-weather cultivation in the Indo-Gangetic plain, however, there can be no doubt. We believe that it will be found to be of universal application in the plains and will lead to material progress of the people.[96]

The idea is simple, and certainly executable within the means of any cultivator: timing is the key. If cultivation is done under the hottest Indian sun, the soil will be sterilized such that phagocytes preying on bacteria that facilitate plant growth will be killed, and the salutary bacteria will multiply at high rates when water finally returns. In their experiments with hot-weather cultivation, Albert and Gabrielle at once help cultivators grow better crops and confirm the theoretical explanation put forth by E. J. Russell for why this cultivation works. Ironically, Russell, longtime director of the Rothamsted Experimental Station, will become after Albert's return to England the target of his organic invectives.[97] In the meantime, Albert and Gabrielle extend the benefits of hot-weather cultivation to include killing weeds, increasing the receptivity of soil to the ensuing monsoon rains, and—most vital—aerating the soil. By the end of their assignment to Pusa, the improvement of crop production in India has been translated into one agricultural practice: soil aeration.

At times, the papers written after World War I sound almost like "soil reductionism" (uncharacteristic for practitioners of a science that is becoming increasingly holistic). Albert and Gabrielle suggest that "in agriculture, the best investment of capital is to maintain the soil in a high condition and to allow nothing to interfere with this." Albert later adds: "The graduate denudation of the soil of the country is the real 'economic drain' from India."[98] Dirt is the linchpin of imperial economic botany—

96. AH and GH, "Milling and Baking Qualities of Indian Wheats," no. 3, 21.

97. AH and GH, *Wheat in India,* 72. An interesting aside anticipates the later battle fought by the organic Albert against the synthetic Rothamsted. I shall suggest shortly that the Howards were, throughout most of the Pusa period, indifferent about the use of natural versus synthetic fertilizers—and, as I noted before, the former are often preferred simply because they are more within the meager economic means of cultivators. Still, in using the strong Indian sun (rare in England!) to test experimentally Russell's theories of "soil weathering," Albert and Gabrielle note that "the intense dryness, heat, and light of an Indian hot weather . . . is similar . . . to that produced by artificial heating and by poisons . . . in England." Nothing ideological or political is drawn from this contrast in 1909, however.

98. AH and GH, "Improvement of Tobacco Cultivation in Bihar," 9; AH, "Soil Erosion and Surface Drainage," 6. Albert's discussion of erosion includes a series of translations that will become mainstays of his work in the 1930s and 40s: "This loss of [soil] fertility reacts on crop production and therefore on the well-being of the people. This impoverishment means debt, increased liability to diseases like malaria and finally rural depopulations." The concerns ad-

holding together the interests of the cultivators (who get more from their crops if they aerate the soil), the commodities market (as we shall see shortly), and botanists (for whom the claim that soil is the primary determinant of plant growth will, if true, be a theoretical principle of considerable significance). The theory of soil aeration holds that roots of plants need oxygen, and that there is a "gaseous interchange" between the soil and the atmosphere above ground. Properly aerated soil will allow carbon dioxide produced by roots to escape into the above-ground atmosphere, where it is consumed in turn by the green part of the plant.[99] Albert and Gabrielle come to believe that "far too much attention has been devoted to the above-ground portion of the plant"—a fault that applies to most of their own work at Pusa—and that more botanical attention should be focused on where root nodules meet the soil, a site of "intense biological activity." Soil aeration is the sine qua non of improved crop production: "Wherever tobacco is grown, the importance of a well-cultivated and well-aerated soil has been emphasized, and it has frequently been laid down that this is the first condition for successful tobacco-growing. Badly cultivated and poorly aerated soil always lead to disaster."[100]

Aeration of the soil can be accomplished through different practices: by cultivation, by adding particulate matter (organic stuff like farmyard manure, inert stuff like potsherds), or by "green manuring" (growing a cover crop typically rich in nitrogen—often something leguminous—and then tilling it under in toto or, if it is harvested for another use, just the lingering stalks and roots). Adding specifically organic manures (including green manuring) has benefits beyond aeration: it also improves soil fertility by adding nitrogen and other needed nutrients and increases the soil's ability to retain water.[101] Soil can achieve a "good heart" if suffi-

dressed here in "Soil Erosion and Surface Drainage" are encapsulated in the title of one of Albert's last writings, *Soil and Health.*

99. See, for example, AH and GH, "Soil Ventilation," 1; AH, "The Manurial Value of Potsherds," *Agricultural Journal of India* 11 (1916): 256; and AH and GH, "Improvement of Tobacco Cultivation in Bihar," 10.

100. AH and GH, "Economic Significance of Root-Development," 17; AH, "Recent Investigations on Soil-Aeration," *Indian Forester* 44 (1918): 193; AH and GH, "Studies in Indian Tobaccos," no. 1, 8. "We are much too inclined to draw conclusions from what can be seen of the crop above the ground and to take no pains to ascertain what is happening down below. We should condemn a judge who gave a verdict after hearing only a portion of the evidence" ("Economic Significance of Root-Development," 17).

101. "Besides cultivation as a means of aerating the soil after the decay of the green crop, there is another possible method which gives certain results. This consists in mixing into the soil fragments of tile or bricks" (AH and GH, "Soil Ventilation," 14). "The fertility and water-holding capacity of the high lands can only be kept up by the application of organic manures"

ciently manured with organic material, and "improved varieties of crops" can be grown "without the slightest fear of exhausting the land." Advantages to cultivators of such scientifically warranted agricultural practices are obvious, but planters "must be willing to place the cultivation of the crop on a proper basis and give up their present methods."[102]

These ideas about the importance of the texture and fertility of soil for successful plant growth are hardly unique to the Howards. What is noteworthy, however, is how vociferously they build on this soil reductionism to launch vigorous attacks on other controversial botanical theories. Albert and Gabrielle register dissatisfaction with a "depletion theory" that attributed wilt disease in indigo crops to a deficiency of phosphorus in the soil, to be remedied by "applications of superphosphate of lime." Their own experiments with green manuring fields prior to the planting of indigo have led to a different explanation: tilling under the cover crop adds organic matter to the soil and averts a "water-logging of the pore-spaces" around the root nodules, allowing the indigo plant to breathe properly. The observed positive effects of dressings with superphosphate are thus traced, not to the chemical nutrients added to the soil, but to the alteration of its physics such that aeration is improved.[103]

Theories of soil fertility and aeration bring Albert and Gabrielle back to one of Albert's earliest interests: disease in plants. Although they worked hard in the lab to make disease resistance a built-in quality of the plant, they put as much effort into arguing that disease stems from the poor condition in which a plant is grown—conditions that prevent it from satisfying normal physiological needs, and making it vulnerable to attack from pathogens (bugs and fungi, for example).[104] In some cases, poor soil conditions mask themselves as "simulated diseases," when "in reality, the

---

(AH, "Soil Denudation by Rainfall and Drainage and Conservation of Soil Moisture," *Quarterly Journal of the Indian Tea Association* [Calcutta] [1914]: 24).

102. AH, "Suggested Improvement in Sugarcane Cultivation in the Indo-Gangetic Plain," 43; AH, "Improvement of Crop Production in India," 577; AH and GH, "Soil Ventilation," 24–25. "After a *shaftal* crop [Persian clover, a green manure], the tilth improves and the soil assumes that elasticity to the foot which is so characteristic of arable land in really good heart" (AH and GH, "Clover and Clover Hay," *Bulletin, Fruit Experiment Station, Quetta,* no. 5 [1915]: 4).

103. AH and GH, "The Continuous Growth of Java Indigo in Pusa Soil," *Agricultural Journal of India* 19 (1924): 611; AH and GH, "A Preliminary Note on the Theory of Phosphatic Depletion in the Soils of Bihar," *Agricultural Journal of India* 18 (1923): 148.

104. "A diseased condition in a plant usually arises from some profound interference with the normal physiological processes after which a pathological phase gradually develops. . . . An invasion of fungus mycelium is usually impossible when the plant is in health as protoplasm is strong enough to resist any attack and the cell-sap is not in a suitable condition to nourish the fungus" (AH, "Application of Botanical Science to Agriculture," 23).

trouble . . . follows from the interference with the supply of air to the roots." This may account, in part, for the spread of wheat rust: "When sown in land badly prepared, [a supposedly rust-resistant] variety is sometimes destroyed by rust. The amount of damage from the fungus appears to increase as the physical texture and aeration of the soil fall off." Similarly, remedy for green flies that attack fruit trees should be sought "elsewhere than in the destruction of the insect," in other words, in the improvement of the soil and root system.[105] Albert and Gabrielle's sustained interest in the problem of indigo wilt—about fifty years after the discovery of aniline dyes for making a synthetic true blue—evinces a wistful agrarian romanticism (which may also be interpreted as chagrin at the loss of a valuable market by Indian cultivators): "There can be no question that the future of Java indigo in Bihar depends on soil ventilation. It is perhaps not too much to say that had attention been paid to the growing of indigo twenty years ago, there would have been no competition from the synthetic product."[106] Proper aeration of the root nodules of the indigo crop would, it is suggested, have cast Liebig's discovery of aniline dyes into the category of curio—not the signal moment in the emergence of a science-based industrial chemistry. As it happens, Albert will battle Liebig again later, blaming him for misusing science to spawn the industrialization of chemical synthetic fertilizers.

As imperial economic botany becomes the scientific study of agricultural practices, it brings the Howards ever closer to the cultivators: ineffectual habits of planting can be improved upon, and because such innovations are both within the means of cultivators and profitable, they will be adopted. But this concern with practice does not take Albert and Gabrielle farther away from botany and its disinterested search for abstract principles to explain the growth of plants. Their experiments with soil aeration and fertility embroil them in what will become (at least from the perspective of Albert's later days) the theoretical fight of their life: natural or artificial manures, organic or synthetic? Which works best for the plant, and under what conditions? Throughout most of the Pusa days, Albert and Gabrielle are indifferent: natural and chemical manures are substitutable or complementary.[107] Near the end of that posting, however, they begin

105. For these examples, see AH and GH, "Soil Ventilation," 23–24; AH, "Disease in Plants," *Agricultural Journal of India* 16 (1921): 628; AH, "The Influence of Soil Factors on Disease Resistance," *Annals of Applied Biology* 7 (1921): 386.

106. AH and GH, "Soil Ventilation," 24–25.

107. Writing about tobacco: "The results suggest that the materials supplied by *sanai* [an organic manure] are not quite sufficient for the needs of the crop. Its value can probably be

a noticeable lean toward the preferability of organic manures—mainly because they improve soil texture (enhancing aeration and water retention) as they improve soil fertility.[108] Albert suggests that the discovery of chemical manures has had undesirable consequences for agricultural science, in that it has steered investigators away from certain research topics vital (in his judgment) for the improvement of crop production in India:

> The discovery of artificial manures has influenced agricultural science just as profoundly as it has revolutionized practice. When most soils are found to respond at once to applications of combined nitrogen, phosphates and potash . . . and when artificial manures are purchasable in any market place of the country, it is natural to regard such soil deficiencies as due to exhaustion and to find in application of artificial manure the natural remedy. Under circumstances such as these no stimulus to the study of factors like soil aeration is likely to occur.[109]

A growing commitment to theory of soil aeration and fertility will, in the years ahead of them, pit Albert and Gabrielle against the accepted wisdom of conventional botany back in England—E. J. Russell and the Rothamsted gang.

The lean toward natural manures will become a free fall as the Howards set up the Institute of Plant Industry at Indore, but even before then

---

increased, either by supplementing it with a light dressing of *seeth* or superphosphate or both" (AH and GH, "Improvement of Tobacco Cultivation in Bihar," 12). And about a rotation of indigo and linseed: "The leaf-mould and potsherds were added to the soil in 1919 once for all when the pits were filled. Sodium nitrate is added every year just before sowing by sprinkling it on the surface and mixing it with the upper soil" (GH and Abdur Rahman Khan, "Studies in Indian Oil Seeds," no. 2, "Linseed," *Memoirs, Department of Agriculture in India,* Botanical Series, 12 [1924]: 141). The next extract suggests that chemical pesticides were welcome in the Howards' arsenal, but frankly I included it because I could not resist sharing the pun: "Something can be accomplished by manuring the surface soil, and in some cases the pests can be kept in check by spraying and other means. At the best these methods are little more than palliatives, and often do not go to the root of the matter" (AH, *Crop-Production in India,* 58).

108. "The limiting factor in the production of wheat is the supply of nitrogen in an available condition. This is best applied in the form of farmyard manure or cowdung which gives better results than saltpetre in the long run on account of the good influence of dung on the tilth and moisture-retaining power of the soil. Saltpetre, although a good manure, should only be occasionally applied as long continued application seems to do harm" (AH and GH, *Wheat in India,* 64). The Howards' growing organicism leads them to see supposed insect pests as beneficial, and not merely something to be exterminated with chemical poisons: "In addition to assisting in the transformation of vegetable matter into humus, [termites] are must useful as aerating agents and their systematic destruction, as is sometimes advocated, would if successful, only lead to great loss to India" (AH and GH, "Soil Ventilation," 20).

109. AH, "The Influence of Soil Factors on Disease Resistance," 374. Note the lovely inversion of "artificial manure" as the "natural remedy."

a new and challenging problem emerges. No scientific improvement in crop production is worth time or effort if it is not within the means of the cultivator to convert into routine practice. The botanical theory of soil aeration and fertility implies that agricultural advance is virtually dependent upon amending the farmer's field with organic matter. But where to get it—in a country that burns almost all of its farmyard dung as fuel?[110] The emphasis placed in green manuring (which I shall consider further in the next section) is now understandable as one possible solution, but Albert finds another in the writings of F. H. King on traditional agricultural practices in Japan and China. It is mainly from King's *Farmers of Forty Centuries* that Albert picks up ideas about "intensive agriculture" accomplished in part by dressing fields liberally with *composted* farmyard waste (undecayed plant waste competes with the crop for oxygen available in the soil).[111] He is not yet convinced that such practices are transferable to India, and even less certain that they represent an agricultural panacea for the cultivators. In 1924 at least, the question of organic versus chemical is for Albert a matter of experimental adjudication. In a classic "truth will out" move, he writes that it would be "exceedingly interesting to compare the cost, results and the nitrogen-balance sheets" in fields grown with the two sources of manure. Ever the botanist, ever the scientist, he concludes: "From these it ought to be possible to formulate *leading principles* which should be followed in the practical solution of this matter."[112] Only a few years later, there would

110. AH, "The Importance of Soil Ventilation on the Alluvium," *Agricultural Journal of India,* Special Indian Science Congress Number, 11 (1916): 48.

111. "Simple methods, within the means of the cultivator, of composting these materials with earth and cow-dung on the lines adopted so successfully in China and Japan must be devised. In these countries, intensive agriculture is practiced, and the soil supports a much larger population than India without any importation of artificial manures. The greatest attention is paid in the Far East to the production of organic matter in the right stage of decomposition before it is applied to the fields. It is never used fresh and undecayed, as is often the case in India. Plant residues, in the hands of the peasantry in China, appear to be much more efficacious than they are in India, and it seems well worth exploring the possibility of adopting many of the practices of the Far East to Indian agriculture" (AH, *Crop-Production in India,* 39; and for a suggestion that Indian cultivators could compost the abundant water hyacinth as a source of organic soil amendment, 140).

112. He sets up the crucial experiment this way: "Before any decision can be arrived at as to the best method of solving the nitrogen problem, a good deal of careful experiment will be necessary. At least two methods of manuring should be tried at these centers. One might usefully follow the lead of China and Japan, and make the fullest use of surface-drainage, nitrogen fixation, green manures and the various forms of organic residues. Another might proceed on the lines suggested by the Sugar Committee, and employ sulphate of ammonia from the coal-fields and synthetic nitrogen compounds like calcium cyanamide" (AH, *Crop-Production in India,* 40–41).

be no experiment around that could convince Albert and Gabrielle that synthetic manures are preferable to natural.

The Market's Standpoint

Imperial economic botany needs one last point of view from which to see the improvement of crop production in India. Albert and Gabrielle's bulging hybrid science is uniquely effective at reaching that goal precisely because it absorbs the standpoints of cultivators, botany, and finally, "the trade." What does the improvement of crop production mean when viewed from the perspective of the market? And who should decide that a crop from India is better than what came before? If Albert and Gabrielle have become anthropologists of indigenous knowledge and practice in order to found them anew on scientific principles, and if they have become systematic geneticists in order to select and hybridize a superior variety of wheat, now they become political economists—knowledgeable about the supply and demand for agricultural commodities, skillful in the arts of marketing and government reform. The goal throughout has been to discern principles that work to improve crop production as interdefined by cultivators, botany, and the market—whose interests together make up the raison d'être of imperial economic botany.

To find out what "the market" wants out of improved Indian crops, Albert and Gabrielle must follow them downstream—on beyond the planting out, cultivating, irrigating, manuring, and harvesting. Whether they are working on wheat, tobacco, fruit, indigo, fiber, fodder, or oilseed, their imperial economic botany assumes additional responsibilities: manufacture of seeds for improved hybrids must be scaled up, and efficient systems for distributing them to cultivators must be devised; packaging and shipping the crops for market must be done in a way that enhances their attractiveness to buyers; the putatively improved varieties must not just be tested by the cultivators for their hardiness and stability under diverse growing conditions, but tested anew by those knowledgeable about how the finished crop will be processed and consumed—millers and bakers for wheat, dyers for indigo.

It takes a big science—a hybridized science—to attend to these matters, while not forgetting about systematic botany or about the indigenous wisdom, practices, and means of Indian cultivators. The Howards tie it all together: "When the crop has been thoroughly surveyed the investigator is able to see how far the present wheats are suitable for local purposes and also for the export trade. He will perceive where improvements can be made with the greatest chance of success and in what directions his science can be most profitably applied." Those improvements might center on the plant, agricultural practices, or—as we shall see—on the preparation of crops for market. Albert and Gabrielle thus link botany and political economy as vital elements of their science: "While . . . those results which bear on the Indian wheat trade will naturally receive most attention in this paper, an attempt will also be made to indicate the scientific methods which have been adopted in prosecuting these investigations. In this way it is hoped that this paper will prove of interest both to those particularly concerned with the trade aspect of wheat and also to members of a scientific society who will not be disappointed to find that the results obtained in applied work must be based on sound biological principles."[113]

Watch how Albert and Gabrielle's science absorbs still more diverse perspectives and interests as they follow the crops downstream. To get hybrid seed out to the cultivators in quantities large enough to monopolize all fields and farms, our botanists consider the relative merits of socialism and capitalism. India simply does not have "seed merchants" as they exist in England, because the poverty of Indian cultivators precludes a profitable market. The Howards suggest the possibility of developing a cooperative credit movement or an "agricultural supply society which is affiliated to the district or central bank"—supported by shareholders to be sure, but "afford[ing] room for cultivators of every grade." They note that such "elimination of profit" in the distribution of hybrid seed "has one great advantage": "It enables the question of the maintenance of the improved variety to be considered on its merits."[114] Stability of an improved pure line of wheat will be achieved not just genetically (breeding true) but economically (in a quasi-socialistic way, by eliminating contamination that would come either from unscrupulous seed merchants or from cultivators' inability to afford the best strains).[115]

Once to market, the crop must look appealing to buyers in order to fetch the best price, and Albert and Gabrielle work out principles for a

113. AH and GH, "Improvement in Yield and Quality of Indian Wheat," 188, 187.

114. Burt, AH, and GH, "Pusa 12 and Pusa 4," 9; AH, *Crop-Production in India,* 69.

115. Or maybe just sloppy seed merchants. Albert and Gabrielle describe the potential for a devastating invasion of flax fields by the parasitic bindweed "dodder" if adulterated seed reaches the country: "Its introduction can be prevented by importing European or American linseed . . . only through well-known and reliable seed merchants and by obtaining a guarantee that such seed is free from dodder seeds" ("On Flax Dodder," *Bulletin, Agricultural Research Institute, Pusa* [1908], 3).

different sort of display—not to persuade cultivators, but brokers. They advise cultivators on the "cleanliness and . . . good appearance" of their wheat, and—with respect to an improved fiber crop—give them persuasive evidence: "The thoroughly retted Pusa sample was valued at £18 a ton, while the local variety, retted by the people, was only worth £8." Some improvements in the appearance of Indian fruit are realized by applying science to packaging: "Experiments were also made in which ventilated wooden boxes were used instead of baskets for the outer package."[116] These experiments were a great success not just because fruit arrived less bruised and battered, but because the packaging system spawned "copies in indigenous materials . . . in all the great markets of India"—another discovery within the means of Indian cultivators. Presentation is everything—it is essential if farmers in Baluchistan are to develop a market in Calcutta for their tomatoes: "The ideal method is to offer each unit of fruit for sale in a suitable gift package, of such character as to encourage the purchaser." From laboratory to field to Madison Avenue, the social organization of the market yields to analysis by imperial economic botany. Albert and Gabrielle extend their science even to crime, siding with the fruit growers who are likely to be ripped-off by dealers who have "attempted to form a ring for the purpose of controlling prices." A solution is to establish agencies at the chief markets "under European supervision" and "outside any rings made by the Indian dealers."[117] (Why only Indian traders are unscrupulous is never made clear.)

But kinder packaging and better policing are not enough: the Howards determine that the present system of consignment on many Indian railways actively discourages growers from using packaging systems that would deliver their fruit in better shape: "It was soon discovered, however, that these boxes, although sound in principle, were quite unsuited to India, on account of the railway rules in force, by which each package is charged for, separately, according to a scale of weights." So Albert and Gabrielle launch into railway reform by proposing two concessions: all parcels sent to one consignee will be grouped, weighed, and priced to-

116. AH, H. M. Leake, and GH, "The Influence of the Environment on the Milling and Baking Qualities of Wheat in India," no. 3, *Memoirs, Department of Agriculture in India,* Botanical Series, 6 (1914): 245; AH and GH, "An Improved Fibre Plant," *Agricultural Journal of India* 10 (1915): 230; AH, "Second Report on the Fruit Experiments at Pusa," 25.

117. AH and GH, "The Cultivation and Transport of Tomatoes in India," *Bulletin, Fruit Experiment Station, Quetta,* no. 1 (1913): 8; AH and GH, "Some Improvements in the Packing and Transport of Fruit in India," *Bulletin, Fruit Experiment Station, Quetta,* no. 2 (1915): 16.

gether; empty boxes will be returned to farms free of charge for reuse.[118] There seems to be no end to the boundaries of imperial economic botany! Albert and Gabrielle write that "experiments are needed by which the economics of this question can be set out in detail." Which question? Should dams and hydroelectric plants be built in the Punjab to provide power not just to major cities but to rural areas, enabling cultivators to use electric tube-wells to extract more water for irrigation? Imperial economic botany must "attack the larger problem of rural development," and Albert and Gabrielle speculate that as more Indians are drawn to the cities by economic opportunities, those left in the countryside will "become more organized" and the "small inefficient holding [of land] will disappear."[119]

Such concern for the long-range future of India may fit into Albert and Gabrielle's expansionist science, but one could also ask from whose standpoint these last predictions make sense. Does the improvement of crop production in India require such sweeping transformations of traditional village life? Are these huge changes—set in motion by the Howards' selection and hybridization of improved varieties of crops and by their efforts to rest agricultural practices on scientific principles—consistent with cultivators' desires for more profit and less labor? Are they consistent with imperial interests of Britain in potentially larger profits from processing Indian agricultural commodities, produced with the cheapest of labor? Are these sets of interests necessarily at odds? Social theories of imperialism would certainly suggest that this is the case: imperialism is economic exploitation, with ideological paternalism to legitimate the cause. Albert and Gabrielle do not read Lenin, but rather cross their own strain of "imperial" economic botany.

118. AH and GH, "Some Improvements in the Packing and Transport of Fruit," 252; AH and GH, "Improvement of Fruit Packing," 52. On Indian railways of the time, see Derbyshire, "Building of Indian Railways."

119. AH, *Crop-Production in India,* 34, 33; AH and GH, "The Irrigation of Alluvial Soils," *Agricultural Journal of India* 12 (1917): 198. Albert elsewhere proposes the complete mapping of drainage in North Bihar, in order to assist cultivators with removing excess water from their fields "in a scientific way." But he then ties the maps to much bigger projects far beyond the immediate needs (or maybe interests) of cultivators: "The utility of such maps . . . does not end with drainage schemes. They would prove of service in many other aspects of the economic development of the country. They could be made use of in laying out new roads and new railways. . . . Other uses suggest themselves, such as town-planning schemes, malarial studies and projects connected to rural sanitation." Now, all those interested in any of these projects are brought within the omnivorous imperial economic botany—aligned with the various interests that define it, so that the solution of drainage problems of cultivators is pursued through the same means as town planning and malaria control—through maps (AH, "Soil Erosion and Surface Drainage," 22–23).

The improved hybrid wheat Pusa 12 has succeeded in aligning the interests of cultivators and botanists. Can it also bend into a consonant shape imperial political or economic forces behind "the market"? Yes, but only if Pusa 12 advances the interests not just of brokers at the colonial market, but—further downstream—millers, bakers, and bread eaters at home. The improved Indian strains must be put to the test: How well will they mill, when compared to the epitome of wheat from Manitoba? How well will they bake in Britain? How will they taste as bread? The market's goal in the improvement of crop production in India is *quality*—not exactly the same as immediate utility (cultivator's standpoint) or purity (botany's standpoint), but not inimical to these other goals either. "Quality" will be the goal attributed to "the market" such that imperial economic botany can bring it into alignment with the interests of Indian cultivators and scientific botany.

Quality in wheat is elusive, and its analysis is perhaps beyond the capabilities of test-tube science: "A good deal of work has been done by chemists to ascertain the ultimate cause or causes of quality in wheat, but . . . they are unable to state with precision, in terms of their own science, the characteristics of wheat which are the ultimate causes of . . . good baking qualities."[120] So Albert and Gabrielle enlist the help of a different authority, A. E. Humphries, Esq., past president of the Incorporated National Association of British and Irish Millers, who is asked to evaluate the quality of new Indian wheats from the standpoint of millers and bakers—by milling it, and baking it, and comparing results against the best wheats in the trade. Albert and Gabrielle do not simply adopt the standpoint of millers and bakers. They literally hand over the task of judging Pusa 12 to Humphries: his verbatim testimony makes up large chunks of articles written by the Howards, as "improvement" is moved to new sites for adjudication—mills and bakehouses, rather than laboratories and fields. All in the name of imperial economic botany.

The miller's tale begins with an announcement of his own discursive position of authority: "I have regarded the whole matter *from the standpoint of* a British miller and am accustomed to buy Indian wheats on a commercial basis for the manufacture of flour to be used in England." Millers at home define high-quality wheat as "absorbing as much water as possible . . . without losing the property of free grinding" and having "a thin and tough skin which will not tear but which is easily removable

120. AH, Leake, and GH, "Influence of Environment on Milling and Baking Qualities of Wheat," no. 3, 246.

from the endosperm"; bakers define a high-quality wheat as capable of producing "large, well-shaped loaves." Humphries reports that the wheats imported from India—before Albert and Gabrielle started their selection and hybridization work at Pusa—were generally soft, inferior to the hard wheats preferred by brokers, millers, and bakers. However, in his milling and baking tests, Humphries vindicates the Howards' hard work in the only court that now really matters: "No housewife would hesitate a moment in her choice between the two Indians [pre-Pusa improvements, and post-] even before she tasted them, a blind man would choose the same loaf of the pair. It will also be seen that the best Indian does not attain the same volume as the Manitoba, but the flours are of the same type and in flavor and appearance . . . there is little to choose between them." The value of the Pusa hybrids is not now measured in profits to cultivators or genetic stability, but rather "the improved produce will find its real value in accordance with the laws of supply and demand."[121]

Maybe this is how "imperialism" takes its place in the titles assigned to Albert and Gabrielle. After all, what really seems to drive the improvement of crop production in India is the market demand (and even judgments of taste) at the imperial capital of London. Indeed, most of the crops worked on by Albert and Gabrielle at Pusa were those with at least the potential for a profitable export market.[122] And one could read Albert's prescient line "Agriculture is, and for many years to come must remain, India's greatest industry" as a prediction for a colony that would not be industrialized in the true sense, but remain economically dependent upon European purchases of its high-quality agricultural commodities at enforced low prices.[123] *Should* the direction of Indian agricul-

121. AH and GH, "Varietal Characters of Indian Wheats," 46 (my emphasis); AH and GH, *Wheat in India,* 148–49; AH and GH, "Milling and Baking Qualities of Indian Wheats," no. 3, 15; AH and GH, "Production and Maintenance of Pure Seed," 174.

122. "Safflower seed oil should become a very valuable economic product if it can only be brought over and utilized on the Home markets" (AH and J. Stewart Remington, "Safflower Oil," *Bulletin, Agricultural Research Institute, Pusa,* no. 124 [1921]: 13). "The crop [hibiscus, a potential source of fiber] is an easy one to grow and is widely distributed in India so that if a good price could be obtained in London there is no reason why an export trade should not be maintained" (AH and GH, "Studies in Indian Fibre Plants," no. 2, *Memoirs, Department of Agriculture in India,* Botanical Series, 4 [1911]: 13). Still, some research was pursued on crops with mainly an internal market: "It is probable that if really good dried vegetables were offered for sale, there would be a steady demand for them all over India" (GH, "The Sun-Drying of Vegetables in Baluchistan," *Bulletin, Fruit Experiment Station, Quetta,* no. 10 [1920]: 2).

123. "For a society that was 'deindustrialized' partly as a consequence of colonial rule, there was more than a ring of truth in [this comment of] Sir Albert Howard's" (Baber, *Science of Empire,* 220).

ture be decided in England—by millers and bakers and consumers of bread?

Or maybe imperial economic botany can align the interests of home and colony, of miller and cultivator—so that the satisfaction of all can be accomplished through scientific principles of improved crop production? "Any improvement in quality, to be of importance, must satisfy both classes of consumers—the people of India on the one hand and the Home millers on the other." Pusa 12 saves the day: "It is fortunate that the class of wheat most liked by the people for food [in India] is that which is worth the most money on the Home market."[124] Not only can British and Indian tastes be brought into alignment, but their purses as well: "Translated into money the improved Pusa wheats are worth from 8 to 10 annas a maund more than the ordinary wheats of commerce. When we consider that the annual export trade in Indian wheat amounted in 1904–05 to over 2,000,000 tons valued at 18 crores of rupees it is clear that a much smaller general improvement than has already been obtained would greatly benefit the cultivator and the merchant. It is bound to be to the advantage of all concerned to deal in an improved product." What British millers and housewives want is high-quality wheat, just what consumers in the colony want as well. Potentially competing interests are translated into harmony by the science of selection and hybridization at the Indian Agricultural Research Institute at Pusa. Here is the same alignment work again, this time with the signature trope of imperial economic botany: "Some of these types are far better wheats from the cultivator's *point of view* and also possess far superior milling and baking qualities to the mixtures, largely composed of inferior soft white wheats, grown for the export trade."[125]

So why, then, have Indian cultivators been growing mainly the inferior soft Muzaffarnagar? Because they lacked exactly what imperial economic botany can uniquely provide: the ability to see the problem of crop production from multiple standpoints. The ascendancy of soft white wheats in India occurred when the principal means of milling flour in England was the stone—for which softness is a virtue. With time, the millstone gave way to the "modern roller mills" that work better with harder wheats—but nobody remembered to tell the Indian cultivator, who persisted in growing the wrong crop for the changed technology.[126] Albert

124. AH and GH, "Improvement of Indian Wheat," 31; AH and GH, "Pusa 12," 2.

125. AH and GH, "Improvement in Yield and Quality of Indian Wheat," 195; AH and GH, *Wheat in India,* 159 (my emphasis).

126. AH and GH, "Varietal Characters of Indian Wheats," 52.

and Gabrielle find a solution—one that works well for all concerned—because they see the problem from the perspectives of both Indian cultivator and British miller (evidently, something neither of them can do by themselves, even though their "interests" both fall within the gaze of imperial economic botany).

This is not to suggest that all concerns of the imperial home and colonial cultivators can easily be brought to alignment. Indeed, Albert and Gabrielle run into a case of what might be called "oil-cake imperialism," and find no easy solution. Here is the rub: the theory of soil aeration and fertility dictates the desirability of adding organic matter to Indian soils (see how botany fits back into imperial economics?). One prime candidate for soil amendment is the oil cake—what is left of various seeds after the oil has been pressed from them. Unfortunately, oil pressing does not take place in India but in Europe, and the "oil-cake does not return to India" where it ought to be available for use as a nitrogen-rich manure. Albert and Gabrielle are frustrated: "Simple as this matter at first sight seems, a solution has not yet been found. It is much easier and cheaper to transport seed in gunny bags than to move oil without loss, deterioration, or contamination in drums or in tanks." And even if such a means for storing and shipping oil were to be devised (something the Howards have done for other crops), the real sticking point would then be confronted—a power apparently beyond their scientific principles: "The European industries engaged in crushing oil-seeds and in utilizing oils must also be considered. They are not likely to remain passive spectators while drastic changes in the trade are in progress."[127] The Archimedean vantage of imperial economic botany allows Albert and Gabrielle to decide which standpoints may be translated into conjoint projects (cultivators' choices, genes inside wheat, railway officials), and which will resist translation.

But the case of oil cakes may, for Albert and Gabrielle, be the exception to the rule. More typical is this: "As regards the utilization of the grain there is therefore no clash of interests. India and Europe regard wheat from the same point of view."[128] A science that stretches from sys-

127. AH, *Crop-Production in India,* 38–39, 148. Tyabji, "Genesis of Chemical-Based Industrialization," discusses the political economy of Indian oilseed during the early twentieth century.

128. Burt, AH, and GH, "Pusa 12 and Pusa 4," 3. Another case where Albert and Gabrielle seek to align diverse interests by means of science takes place during World War I. The development of a particular fodder crop "is of importance from *two points of view*—that of the zamindars and of Government" (my emphasis). The Howards also adopt the standpoint of the soil, arguing that it wants organic matter so that it can retain water better and be more fertile. *Shaftal*

tematic botany to agricultural practices to fruit packing to railway reform to milling-and-baking tests succeeds in aligning the interests of home and colony in the improvement of crop production in India. The Howards' hybridized "imperial economic botany" becomes a power without a standpoint of its own: they do not side with either science or indigenous wisdom, plant or conditions, gene or environment, pure or applied, quantitative or qualitative, agriculture or botany, plant or practices, cultivator or miller, India or England—but with them all. There is nothing in their writings about Indian nationalism, despite the fact that there were revolutionary stirrings throughout the period of their work at Pusa. Neither do their papers deal with improvements in crop production that favor only home interests at the immediate expense of Indian cultivators. With disinterested Archimedean science, everybody wins.

Chafing under Science

Maybe not everybody: do Albert and Gabrielle consider *themselves* winners? Because they submerge their own personal and professional interests beneath the interests of everybody and everything else connected to the improvement of crop production in India, it would be difficult to know exactly what success might mean for them. Near the end of their stay at Pusa, the Howards express opinions reminiscent of John Tyndall (chapter 1): to solve the problems of the world, scientists need "freedom for the investigator and proper control of the money provided for their work."[129] They give hints that their accomplishments have been underappreciated, because they are doing "applied" research or perhaps because they are at Pusa and not Cambridge: "The man who makes two blades of

(the crop) keeps everybody happy: it can be baled up and sold to the army to feed their hungry horses, returning a nice profit to the zamindar, while the stubble can be turned into the hungry soil as a nitrogen-fixing green manure (AH and GH, "Clover and Clover Hay," *Agricultural Journal of India* 11 [1916]: 77; AH and GH, "Leguminous Crops in Desert Agriculture," *Agricultural Journal of India* 12 [1917]: 28, 34, 42–43). A report from the commandant of the Seventy-second Heavy Battery, R.G.A, Colonel M. H. Courtney, R. A., attests to the suitability of the finished fodder: "I found 3 lbs. *bhusa* to one of *shaftal* made an excellent chop, and the horses really throve well on it and the *shaftal* made even the shyest feeders eat *bhusa* freely" ("Clover and Clover Hay," 6). I include this example of a "testimonial" because it anticipates the kind of authority Albert will increasingly draw upon after his permanent return to England. Other witnesses are called in to testify to the better taste of a vegetable grown organically; experimental data is used less often after the Pusa and Indore years. I shall argue that Albert expands the authority of "science" to include both sources of evidence—credible witnesses and experiment—as he seeks to defend his fact (the necessity of composting organic waste).

129. AH, *Crop-Production in India,* 184.

grass grow where one grew before, deserves well of his country, and must be promptly and adequately remunerated." Moreover, "while [those working toward practical applications of their science] have to sacrifice the rewards open to workers in pure science, they often make discoveries from which the country reaps benefits which may run into large sums of money. If some method of reward could be devised for applied work of this character, a great step forward would be made."[130]

But, ironically, their biggest gripe is with science per se: not the promiscuous science they have fashioned in order to succeed as imperial economic botanists, but the science that came from England to found the Imperial Agricultural Research Institute. Albert rails against "organization," taking sides with those who believe that genuine science cannot be managed (especially not by government): "All notable advances in agriculture up to the present time have been initiated by individuals and not by systems of organization. This applies to creative work of every kind. The individual has always triumphed over the committee or the organization." Later, he adds: "in all research work the pioneer, the man who had to find his way into the unknown, should be unfettered and allowed to work out his own salvation." What is wrong with organization? How does it stifle the kinds of creative solutions to problems like those Albert and Gabrielle confronted at Pusa? "All organizations sooner or later become affected by disease. In India, this often takes the form of acute departmentalism."[131]

The holistic science Albert and Gabrielle have constructed in order to improve crop production in India is impossible in a bureaucratic machine—for the education of students, for the administration of research projects—that is founded on the necessity of specialization. By now it should come as no surprise to hear Albert say: "The various problems which arise with regard to crops should not be viewed from too narrow a standpoint, and that in their investigation, knowledge from as many different sides as possible should be brought to bear." But this is not how others conceive of science: "One drawback of even the best scientific training [is] the artificial subdivision of science rendered necessary by

130. AH, "Application of Botanical Science to Agriculture," 26; AH, *Crop-Production in India,* 194.

131. AH, "Improvement of Crop Production in India," 573, 577. Albert draws a line between what science needs and what government depends on: "There is therefore, no room for individual opinion in official pronouncements, and it naturally follows that, in official publications, it becomes a very safe rule to play for safety and to take the middle course. Very different is the work of a scientific investigator. Here the man is everything, the system nothing" (*Crop-Production in India,* 192).

the exigencies of teaching and examinations. The advanced student should be made to realize that such subdivision, and the consequent existence of schools of botany, chemistry, physics, and so on, is after all an artificial thing which is not recognized by the living plant." For all their achievements in improving crop production, Albert and Gabrielle increasingly chafe at the bureaucratic specialization of others' science, which seems to them to preclude any further improvement.[132] It will take a new science, built from scratch in India, to take the next steps.

EXPERIMENTING ON SCIENCE: INDORE, 1924–1931

It is difficult to judge how much Albert and Gabrielle were transformed by their first two decades in India—psychologically, spiritually, emotionally, or politically. Such matters do not emerge from their published writings, and in the absence of more immediate personal data from diaries or letters, we can only guess that the chasm between Occident and Orient might have compelled them to look upon life from a slightly different point of view. Clues to a sea change can be found. In 1924, Albert and Gabrielle move from the Imperial Agricultural Research Institute at Pusa to the Institute of Plant Industry at Indore—created by them to carry forth their efforts at improving crop production in India. Like the several transits earlier in their lives, this one is accompanied by a redirection of professional goals and strategies—and it is also marked by a different discourse. One phrase turns up several times in the writings from Indore, though I never found it once in the far more voluminous books and papers from Pusa: "wheel of life." The concept is Buddhist (also called the "wheel of the law"), and suggests that "everything is subject to causations" so that "the moon beheld in water is neither removable nor self-subsistent." Ignoring several millennia of metaphysical niceties, one can say that the "wheel of life" implies the interconnectedness of everything, the necessity of seeing the whole while avoiding distractions of the parts.

Albert and Gabrielle never cite the Buddha, nor is their use of "wheel of life" an especially spiritual one. It turns up first in a discussion of the

132. AH, "Disease in Plants," 627; AH, *Crop-Production in India,* 187. "It is no easy task for the student to appreciate fully the many-sided aspects of the living plant and to master the manifold details of a science. . . . The investigator too is hampered in this direction by the necessity of specialization and of narrowing down the conditions of a problem, so that the ordinary clear cut methods of academic research can be applied" (AH, "Application of Botanical Science to Agriculture," 17).

disastrous effects of canal irrigation on the fertility of the soil: "Canal irrigation in the hands of the cultivator seems to put a brake on the wheel of life. In some places . . . the wheel of life is brought to a standstill altogether by the land becoming a wilderness of alkali on which nothing can grow. Here the canal has produced dead soil."[133] Cultivators believe that water brings life to their plants but they do not see that it brings death to the soil on which their crops also depend. "Wheel of life" also appears in a discussion of the salutary effects of leaving plant roots in the farmer's field after harvest (soil fertility is maintained somewhat even in the absence of any intentional manuring): "The organic matter of a soil, constantly cropped without manure, does not disappear altogether. The wheel of life slows down. It does not stop."[134] In this case, cultivators believe that they have carted off to market the only valuable portion of their crop, failing to appreciate that by leaving the roots to decompose into nutritious organic matter in the soil they have unwittingly enabled the next crop to be almost as successful.

I doubt very much that Albert and Gabrielle began their days at Indore with a chant, but the kind of holistic perspective conveyed by "the wheel of life" describes perfectly their appreciation of nature and of society—and their prescriptions for how both must be examined scientifically. If indeed everything is causally interconnected with everything else, a science fragmented by disciplinary specialization will only capture the distracting bits and pieces and—because of this—can never solve the big practical problems of crop production in India. "A successful biological investigator must be able to visualize the complicated processes taking place in a plant, and to realize that such processes do not take place independently but that each influences the other." The interconnections, causalities, and dependencies fan outward from the plant in every direction, so that an effective agricultural science "embraces not only the health and well-being of the people but also the main facts of rural economy as well as the problem of the maintenance of the fertility

133. AH, "Agriculture and Science," *Agricultural Journal of India* 21 (1926): 175. Albert explains that anaerobic bacteria that thrive in oxygen-starved soil are "probably the real agents which give rise to harmful salts which occur in alkali tracts" (AH, "The Origin of Alkali Land," *Agricultural Journal of India* 20 [1925]: 461). The Buddhist phrase is also found in a later paper on soil aeration: "An inadequate supply of oxygen in the soil puts a brake on the wheel of life" (AH and GH, *The Development of Indian Agriculture,* 2d ed. (Humphrey Milford: Oxford University Press, 1929), 16.

134. AH and Yeshwant D. Wad, *The Waste Products of Agriculture: Their Utilization as Humus* (Humphrey Milford: Oxford University Press, 1931), 38.

of the soil."[135] Plants, people, economy, soil—no single conventional science has the breadth or depth to solve problems that inevitably stretch across all these ontological domains.[136] The new institute at Indore will be designed with the wheel of life in mind: its science will be holistic rather than fragmented, its scientists will look out from a "broad standpoint" and be "capable of bringing several sciences simultaneously to bear on these problems."[137] The multiple and varied points of view of crop production at Pusa must be amalgamated into a single encompassing standpoint—a standpoint that cannot be distributed throughout a network of bureaus or agencies but must be crammed into the lonely choices of Albert and Gabrielle.

One problem especially is still unsolved after twenty years of work at Pusa: what Albert and Gabrielle now describe as "the human factor." All of the genetics that have yielded Pusa 12, all of the experiments that have proved the effectiveness of surface drainage or green manuring, all of the milling and baking tests, all of railway reform and packaging design—all of that will not add up even to a hill of beans unless improved agricultural practices can be widely and consistently adopted by cultivators, zamindars, and plantation owners. Will they plant the hybrid best for their local soil and water conditions? Will they scratch out furrows for irrigation according to Pusa's precise plan? Will they aerate the soil at just the proper time, at just the proper depth? That challenge is still ahead for the Howards: "Everywhere it is the human factor which stands in the way of progress. . . . Till the inhabitants of the villages of India can be awakened and till a general desire for rural uplift can be implanted in the people themselves, it must take centuries to effect any real and lasting development." Despite the cultivators' innate and timeless wisdom, de-

135. AH, "Agriculture and Science," 182, 178.

136. "These problems do not fall within the limits of any single branch of science. It naturally follows therefore that the conventional method of attack cannot hope to be completely successful" (Ibid., 174). Later, Albert adds: "We have been impressed by the evils inseparable from the present fragmentation of any large agricultural problem and its attack by way of the separate sciences. All this seems to follow from the excessive specialization. . . . The tendency of knowing more and more about less and less is every year becoming more marked" (AH and Wad, *Waste Products of Agriculture,* 17 n. 1). Blame for the fragmentation of science is laid at the feet of Liebig, who emerges from these Indore days as villain several times over (he will also be blamed for Western farmers' embrace of synthetic fertilizers—misguided and dangerous, says Albert): "The approach by way of a single science is really an inheritance of the Liebig phase when agricultural chemistry and agricultural science were synonymous" (AH, "Agriculture and Science," 180).

137. AH, "Agriculture and Science," 178–79.

spite attempts at Pusa and Quetta to display the advantages of improved varieties and practices in carefully staged demonstrations, Albert and Gabrielle admit that "the response of the cultivator to these efforts has not been as promising as the results yielded by the soil." Perhaps they had chosen to attack the easy problems first: hybridization, vicinism, irrigation, drainage, railway reform. Or perhaps the conventional specialized science they practiced at Pusa—eagerly at first, reluctantly later—blinded them to the human factor: "Much more attention should have been paid from the very beginning to the village as a whole, to its people, to their ideas, and to their general condition and outlook."[138]

The science at Indore must recognize "the supreme importance of dealing with the Indian village and its fields as a single subject."[139] It cannot be enough to apply science to plants, environmental conditions, and agricultural practices, nor is it enough "to bring the results to the notice of the people. This is only half the battle. The people themselves must desire to make effective use of the results and to improve their general condition." A different set of scientific principles must be discovered to guide programs of education, training, "propaganda" because "the new methods must be welded permanently into the rural economy."[140] Indian cultivators must come to view their cooperative societies not just as sources of credit but as sources of accurate scientific information about improved varieties and practices—and as organizational bases for community development projects (such as the realignment of holdings to enable large-scale improvements to surface drainage) and for marketing commodities (so that the "small grower" will no longer "be at a disadvantage in disposing of his produce").[141] The key is to get the culti-

138. AH and GH, *Development of Indian Agriculture,* 58, 84, 31. "The most formidable obstacle encountered in making practical use of the results obtained at the experiment stations is the unfavorable economic and educational condition of the Indian villages." The Howards describe a "mentality enslaved by superstition," where "the men and women, on whom all developments in Indian agriculture must depend, can neither read nor write and therefore cannot be reached by any form of literature" (57).

139. Ibid., 83. "In the development of rural India not only its soil, crops and cattle but also its people must be considered" (21). The only hint of the existence of social science comes not from Albert and Gabrielle's pens, but from a paper by G. Clarke, which they reprint as a friendly appendix to *Waste Products of Agriculture:* "This opens up a wide sociological study" (140).

140. AH and GH, *Development of Indian Agriculture,* 58–59. "This cannot be done effectively unless the support of the present adult population is enlisted and until they are made willing partners in the enterprise. . . . The people must be taught to desire better education for their children and better villages for themselves and they must also contribute a portion of the cost. Unless all this is accomplished, there can be no real progress and the tree will not take firm root in village life" (62).

141. Ibid., 73, 75, 80.

vator to want to embrace these scientific principles for improving his life, and it is toward these persuasions that the new Indore Institute will direct much of its energy.

What will do the job? "If the cultivator is to be made a willing partner in the new scheme he will have to be handled from the outset by men who are in sympathy with him, who understand his point of view, who speak his language, wear his dress and who can live in his village. One of the greatest difficulties will be to find and train an adequate supply of raw material for dealing with the people." Getting the word out—by training those who will in turn take the message of scientific agriculture to the people in a familiar and comfortable garb—will be a central organizing principle at the new institute. Its other raison d'être pertains to a message that may be more vital than all others to get out: "The problem is to show the people how to make the most of the organic matter now available and how to improve the supply." It will take a holistic science indeed to stretch from propaganda to composting. It is impossible to say whether Albert and Gabrielle's holism stems from twenty years in India or from their struggles to solve the multifaceted problems of crop production, or both. But I think that the wheel of life inspired their experiment on science at Indore, and led them to create a place where agricultural science is returned to its human context (getting the word out, to change cultivators' minds) and where plants are returned to their natural context (composting, to increase soil fertility organically).[142]

Institution Building

The Institute of Plant Industry affords Albert and Gabrielle an opportunity to mold their principles of holistic science into the walls and fields of an experiment station, and into its bureaucracy and programs as well—a chance to build a new science from scratch. It has three goals: to do fundamental research into the growing of cotton on the black soils of Central India and Rajputana; to train postgraduate students and others who will effect agricultural improvements throughout the countryside; to serve as an "object lesson" for cultivators and political officials, an exemplification of how best to grow cotton and structure rural villages. Clearly, Indore will not be a clone of Pusa, but rather a different place

142. Ibid., 84–85, 16. Albert would later write that two things were accomplished at Indore: "(1) the obsolete character of the present-day organization of agricultural research was demonstrated; (2) a practical method of manufacturing humus was devised" (AH, *An Agricultural Testament* [1940; London: Oxford University Press, 1943], 41).

"where the sheltered corners created by universities and by the endowed experiment station do not exist and where the working conditions can be summed up briefly in three words—payment by results."[143] Albert and Gabrielle resort to cultural cartography to compare the local agricultural research institute they plan to set up at Indore and the central institute they have left behind at Pusa. They reject any attempt to distinguish the two along lines of "fundamental" versus "local" research, for that is "a distinction without any real difference." The Indore Institute must embrace both, and so enable investigators to "work out their own salvation, and to follow the gleam untrammeled in whatever direction it may lead." An exclusively fundamental research "may lead to nothing," while those who are restricted to local research "may solve their problems at the expense of their own interests." An agricultural research institute that respects this pseudodistinction between fundamental and local research "erects walls where . . . the rule should be—no walls."[144]

But walls do go up at Indore, metaphoric walls as well as those of clay. If the distinction between fundamental ("long-range") and local problems is dismissed as an "artificial division of the subject," it is only to be replaced by a "more natural one" between research and demonstration. The Indore Institute will house research and provide training for those who are to carry out demonstrations of the improved practices that it discovers. Still another kind of institute or center is needed to finish the job of agricultural improvement: "Two branches—research and demonstration—which are both equally important should be developed in every agricultural department." Walls, or better still, distance will separate the two: "Such demonstration farms [will] require no scientific equipment and [will] be almost self-supporting. It should be clearly understood that no research work is [to be] attempted by the district staff. The art of demonstration and of inducing cultivators to adopt improvements is as important as that of research and every endeavor should be made to develop this branch of the subject as a separate and as an honored profession."[145] Indore science will comprise both fundamental and local research, and although it will pursue principles of inheritance, agricultural practice (like composting), training, and propagandizing, the actual task of face-to-face demonstration is removed from this cultural (and

143. AH, "Agriculture and Science," 181. See also AH and GH, *The Application of Science to Crop-Production: An Experiment Carried out at the Institute of Plant Industry, Indore* (Humphrey Milford: Oxford University Press, 1929), 4.

144. AH and GH, *Application of Science*, 60.

145. Ibid., 61, 60.

physical) space. The Howards' rationale for this wall will become clear during their description of the physical arrangement of buildings at the institute—but first something must be said about how the wherewithal for the place is cobbled together.

Original plans for a new agricultural research institute at Indore had been proposed in 1919, but the project had to be scuttled for want of financial backing. Five years later, that problem is solved through a union of the Indian Central Cotton Committee—described by Albert and Gabrielle as a "republic of cotton," financed by cultivators through a small "cess" on each bale of cotton they produced—and eight contributing states.[146] The Cotton Committee furnishes the capital cost of the building and equipment, while the Central States and Rajputana contribute to annual operating expenses (along with individual benefactors, who supplement revenues generated from the sale of publications and equipment). "Payment by results" now takes on flesh: whatever else it might do, the institute must address cotton growing in the black soil region of the Indian Central States.[147] When I get around to describing the agricultural science of the institute, it will be interesting to see how Albert and Gabrielle translate this specific and local mandate into a research program with universal ambitions—well beyond cotton in the Central States of India. Also, it is perhaps ironic that the Howards use their successes at Pusa to create a rival research center that will challenge its conventional sense of "science." Their subscribers (Cotton Committee, Central States, and Rajputana) may have been impressed that the Pusa wheat hybrids were "increasing wealth" by £20 million per year—a return on investment "many times greater than that yielded by the most successful industrial enterprise."[148] The Indore Institute for Plant Industry is brought into being on the hunch that similar returns are possible in the Indian cotton industry. Its subscribers evidently have confidence that Albert and Gabrielle—if given free reign—can work their scientific magic again.

The Darbar of Indore provides three hundred acres at nominal rent for ninety-nine years: the site has pros and cons. It is located near the

146. AH and GH, *Development of Indian Agriculture,* 28. For a brief history of British efforts to improve Indian cotton growing, see Henry, "Technology Transfer," 62–64.

147. "As the Indore Institute owes its existence to the grants made by the Central Cotton Committee and was founded as an agricultural research institute for work on cotton, it follows that the investigations on this crop take up a great deal of the energies of the staff" (AH and GH, *Application of Science,* 17).

148. AH, "Agriculture and Science," 173.

city of Indore (population 125,000), with its good communication and transportation infrastructure, and ready supplies of labor, water, and urban amenities for the staff. But the land itself is waterlogged to the point where cultivators have abandoned it to the weeds. The new science must be built from the ground up! Given their theoretical conviction that environmental conditions play a huge role in plant performance, we should not be surprised that Albert and Gabrielle invest several years preparing the Indore acreage for experimental crops: they install a system of surface drainage; level small hills and fill ravines; they construct metaled roads, culverts, and bridges. Such improvements to the land are justified theoretically, in terms of scientific aims: the Howards point out that experimental errors in variety trials are typically dealt with by "repeating the trials a number of times and by subjecting the figures to mathematical treatment." They have a better way: because "soil-erosion and poor surface drainage are very important factors in the production of irregular yields," it is obvious that "the improvement of drainage is the first step in dealing with experimental error."[149] In effect, much early work at Indore centers on bringing the conditions of the land under experimental control, so that drainage and erosion will not be factors in subsequent efforts to evaluate varieties of cotton. Moreover, these improvements to the land will provide object lessons for cultivators and larger landowners: prima facie evidence that even the worst fields can be made most productive. Perhaps the detail with which Albert and Gabrielle explain all this needed infrastructural work hints at subscribers' impatience, or maybe it signals that *nothing* about Indore—even the width of the roads is considered at length—is outside these experiments on a new science.

Even more is said to describe, explain, and justify the design and physical arrangement of the institute's buildings. Albert and Gabrielle are plainly committed to the idea that research has a spatial dimension, and that experiments can and should be done on the architecture of science. Holistic science is here built-in. On the site, they build laboratories, farm buildings, a model village, and a ginning factory. Of greatest sociological interest is their account of the positioning of laboratory buildings (offices, labs, seed stores, pot-culture house, seed-drying house, library, balance room, stores) vis-à-vis farm buildings (cattle sheds, godowns [storage], threshing yard, threshing floor, bullock gear, fodder barn, seed store, cart shed, mortar mill, water tank, seed-drying shed, and "perhaps the most

149. AH and GH, *Application of Science,* 35–36; on the conditions that the Howards met at Indore, see 7.

important item of the farm buildings, namely the compost factory and the pits for silage"). Albert and Gabrielle design from the following principle of holistic science: "It is most essential that the scientific and agricultural aspects of the work on crops should be as closely coordinated as possible." Perhaps they recall their days at Pusa and Quetta: "In all institutions of this character, a cleavage tends to develop and opposition between the staff employed on the two aspects of the work often arises." Such conflicts can be overcome spatially: "One of the tasks of the Director is to weld these two aspects of one subject into a real working unit instead of a combination in name only. For this reason, we wished to place both laboratories and farm-buildings together." However, their good principles must take into account the "noise and dust" generated by the farmyard, which is "not conducive to quiet work" of the lab. Such "practical difficulties" require that they put lab and farm on opposite sides of the main road, "farm buildings being to the windward." There is another reason why such a spatial demarcation has to be effected, albeit reluctantly. The farm buildings have been designed (as everything else) as a model for the demonstration farms being established in the separate states, and for the farms of cultivators as well. The Howards can therefore expect many visitors to Indore, and "the presence of laboratories next to the farmyard is apt to discourage the purely agricultural visitor."[150] Albert and Gabrielle do not elaborate: Will cultivators be intimidated by white coats and test tubes? Will they shrink from the purity and cleanliness required by a laboratory of science but virtually impossible in a stable for cows? No doubt this line of thinking also led Albert and Gabrielle to conclude that demonstration farms were better built on spots at a distance from agricultural research stations.

The bureaucracy and personnel at Indore differ from the organization needed for conventional science at Pusa. Given the Howards' view that in science the individual is everything and the organization nothing, one might a expect lean and mean organizational chart. So it is at Indore: although the institute will employ more than 150 people, almost all of them are farm laborers or trainees. The professional and clerical staff is minuscule compared to the elaborate system of section heads at Pusa (admittedly, a much bigger enterprise overall): "Director of the Institute and Agricultural Adviser to States in Central India and Rajputana [Albert, of course—Gabrielle remained on deputation from Pusa], Cotton and Physiological Botanist, five Assistants on Provincial Service pay (As-

150. Ibid., 14, 9.

sistant for demonstration work in the States, Personal Assistant, Assistant in charge of field experiments and farm, Chemical Assistant and Botanical Assistant), five Junior Assistants (Plant-Breeding, Farm and Botanical), Librarian, Artist, two Clerks, three Fieldmen, Store-keeper, two blacksmiths for the farm, one blacksmith for extension work and the usual menial staff."[151] By my count, there are two senior scientists at the institute, and only one of them identified by disciplinary specialty. This is surely in keeping with Albert and Gabrielle's desire to transcend the fragmented character of knowledge making: "In place of the conventional approach by way of the separate science, the plant will be regarded as the center of the subject. A knowledge of several sciences, of practical agriculture and of the requirements of the trade will be brought to bear simultaneously on the chief problems presented by cotton and other related crops." Unlike Pusa, Indore will not have rafts of specialists in bacteriology, mycology, veterinary science, entomology, chemistry, or even in economic botany. "It became increasingly evident that the organization of an agricultural research institute, on the basis of practical agriculture on one hand and on the separate sciences on the other, was by no means the ideal arrangement. . . . It was soon discovered that the problems, presented in the improvement of the crop, cannot be split up into a number of parts without grave detriment to the *whole.*"[152] This is as close to an admission of "holism" as I could find.

Perhaps because the bulk of employees at the Indore Institute are farmworkers, Albert the director must consider potential problems of "labor-management relations"—of course, he simply adds it to his scientific repertoire by publishing a paper in the respected journal *International Labor Review* entitled "An Experiment in the Management of Indian Labor." His specific findings are less interesting for us than the choice by this trained fungus expert to bring such issues within the sphere of his ballooning science. Albert faces an immediate problem: the city of Indore provides many good employment opportunities, and if the institute is to procure a reliable labor force for its farms and fields, it will have to discover "the principles" behind effective incentives.[153] The key, writes Albert, is to make "the laborer feel that he is receiving a

151. Ibid., 16,

152. AH and GH, *Development of Indian Agriculture,* 29; AH and GH, *Application of Science,* 1 (my emphasis).

153. "The Director would have to pay attention to the labor problem and devise means by which an efficient and contented body of men, women and children could be attracted and retained for reasonable periods" (AH and Wad, *Waste Products of Agriculture,* 159).

square deal." What constitutes "square" then and there measures benevolent colonialism: the work day will be reduced from 10 to 7.5 hours, payment will be direct and regular (no middlemen to siphon off their share), no company stores to tempt workers to put their wages back in the institute's coffers, free medical care, a one-room lockable earthen-floor cottage with a veranda and drinking water, and a "certificate plan" attesting to the laborer's acquired skills. Accumulation of these certificates is to be tied to rising wages, a kind of incentive program that "enables an ambitious laborer to save enough money in a few years to purchase a holding and to become a cultivator." Nothing is overlooked in Albert's characteristically exhaustive foray into administrative science: "It is equally important to allow laborers to reach their homes by sundown, particularly during the rains when snakes abound." Albert's measure of success would surely have pleased his contemporary Frederick Winslow Taylor: "To everyone's surprise, it was found possible to speed up the work very considerably."[154]

Apart from its research on cotton, the institute developed two other programs: encouraging visitors and training students. A prior investment in good roads now pays off, as they "have enabled important visitors, who are more frequently pressed for time, to gain a good idea of the Institute and of the work in progress in little more than an hour." Less-hurried visitors have also been drawn in from the ranks of cultivators.[155] In effect, the Indore Institute becomes a demonstration of demonstration: the grand plan calls for the separation of research institutes from demonstration centers in each of the constituent states. The farms at Indore—a place for research—provide the model for the proposed demonstration centers. But because most of these other demonstration centers remain on the drawing board, Albert and Gabrielle welcome to their institute the ultimate end-user of their science: local cultivators and zamindars. No longer do the Howards assume that high-yield varieties or experimentally proven improvements in drainage or tilling systems can sell themselves to the people. The institute works hard at getting the word out by sponsoring regular demonstrations for villagers on cultivation, well irrigation, crop selection, manures, and cattle food, and by opening the institute's library as a center of agricultural information: "Very few

154. Ibid., 165, 162, 161.

155. AH and GH, *Application of Science,* 8. "There has been a growing stream of visitors, many of whom are either local notables or well-to-do men who are beginning to take up the improvement of their land. Actual cultivators from the villages are now coming in larger numbers, many of whom work for a time to learn new methods" (55).

visitors leave the Institute without purchasing some book. The interest aroused by a visit to the Institute is thus crystallized and made more permanent." The manure factory "in which indigenous materials only are employed . . . attracts many visitors," and Albert and Gabrielle seem to believe that they have satisfied their subscribers by establishing a place that "exports ideas and information on rural reconstruction as well as improved varieties of crops and new methods of cultivation."[156]

But ideas and information are not the only product of the Indore Institute: training programs for "sons of cultivators," supported by a generous plan of studentships, allow the institute "to export every year a number of trained workmen." The goal of such training is "to stimulate interest in the development of the countryside rather than to teach the principles of agricultural science," and many of the graduates will go on to become agricultural officers in the larger states.[157] The curriculum will be as holistic as the science on which it is based: students will learn about plants, soils, the arrangement of farm buildings, cultivating and drainage practices, packaging and the market, cooperative credit, rural redevelopment, and propagandizing (so that they in turn might help overcome the "human factor" in the rural development of India).

Intensive Agriculture

The Indore Institute is only in part a place for the dissemination and demonstration of agricultural science: it is also to be a place for the discovery of new knowledge about better ways to grow cotton on the black soils of the Central States. Albert and Gabrielle must keep their subscribers happy by enhancing the profitability of the cotton crop in central India, while at the same time pursuing their own quest for a holistic agricultural science that might yield universalizable principles potentially transportable to other crops, other growing conditions, and even other cultures. The centerpiece of their vision is "intensive agriculture," the idea picked up from King's *Farmers of Forty Centuries,* and it will hold tight the interests of subscribers and those of two holistic scientists. "Intensive agriculture" will translate the economic ambitions of the institute's backers into the sorts of experiments that Albert and Gabrielle left the constricting conventions of Pusa to pursue.[158] These new experiments

156. Ibid., 11, 41, 56.

157. Ibid., 51–53.

158. King's book "should be prescribed as a textbook in every agricultural school and college in the world" (AH and Wad, *Waste Products of Agriculture,* 12). "The great problem of agricul-

into intensive agriculture will tip the balances the Howards had kept level at Pusa: now, India's differences from the West will become more important than its similarities; the conditions in which a plant is grown will become more important than its genes; and organic manures will become a far preferable means for enriching soil than artificial (chemical) ones. The result? New facts: the Indore Institute is all about "doing science in a new key," to produce principled and useful understandings of the world that are undiscoverable (or rejected) by conventional scientists and their fragmented methods.

East Is East

Albert and Gabrielle begin their efforts to improve cotton growing on the black soils of central India with a question of global proportion: Can scientific advances that have made agriculture in Europe and North America the envy of the world be transported to India? Twenty years of hindsight now give them a decisive answer: No. The agriculture of the West is "extensive" and "essentially modern": large landholdings; machine-based, expensive labor costs; abundant capital for improvements; sophisticated food preservation; efficient systems of transportation and marketing; and dependence upon artificial manures to maintain soil fertility. The contrast to India could not be crisper: landholdings are minute; labor is cheap and abundant but capital is not; machines and chemical fertilizers are out of economic reach; transportation and marketing systems are risky at best.[159]

Because of such ineradicable differences, India should not be envious of the West and it would be foolish to try to emulate Western means of success: "The systems of agriculture of the west and of the east are very different and the two have little or nothing in common." Extensive methods of agriculture in Indian conditions "neither utilize the full energies of the workers nor the potential fertility of the soil. Such a system of agriculture is bound to prove un-economic and to result in poverty."[160] Indeed, extensive agriculture might not even be well suited to the envi-

---

ture of the near future is the *intensive cultivation* of improved varieties by which the present production can be vastly increased in quantity as well as improved in quality" (AH, "Agriculture and Science," 174; my emphasis).

159. AH and Wad, *Waste Products of Agriculture,* 5, 2. The Indian landholdings are so small because "every male child inherits an equal share of every description of land. . . . The plots get smaller and smaller and in some cases so narrow that cross-plowing is impossible" (AH and GH, *Development of Indian Agriculture,* 73).

160. AH and Wad, *Waste Products of Agriculture,* 16; AH and GH, *Development of Indian Agriculture,* 3.

ronmental conditions of the West—or anywhere else, for that matter. The jury is still out on modernity: "This system of agriculture lies in the fact that it is new and has not yet received the support which centuries of successful experience alone can provide." Rarely do Albert and Gabrielle invoke a discourse that would today be labeled "environmentalist," but here is one instance—as they criticize the possible shortsightedness of Western extensive agriculture: "Mother Earth is provided with an abundant store of reserve fertility which can always be exploited for a time. Every really successful system of agriculture however must be based on the long view, otherwise the day of reckoning is certain."[161] The wheel of life may yet turn against the modern agriculture of the West.

However cotton growing in the Central States and Rajputana might be improved, it will not be done by importing the modern extensive agriculture of the West. The Howards' holistic science allows them to seek models elsewhere (even outside conventional science), and they find a more workable one further to the east—in an agriculture that has indeed stood the test of time. King's account of the intensive farmers of Japan, China, and Korea situates their agricultural practices in contexts not unlike those of India: small landholdings, intense population pressures, little capital for investment in machinery and other modern improvements. Albert and Gabrielle are obviously impressed by King's assertion that "fertility has for centuries been maintained at a high level without the importation of artificial manures," and inspired by his vivid descriptions of how all agricultural residues are piled up into heaps until they become a "finely divided organic matter" then spread on the fields and paddies.[162] On a hunch that Indian agriculture is best served by importing intensive agriculture from the East, Albert and Gabrielle map out their experimental program for the Indore Institute: "The obvious line of advance is the gradual introduction of more intensive methods, for which the supply of suitable manure, within the means of the average cultivators, is bound to be an important factor."[163]

The Howards' growing conviction that the route to improved cotton growing in India goes through Chinese farmers rather than through Cambridge laboratories is not, however, a retreat from science, but an expansion of it. A properly holistic science will transcend differences between East and West and translate the practices and beliefs of each into universal principles enabling improvements tailored to the specifics of

161. AH and Wad, *Waste Products of Agriculture,* 2, 5.
162. Ibid., 11; AH and GH, *Development of Indian Agriculture,* 49.
163. AH and Wad, *Waste Products of Agriculture,* 7.

place. Some bits of conventional-looking science will be useful for the tasks at Indore. For example, Albert and Gabrielle lay out (as they did earlier for wheat) the classification of cottons from the standpoint of systematic botany, its irrelevance to the agricultural conditions of India, the possibility of improved varieties via selection and maybe hybridization. As at Pusa, the pursuit of improved cotton will inevitably draw on scientific principles of genetics developed in the West—but applied in the distinctively Indian context. "In the solution of these questions, the inheritance of characters will have to be studied," and in the nomothetic spirit of science that Albert and Gabrielle endorse: "The results will apply not only locally but will also be of general interest." They report early successes: "The selections made show great promise and the isolation of an improved malvi cotton which will ripen early, yield well and produce a much better fiber than the mixture now grown is only a question of a very short period of time." And, remembering what happened to the demand for indigo, Albert and Gabrielle express some urgency that such work continue: "In the world of cotton, artificial fibers are already on the horizon." But that future may be staved off if "science is effectively brought to bear on the production of the raw material."[164] Science remains the hope for a profitable cotton industry in India, but only a science as comfortable with Chinese dung heaps as with systematic botany. After all, for the holistic Howards, "The one supplements the other: each can be regarded as a part of one great whole."[165]

A Hundred Percent Conditions, Ten Percent Plant

Looking back on the papers written during the first decade or so at Pusa, we find a majority of them center on questions of systematic botany and the genetics of inheritance. By contrast, among the Indore writings from 1924–31, explicit discussions of the improvement of the cotton plant oc-

164. AH and GH, *Application of Science,* 80, 22, 28.

165. AH and Wad, *Waste Products of Agriculture,* 16. On a continued role for science in agriculture, see AH, "Agriculture and Science": "Science is certain to play an important part" in the solution of "some of the agricultural problems of to-day" (171). "The problems . . . now await solution by the man of science" (180). The Howard team will draw on the research of Selman Waksman at Rutgers in New Jersey, who is engaged in experiments (well before his more famous research on polio vaccines) that will culminate in his 1936 book *Humus: Origin, Chemical Composition, and Importance in Nature.* Holism makes possible this massive cultural hybridization: "The latest scientific work of the Occident and particularly that recently accomplished at the experiment station of New Jersey, together with the practices in vogue in India and the Far East, have been welded together and synthesized into a system for the continuous manufacture of manure throughout the year so that it forms an integral part of the industry of agriculture" (AH and Wad, *Waste Products of Agriculture,* 17).

cupy only one ten-page chapter in *The Application of Science to Crop-Production,* one brief note in the *Agricultural Journal of India,* several pages in *The Development of Indian Agriculture,* and no pages at all in *The Waste Products of Agriculture.* This signals a shift: the Indore Institute intends to make a science of intensive agriculture (not just genetics).[166] "The plant" is not examined so much for its own sake (whether to reveal patterns of inheritance or to stabilize improved characters of market significance), but as a measure of environmental conditions: "The plant itself [is used] to indicate the general soil conditions and its deficiencies. . . . In this way soil studies resolve themselves into problems of adaptation—the relation of the plant to its environment." When Albert and Gabrielle first started their work on wheat at Pusa, progress "lay in the improvement of the variety," so that "the possibility that various soil factors might limit the growth of the . . . plant naturally received little attention."[167] But there has been progress in the twenty years since: upon completion of the "work of replacing the indigenous crops of India by higher yielding varieties, it was soon realized that the full possibilities in plant breeding could only be achieved when the soil in which the improved types are grown is provided with an adequate supply of organic matter in the right condition." The Howards take this lesson to heart in their work at Indore, and looking back later on their successes, Albert recalls: "Improved varieties by themselves could be relied on to give an increased yield in the neighborhood of *ten per cent.* Improved varieties plus better soil conditions were found to produce an increment up to a *hundred per cent* or more."[168]

Still, genetics will not completely disappear from Indore. Studies of the inheritance of plant characters will fit into intensive agriculture by looking at roots (which, of course, come most directly into contact with

166. You can forgive Albert and Gabrielle another pun: "The best results on cotton will therefore be obtained, not by following the methods of the past, but by breaking new ground. . . . The cotton work of the future must be a well-balanced combination of agronomy and genetics with soil science" (AH and GH, *Application of Science,* 27).

167. AH and GH, *Development of Indian Agriculture,* 45; AH and GH, *Application of Science,* 27.

168. AH and Wad, *Waste Products of Agriculture,* vii–viii. Once again: "This amount of cotton per unit of area is limited by the conditions under which the cotton plant grows. Improved varieties give some increase in the total yield but such results are small compared with the enormous increment made possible by better agricultural conditions. By improving the agronomy of cotton on the black soils, it is possible to double the acre yield. By merely changing the variety, the increase in the crop is often not more than ten per cent" (AH and GH, *Application of Science,* 27).

soil conditions): "There can be little doubt that much of the plant-breeding work of the future on cotton will concern the life history of the root-system and its relation to the soil types quite as much as the growth of the shoot and the character and amount of fiber." Gabrielle considers the future of genetics in her only solo paper after 1924, the Presidential Address to the Agricultural Section of the Sixteenth Indian Science Congress meeting in Madras just one year before her death. She makes the case for a chair of genetics at some Indian university (there were none in 1929). Although recognizing that "little or none of the fundamental work on the theory of heredity has been carried out in India," she argues that "it will be increasingly difficult to maintain the economic [agricultural] work at its present level unless it is stabilized by a school of pure research in the country itself."[169] At first, such a message would seem inconsistent with the principles guiding the organization of the Indore Institute: wouldn't that sort of genetics be one of those fragmented sciences lacking the breadth and depth needed to solve problems of agriculture and rural development?

Not really: Gabrielle fashions a recent history of genetics in which its next phase becomes the holistic science she and Albert are building at Indore. Once upon a time, it was thought that Mendelian laws of inheritance could be ridden smoothly into agricultural nirvana: select, isolate, hybridize, stabilize, scale up, grow—and enjoy. But "after twenty-five years, plant breeders are sadder and wiser." Without denying demonstrable accomplishments such as Pusa 12, Gabrielle suggests that genetics is "still a long way from the perfect variety in any crop and the way of improvement is long and arduous." Things are more complicated than Mendel had thought: what were unit characters for him are not readily distinguished and isolated. "Each character we see, such as flower color, rust resistant and so on, is not produced by one simple gene but is controlled by a large number of factors." Moreover, "certain genes do not affect one character only but produce an impression on several organs of the plant—that is, their effect is manifold." The implication is that "the direct application of simple Mendelian rules [is] impracticable." Further complexities abound: Gabrielle cites studies seeking to resuscitate interest in once lifeless Lamarckian thoughts about the inheritance of acquired characters. One study suggests that "if differences arise in the

169. AH and GH, *Application of Science,* 18; GH, "The Improvement of Plants," *Agricultural Journal of India* 24 (1929): 158.

gametes under varying growth conditions then the collection of seeds from different parts of the plant must affect the statistical data and Mendelian ratios." Another is reported to have found "something like the inheritance of acquired characters although . . . not of the Lamarckian type but [which] illustrates a new evolutionary principle that heritable variations may be induced by means of the food supplied." Genetics (as well as agriculture) needs a more holistic approach: it "is entering upon a new phase in which the effect of the environment on inheritance is the main theme."[170] That kind of genetics sits comfortably in the labs and farms of Indore.

If the key to crop development, rural uplift, and even future genetics lies in the conditions of plant growth, then what specifically about those conditions looms largest? It boils down to soil fertility, which becomes, in the more confidently reductionist language of Indore, a function of the organic matter that it contains: "The ancients and the moderns are in the completest agreement as to the importance of organic matter in maintaining the fertility of the soil." A problem encountered late at Pusa, but set aside then, now moves to center stage: "What is needed is to show the cultivators where to find the organic matter required and how to prepare it the best way."[171]

Going Organic

Albert and Gabrielle have so far effected an impressive string of equivalences: improvement of cotton growing on the black soils of central India equals adoption of the intensive agriculture of the Far East as a model, equals primary attention to the conditions in which plants are grown, equals maintenance of soil fertility by amendments of organic matter—and now, the coup de grâce—equals *composting*. Enlarged profits from Indian cotton lay in the compost heap! The judicious position taken at Pusa (either organic or artificial manures could provide everything that plants needed to thrive, and the only liability of the latter was cost) has given way to insistence that organic matter provides both soil and plant with things that synthetic chemicals cannot: "Artificial ma-

170. GH, "Improvement of Plants," 155, 157, 158.

171. AH and Wad, *Waste Products of Agriculture*, 20; AH and GH, *Application of Science*, 38. There is much room for improvement: "Most of this vegetable refuse is at present wasted or is misused. It is not uncommon to see dried leaves, stalks and other vegetable refuse burnt by the municipalities to save the expense of removal. Villagers often burn the refuse on the fields. . . . It is a distressing sight to see India's potential wealth going up in smoke" (39).

nures do not supply what Indian soils really need, namely, fermented organic matter in a finely divided condition," and because of this "all attempts to solve the manurial problems of India by means of the conventional [synthetic] methods of the West have proved a failure." The preferability of natural over artificial manures is no longer an issue for scientific adjudication: "No experiment is needed to demonstrate the effectiveness of more organic matter for cotton and other crops."[172]

I wonder whether Albert and Gabrielle ever hesitated at this moment, to ponder the following dilemma: If organic intensive agriculture is so wonderful, why is its current practice restricted to farmers of the Far East? Why is it not widely pursued, either in India, where cultural conditions are similar, or elsewhere—don't the extensive farms of North America and Europe also need a fertile soil? Blame is placed on Liebig and his fragmented science (he reduced everything agricultural to chemistry). Albert and Gabrielle frame organic agriculture as the naturally enduring human means for growing crops effectively, a practice of the ages interrupted only for the last century by the misdirected science of Liebig and his epigones at Rothamsted, Lawes and Gilbert: "There has occurred only one brief period during which the role of organic matter was to some extent forgotten. This took place after Liebig's *Chemistry in its Application to Agriculture and Physiology* first appeared in 1840."[173] According to the Howards, once Liebig reported in 1840 that plants got the carbon they needed from the carbon dioxide of the atmosphere and not from the soil, scientists and modern farmers were persuaded that the only ingredients in the soil that needed to be replaced were the minerals that were extracted by the plant: N-P-K.[174] They embraced the Rothamsted finding of 1843 that plants could be grown effectively forever if only the proper replacement mix of artificial manures was spread on the soil and if the roots of crops were left in the fields to rot—no natural manures need be added.

To get human history back on track—to return us all to our primordial organic and intensive agriculture—we need not less science but more

172. AH and GH, *Development of Indian Agriculture,* 48; AH and GH, *Application of Science,* 38.

173. AH and Wad, *Waste Products of Agriculture,* 21. For an account that makes Liebig, Lawes, and Gilbert into heroes, by one who swims in their wake, see Russell, *History of Agricultural Science.*

174. "All that mattered in obtaining maximum yields was the addition of so many pounds of nitrogen, phosphorous and potassium" (AH and Wad, *Waste Products of Agriculture,* 22).

better science: the holistic kind from Indore (or so Albert and Gabrielle would have us believe). Neither King's book nor the reported successes of farmers in the Far East nor the obvious similarities between those countries and India can in themselves compel cultivators in the Central States to start composting or adding finely shredded agricultural residues to their fields as a routine practice: Albert and Gabrielle will not again forget the "human factor." Only an improved holistic science can shake lethargic Indian cultivators from their wasteful practices and also eclipse the blinded science of Liebig, Rothamsted, and the "great . . . artificial manure industry [that] follow[s] as a matter of course."[175] In order for intensive agriculture to spread beyond the Far East, the traditional practices described by King must be translated into principles, moved from indigenous wisdom into science, given the authority and credibility of experimental test, and transposed into formulaic methods that will make it work time after time. This is what the institute at Indore is designed to do.

The Howards first have to figure out why certain practices that resemble Chinese methods of intensive agriculture have not always improved crop production when tried in India. In lieu of animal dung, unavailable as a manure because it has already been claimed for fuel, they test green manuring as a viable alternative source of organic matter. Indeed, it seems simple and cheap enough to implement: grow a crop of leguminous vegetables (which are able to fix atmospheric nitrogen) and then till it under (or at least leave its stalks and roots behind after harvesting fodder); the accumulated nitrogen will be returned to the soil (along with the decayable stuff that improves soil aeration and moisture retention), ready for use by the next crop of wheat or cotton. But when Indian cultivators have tried this, sometimes their harvests have been worse than if they had planted in unmanured fields—a discovery confirmed at Pusa.[176] Evidently, the Chinese figured out long ago that this style of green manuring does not work, and Albert and Gabrielle call their confirmation of this fact their "master idea": "The soil cannot ferment raw organic matter and grow a crop at the same time." Chinese farmers first compost fresh green matter and allow it to ferment or decompose before it is added to their fields. Their intensive agriculture "involves two separate processes, the preparation of food materials from vegetable and ani-

175. Ibid.

176. "There is at present a serious waste of manure going on in India due to the fact that raw unfermented materials are applied to the soil. This practice is incorrect." [AH?], "The Preparation of Organic Matter for the Cotton Crop," *Agricultural Journal of India* 21 (1926): 485.

mal wastes which must be done outside the field [in the compost heap] and the actual growing of the crop."[177]

But like their counterparts in India (who also sometimes intuitively got it right), Chinese farmers could not explain why the composted green manure enriched their soil in a way that the raw stuff did not. Their practices had to be understood scientifically before they could be converted into principles that would help Indian cultivators find a better way to use the residues of leguminous plants to fertilize their soil. Holistic science was up to the challenge: "The great value of this broadening of the basis of agricultural science was to afford an explanation of *practices* which had been arrived at on the basis of experience and to add a number of important *principles* to the subject." As it happens, the fungi and bacteria responsible for breaking down leaves, stems, and twigs into "finely divided fermented organic matter" actually consume nitrogen in the process—and if this process goes on in the same soil where crops are being grown, they will rob vital nitrogen from those crops and produce a disappointing yield. The traditional Chinese practice of composting fresh green agricultural waste becomes a "balance sheet on nitrogen," in which Albert and Gabrielle take scientific readings of the amount of nitrogen produced and consumed at each stage of the process. The first principle of intensive agriculture, now experimentally confirmed and theoretically explained, is this: "There must be no competition between the growth of the plant and the preparation of its food materials."[178]

This is, of course, just the start of the Howards' scientification of Chinese composting. King's account would seem to provide a sufficiently detailed recipe for anyone to cook up a hot pile: all sorts of agricultural residues are mixed in a fresh condition with dirt, a little cow dung, wood ashes, then heaped up into a rectangular pile and moistened and later turned—"in the Chinese fashion."[179] Two months later, fermentation is complete and stuff resembling moist leaf mould (humus) can be returned to the fields, with all good results for subsequent crops. Things are not that simple, as the early failures with green manuring attest: "The decay and incorporation of green-manure in the soil has been shown to be a very *complex process* depending on . . . the chemical composition of

177. AH and GH, *Development of Indian Agriculture,* 45, 49; AH and Wad, *Waste Products of Agriculture,* 27.

178. AH, "Agriculture and Science," 173; AH and Wad, *Waste Products of Agriculture,* 27; AH and GH, *Application of Science,* 39.

179. AH, "The Water-Hyacinth and Its Utilization" *Agricultural Journal of India* 20 (1925): 395.

the plants, . . . nature of the decomposition [water, air], metabolism of microorganisms taking part in the decay." Successful composting depends on the exact mix of different kinds of organic waste, on the dimensions of the heap, on the amount of water provided, on the timing of turnings, on wind and ambient air temperature: "Like all manufactured articles, [humus] must be properly made if it is to be really effective," and "the various steps from the raw material to the finished product follow a definite plan." The "Indore method of composting" is born: "The object of the process is to bring these changes under strict control and then to intensify them. A knowledge of the chemical processes involved and of their relative importance is therefore essential in applying the process to other conditions." Indian cultivators (and everyone else) need a "method" of composting, something more than the traditional knowledge provided by King and his wise Chinese farmers. "The agriculturalist of the future must be shown how to become a chemical manufacturer: the method finally adopted must be so elastic that it can be introduced into almost any system of agriculture. . . . It must be simple, safe and must yield a continuous and uniform product, capable of being instantly utilized by the crop."[180] Plainly, Albert and Gabrielle are looking beyond the cotton crops of central India.

It is impossible for me to give a brief synopsis of the "Indore method" if indeed its raison d'être is to explicate scientifically the elaborate and complicated process of making humus in a compost pile. In its rudiments, the Indore pile is not all that different from what King described in China, except the raw materials may be any agricultural residue (although they should be first used as bedding for cattle, where they will naturally get mixed with some dung). To break up large stalks and stems, the raw stuff should be removed from the cattle pens and placed on a road where the cattle can trample it further. Green manures must be withered for at least a day before being added in, and the ratio of carbon (brown stuff) to nitrogen (green stuff) should be about thirty-three to one. The heap must measure thirty by fourteen feet precisely, and be two feet deep (to retain moisture). Urine-impregnated earth and/or wood ashes should be layered in to serve as a base (on its own, humus will turn the pH of most soils in the acidic direction). A scoop of extant compost (one month cooked) should be added to the pile as a "starter" or "charge," which inoculates the new pile with needed fungi and bacte-

180. AH and Wad, *Waste Products of Agriculture*, 42, 26, 77, 57.

ria. The pile should be periodically moistened (except during the monsoon, when it should not be built in a pit at all but on level ground), and it should be turned precisely sixteen days, one month, and finally two months after it is charged by the inoculants. After three months, the finished humus may be put into the fields where it is certain to improve soil physics (particles not fully decomposed will enhance aeration and water retention), soil chemistry (nitrogen is returned to the soil with interest), and soil biology (microorganisms continue to do good things for the plant—a thread that Albert will pick up after his final return to England).[181] Brief descriptions of some of the topics covered in the fifteen statistical tables included in *Waste Products of Agriculture* can only suggest the immense range of experimental variables examined at Indore: composition of raw materials (measuring organic matter, ash, proteins, fats, fiber, soluble carbohydrates, nitrogen) for twenty-one likely agricultural residues; losses of nitrogen resulting from the close texture of the mass (the loss is higher for single-ingredient piles, leading Albert and Gabrielle to recommend a promiscuous mix of raw materials—purity comes later, from the fungi and bacteria!); results with reduced and full dung proportions (the process works just as well with reduced dung so that "the cultivator really requires only a fraction of his cow-dung for converting all his vegetable wastes into humus"); effect of heavy rainfall on pH and temperature in the piles; nitrogen balance sheet in pits and heaps (shows a net increase in nitrogen).[182] But does this methodically manufactured and scientifically known humus really improve crop production—better than no manure at all, or something artificial? When compared, for example, to the elaborate milling and baking tests performed on the many varieties of Pusa wheat, Albert and Gabrielle's QED from Indore is almost cryptic (perhaps these tests are still *in medias res* when Albert packs up for Britain in 1931): "This material when added to the soil stimulates growth in a remarkable manner and is proving a valuable manure."[183] A photograph compares millet grown with Indore compost (obviously fuller and taller) and that grown with "Adco," a patented

181. Albert and Gabrielle are a little fuzzy on the *biological* processes that distinguish soil rich in humus from soil where humus is absent. The answer will not come until the late 1930s, in the form of the 'mycorrhizal association' (a symbiosis between fungus and root hairs). This idea will be key in Albert's debates with Rothamsted scientists, who quite agree with him on the chemical benefits of humus (N-P-K is added) and on the physical benefits (particles improve aeration and soil retention).

182. AH and Wad, *Waste Products of Agriculture*, 86

183. AH, "Water-Hyacinth," 395. No additional details are provided.

artificial or synthetic compost additive that provides food for busy microorganisms and a base for neutralizing the pile—developed and marketed, as it happens, by the Rothamsted Experimental Station. It no doubt gives Albert and Gabrielle great satisfaction to conclude that "far more satisfactory results have been obtained with the indigenous materials."[184]

With Chinese composting practices successfully converted into the "Indore method," Albert and Gabrielle repay their debt to the Central Cotton Committee and their subscribing Indian states. New facts about growing cotton have been experimentally proven and scientifically understood: any variety of crop will perform better in fertile soil; soil fertility is uniquely enhanced by the addition of organic matter; raw agricultural residues must be properly composted if advantages to crop yield and quality are to be realized. Have they also discovered principles applicable to other crops, to other conditions, to other parts of the world—principles that might reconnect the chemically dependent agriculture of the West to its intensive organic roots in the East? "How rapidly the method can be incorporated into the large-scale agriculture of the west is a question which experience alone can answer."[185] It may be telling that Albert and Gabrielle write "experience" here rather than "experiment." Throughout the writings from India, "experience" is what Indian cultivators and Chinese farmers acquire (empiricist intuitionism), in contrast to the principles unearthed only by "experiment." Perhaps Albert, on the verge of retirement, facing the loss of his collaborator and spouse, sees his days for experiment as numbered. Or is he getting ready for another epistemic transit upon his return home—an authorization of experience (and personal testimony) as a means for producing credible scientific knowledge?

I do not know how Gabrielle died. Her sister implies that it may have been sheer exhaustion, from the painstaking job of sorting out those tiny nicotiana seeds under demanding Indian conditions—all the while keeping house. Or it might have been cancer—a possibility suggested by mention of it three times in the publications from Indore.[186] How charac-

184. AH and Wad, *Waste Products of Agriculture,* 87.

185. Ibid., ix.

186. Speaking of agricultural pests, Albert writes: "This and allied results follow alterations in the chemical composition of the cell-sap brought about by changes in the general metabolism induced by the altered methods of cultivation. The pest will only become acute if certain substances are present in the cell in quantity—a close parallel with the recent work on cancer in

teristic it would be of Albert to consider Gabrielle's illness (if I am right) as he had considered disease in plants. Food—and its connection to health—runs as a leitmotif through these papers and books: food for the microorganisms that feed the soil that feeds the plants that feed us. Albert and Gabrielle's "resistance theory" suggests that disease results from pathologies in a plant that occur because its growing conditions are inadequate: if fed properly by a healthy soil, plants are more likely to better resist nasty fungi and insects. Writing about cotton, they note how "unfavorable soil conditions lead to changes in the acidity and other characters of the sap and so prepare suitable food for the insect or fungus which thrives just as long as this food supply is available. Favorable soil conditions, on the other hand, bring about a marked increase in the resistance of the plant." They add that "the wider aspects of disease and immunity are being taken up at Indore," in particular with respect to an outbreak of hoof-and-mouth disease among the institute's cattle. The same theory solves this problem as well: "These simple and obvious improvements in the food supply of the hot months have worked wonders. The animals are now in first class condition during the hot weather and . . . can now be shown to visitors."[187]

Surely Albert and Gabrielle's new science is holistic enough for them to see human health as well in terms of resistance theory. Albert writes approvingly about research by McCarrison and Bentley (also in India) on the connection between, on the one hand, nutritiousness of human food and soil fertility, and on the other, poor harvests and the incidence of malaria. "A definite relationship between the incidence of malaria and the way crops are grown is suggested by the observations of Bentley in Bengal. . . . The fertility of the soil is diminished, the area of wasted land increased, population declined and with it malaria increased to an appalling extent. Where rice is grown to perfection there is little malaria; where the crop is cut off from the necessary inundation, malaria is rife."[188] A link between soil fertility—a function of humus—and the health of ev-

Great Britain" (AH, "Agriculture and Science," 180). Speaking of erosion, waterlogging, and denitrification of the soil, Albert and Gabrielle again draw on the image of disease: "Nature's system of drainage in the black soil areas of India resembles a huge cancer, which slowly absorbs the life-blood of the countryside" (AH and GH, *Application of Science,* 31). "Loss of soil by erosion follows as a matter of course which, if unchecked, leads to the formation of a deep nullah with its attendance antennae. The agricultural cancer of Central India is then established which proceeds to feed on the natural fertility of the countryside" (34).

187. AH and GH, *Application of Science,* 20, 48.

188. AH, "Agriculture and Science," 176.

erything, including humans, is the centerpiece of Albert's efforts after 1931, but whether this turn had anything to do with Gabrielle's death cannot be decided.

YOU CAN'T GO HOME AGAIN: ENGLAND, 1931–1947

Albert retires from his post at Indore and after twenty-five years in India returns home to England, to face the Great Depression and then World War II. He is alone—without Gabrielle "who is no more." But there is no time for rest. Albert looks upon his retirement as an escape from the "strait-jacket" he has worn as an "independent" scientist working in research organizations whose fragmentation and bureaucratization made it impossible for him to pursue genuinely innovative solutions to perennial problems.[189] Even the institute at Indore—the embodiment of holistic science—carried its restrictions: subscribers among cotton cultivators and governments of the Central States and Rajputana had to be placated, and inquiries that strayed too far from cotton or central India had sometimes been perceived as tangential to their interests. Now in London, free from immediate responsibilities to any organization, out from under the narrow interests of others, honored by knighthood and courtesy appointments at scientific institutions, Albert embarks on "a new field of work . . . in which the research experience of a lifetime [can] be fully utilized."[190]

Having found the truth in India—crop production and resistance to disease depends on the fertility of the soil, which must be enriched by the addition of organic waste prepared in scientific heaps according to the Indore process—Albert now sets out to convert the rest of humanity to his fact. He begins a "national campaign" for the "reform of agriculture," enlists an army of "compost-minded crusaders," eager "to convert a nation of gardeners into a nation of compost gardeners."[191] The effort

189. The reference to Gabrielle's death is from Albert's dedication in *An Agricultural Testament* (1940; reprinted, London: Oxford University Press, 1943). Albert speaks of his straitjacket in *The Soil and Health* (New York: Devin-Adair, 1952), 105, a work originally published in England as *Farming and Gardening for Health or Disease.* Albert refers specifically to his time in the West Indies as confining, but he elsewhere has equally unkind things to say about life at the Indian Agricultural Research Institute at Pusa.

190. AH, *Agricultural Testament,* 55. "When this day of retirement came, all those obstacles vanished and the delights of complete freedom were enjoyed" (AH, *Soil and Health,* 14).

191. AH, "The Waste Products of Agriculture: Their Utilization as Humus," *Journal of the Royal Society of Arts* 82 (1933): 110; AH, *Soil and Health,* v; AH, "Editorial [Origin and Purpose]," *Soil and Health* 1 (1946): 3; AH, *The War in the Soil* (Emmaus, Pa.: Rodale Press/Organic Gardening, 1946), 22.

cannot stop at England's shores: "A new civilization will have to be created," and it will have to be created "on a fresh basis—on the full utilization of the earth's green carpet." Nothing less than the survival of the species depends upon his success: "Can mankind regulate its affairs so that its chief possession—the fertility of the soil—is preserved? On the answer to this question the future of civilization depends."[192] Albert pursues the greening of the world with abandon. His second wife and Gabrielle's sister, Louise, describes Albert in his later years this way: "The distractions and amusements of retirement wholly foregone, as though hurrying, hurrying himself to fulfill, and to call on the future to fulfill, that great design which in his eyes illuminated the universe—man perfectly adjusted to his environment, understanding natural law and obedient to it." Among the encomia to arrive after his death, one seems particularly apt: "He was the most single-minded creature I think I have ever met."[193]

What kind of science works best for this last ambition—to change the world? Perhaps Albert need only exploit the authority he has built up during three decades of accumulated expertise and skill, patient experimentation and demonstrated practical results? Will humanity listen to more research reports and monographs like those he and Gabrielle issued from Pusa, Quetta, and Indore—and trust him because of these obvious credentials as "scientist"? In his exhortations from London—three major books, innumerable articles in the scientific and popular press—Albert does nothing to deny his scientific qualifications for recommending humus as the world's panacea. He advertises the scientific base on which his message rests: "I began such an investigation of the plant and animal disease section of agricultural science in 1899 and have steadily pursued it since. After forty years' work I feel sufficiently confident of my general conclusions to place them on record, and to ask for them to be considered on their merits." The Indore process offered to the world bears the stamp of scientific legitimacy: "It claims the surety and detailed exactness which attaches to a method conforming to scientific principles."[194] Moreover, Albert enlists experimental findings of other scientists who confirm his theories about humus, soil fertility, and

192. AH, *Soil and Health,* 193, 261; AH, *Agricultural Testament,* 20.

193. Louise E. Howard, "Sir Albert Howard's Career," *Soil and Health,* Memorial Number (1948): 24; Lionel J. Picton, "Sir Albert Howard's Association with the Cheshire Doctors," *Soil and Health* (spring 1948): 50.

194. AH, *Agricultural Testament,* 159; AH, "Soil Fertility," in *England and the Farmer: A Symposium,* ed. H. J. Massingham (London: Batsford, 1941), 39.

disease: "Waksman's insistence on the role of microorganisms in the formation of humus as well as on the paramount importance of the correct composition of the wastes to be converted has done much to lift the subject from a morass of chemical detail and empiricism on to the broad plane of biology to which it rightly belongs." He is at pains to distinguish his scientifically grounded program for organic agriculture and gardening from collateral efforts by seeming fellow travelers who lack science—such as the biodynamics school on the Continent: "I remain unconvinced that the disciples of Rudolf Steiner can offer any real explanation of natural laws or have yet provided any practical examples which demonstrate the value of their theories."[195] Albert's *faith* in humus-as-salvation is born of his three decades of *science* in India—his hybridization of culture does not cease with retirement.

Albert finds that he cannot change the world by resting on his laurels. The "science" behind his claims is challenged vigorously by scientists with their own insignia of credibility—notably, researchers at the Rothamsted Experimental Station for Agricultural Research. The essential question—how do organic and artificial manures compare in their effects on crop production?—had been under systematic investigation at Rothamsted almost since Liebig's N-P-K discovery in the 1840s. The Broadbalk plots consisted of side-by-side fields of crops raised with no manuring, farmyard manures, and artificial manures—compared over decades of growing seasons. For most Rothamsted scientists, the absence of noteworthy differences between the artificially and organically manured plots constituted convincing evidence that the two soil amendments were substitutable. For them, Liebig had it right: the soil necessary for productive crops needed only the replacement of minerals that growing plants had extracted—phosphorous and potassium—plus a shot of nitrogen. Moreover, the Rothamsted school argued that the raw materials (animal and vegetable residues) needed for Albert's Indore composting were not available in England in the quantities necessary for farmers to get by without chemical N-P-K. Preparation of humus would take scarce time and expensive labor away from farmers' primary task of growing food (for humans). Albert finds it challenging to promote humus-as-fact on scientific grounds when the credibility of his claims about the unique benefits of organic manures has been dismissed by equally authoritative scientists at Rothamsted.

195. AH, *Agricultural Testament,* 51, ix.

The "war in the soil" is on![196] Albert seems to enjoy being cast out by the hyperspecialized head-in-the-clouds scientists at Rothamsted and other centers of conventional science. He admits with delight that reactions to his "unorthodox views" are "definitely hostile and even obstructive," and that he has been "given a most unfortunate cold shouldering by the leaders of agricultural education and research": "I was thought by many to be mad."[197] He discovers during a talk to the School of Agriculture at his alma mater that the Indore process and organic manuring is "up against a solid armor-plate of fear." Albert reports that "the students . . . were not only deeply interested in the subject, but vastly amused at finding their teachers on the defensive and vainly endeavoring to bolster up the tottering pillars supporting their temple. . . . It was obvious from this meeting that little or no support for organic farming would be obtained from the agricultural colleges and research institutes of Great Britain."[198]

Perhaps the hostile reactions of many agricultural scientists to Albert's organicism stems from their weariness at being forced to reconsider an issue that has for them been settled for nearly a century. In his definitive insiders' *History of Agricultural Science in Britain,* Sir E. John Russell—once Albert's colleague at Wye, since then director of Rothamsted—makes only one reference to Howard. It comes in a section labeled "Neovitalism," where he embeds Albert's speculations in a "Buddhist phrase"—the wheel of life—and implies that he is a throwback to the pre-Leibigian era, when it was believed that the soil contained some "vital principle" in addition to the minerals now provided so well by chemical fertilizers.[199] Or maybe the hostile reaction was due to the maverick Albert's relentless attack on conventional science, on the supposed inability of research cooped up in tiny disciplinary boxes to see outside. Either way, Albert now finds the authority of science turned against him: "The magic word Science will be freely employed to bludgeon the iconoclast. It will be dinned into our ears that the scientific foundations of chemical

196. This is the title of a book published in America by the founder of this country's organic farming movement, J. I. Rodale. It is, I believe, an assemblage of papers and chapters that Albert had published in England in various books and magazines.

197. AH, *Soil and Health,* 17, 245, 211; AH, "Humus: The Key to Prosperity," *Revista del Instituto de Defensa del Cafe de Costa Rica* (San Jose,1939), 7.

198. AH, *Soil and Health,* 111, 249. Albert's talk refers specifically to his theory that disease is not caused by pests but by low soil fertility that reduces the resistance of plants to invasion.

199. Russell, *History of Agricultural Science,* 467–68.

manuring have been well and truly laid: the Rothamsted plots will be quoted."[200] Did Albert spend too many years in India? Has he gone native? "Opponents of compost accuse its supporters of evading proof and taking refuge in mysticism."[201]

But Albert clearly feels that the tide is turning in his favor, and that the disciples of Liebig are on the run. He uses the scorn heaped upon his organicism as evidence that other scientific authorities at least consider his theories a serious target: "That such a precious commodity as abuse should have been poured out so lavishly suggests that the chemical farmers are by no means certain about their position." He notes that the Rothamsted types have retreated from their original stronghold—artificial fertilizers do everything for the soil and plant that organic manures do—to a more conciliatory position of complementarity: "A combination of humus and artificials will give the best results."[202] Indeed, from the few scientific papers published by Albert's opponents that I have examined, it appears that their authors are comfortable with the physical benefits of humus (improves tilth, porosity, aeration and water retention) and admit even to its chemical benefits (a chemical analysis of humus will reveal N-P-K, in widely varying concentrations). So Albert needs his own stronghold, a bulletproof scientific argument for why organic matter is *uniquely* capable of satisfying the needs of the soil, of plants, and, in turn, of animals and humans—ideally, an argument that will also show artificial manures to be harmful.

Albert's last stand against those other scientists is the "mycorrhizal association," an argument altogether absent from his research and writings prior to 1931. The mycorrhiza is a symbiotic relationship between a fungus and the roots of many plants. The fungus lives in soil, but only if it has been enriched with humus. It insinuates itself into the active root cells, where it is eventually digested by the plant into constituent carbohydrates and proteins. The benefits of the mycorrhiza are then passed upward in the sap to the stems and leaves. Albert suggests that the mycorrhiza are a "living fungus bridge between humus in the soil and the sap of plants," providing crops with a second vital source of nutrients (the other source is the root hairs, which take up minerals [N-P-K] and nutrients from the watery film that lines pores in a friable

200. AH, "Editorial [Origin and Purpose]," 3.

201. A. G. Badenoch, "Organic Cultivation and Disease," *Soil and Health,* Memorial Number (1948): 52

202. AH, *War in the Soil,* 11, 18.

soil).[203] He searches for evidence that the mycorrhizal association is so important that some plants cannot be grown effectively without it.[204] In terms of the various benefits of humus, Albert's discovery of mycorrhiza tips the balance in favor of the biological improvements—especially over the chemical ones. "The chemical aspect is likely to be the least important, . . . the biological side is probably the most important, and . . . adequate attention should also be paid to the physical factors involved."[205] This is why chemical N-P-K alone is not enough: something more is needed—the mycorrhiza. And now the fatal blow to Rothamsted: artificial manures kill mycorrhiza. Albert's work with sugar plants offers proof: "Artificials tend either to eliminate the association altogether or to prevent the digestion of the fungus by the roots of the cane." In short, the soil itself is a living thing—mycorrhizal fungi—and so with humus "we are dealing not with simple dead matter like a sack of sulphate of ammonia which can be analyzed and valued according to its chemical composition, but with a vast organic complex. . . . How profoundly it differs from a chemical manure."[206]

Still, even if Albert can win the battle over artificial versus natural manures on theoretical and empirical grounds, he still faces the "human factor" in trying to convince the rest of humanity that he is right. He learned in India that the cultivator will not readily adopt innovations even when science proves that they serve his best interests: demonstration and propaganda are needed as well. The easy part of improving crop production was hybridizing high-yield varieties and working out the best recipe for making compost: the real challenge was getting farmers on board. Albert beats the same drum as he now considers the possibility of using composted municipal waste from urban areas as a source of organic manure: "The problem of getting the town wastes back into the land is not difficult. The task of demonstrating a working alternative to water-borne sewage and getting it adopted in practice is, however, stupendous."[207] Convincing the world of the truth about soil fertility, about the workabil-

203. AH, *Agricultural Testament,* x. Ironically, Albert began his mycological career by studying fungi that messed up sugar cane, and he now ends up with fungi that bring life itself. This shift mirrors his general turn from the study of disease to the study of health.

204. "The role of the mycorrhizal relationship in tea helps to provide a scientific explanation of these results" (AH, *Agricultural Testament,* 66).

205. AH, *War in the Soil,* 84. "The problems of the farm and garden are biological rather than chemical" (AH, *Soil and Health,* v).

206. AH, *Agricultural Testament,* 68, 27.

207. Ibid., 105.

ity of the Indore process, about the consequentiality of humus for the quality of human life—and getting the world's farmers to compost—requires Albert to reinvent science one more time.

Externalizing Nature

If scientists disagree, perhaps the world would be more easily persuaded by Nature herself. The truth about soil fertility, disease, and salvation will—in the end—not be found in laboratories, experimental plots, professional journals, or even at holistic science centers like Indore. Only by listening to Nature, learning from her lessons, obeying her laws, can agriculture or any other human endeavor succeed over the long haul. During their stay at Pusa and Quetta, Albert and Gabrielle developed a science that displaced their own interests by sequentially adopting the standpoints of others (cultivators, botany, the market). A different displacement is needed here to change the world—in effect, a displacement of the scientific observer. The interests of Nature reign, and she alone will arbitrate between rival representations of her laws and competing projects to intervene in her workings. "The first duty of the agriculturist[s] must always be to understand that [they are] part of Nature and cannot escape from [their] environment. . . . They must on no account flout the underlying principles of natural law." Nature is thus cast as the "supreme farmer," who can "be no longer defied."[208]

Putting Nature outside and beyond science is just the first step: Albert then realigns his science with Nature by arguing that the Indore process "merely copies what goes on on the floor of every wood and forest." It is true not because experiments prove it or because Albert has the proper credentials: it is true because it is *in Nature* (beyond science). "Nature makes humus," or more specifically, earthworms do, along with termites, fungi and other microorganisms: "Living organisms and not human beings are the agents which make compost."[209] The best that scientists or

208. Ibid., 194, 2.

209. AH, *Soil and Health,* 26, 212. Worms are the "farmer's invisible labor force—the organisms which carry on the work of the soil" (AH, *Agricultural Testament,* 27). "The earthworm was proved to be the farmer's natural manure factory" (AH, *War in the Soil,* 16). "We, with our clumsy two feet and our even clumsier mechanical transport, have nothing like the million-footed labor force—the earthworms, termites, insects—which the forest can command to do its work" (AH, "Soil Fertility," 41). "On each acre of land in good heart—at least 25 tons of worm casts are produced and distributed each year. These casts contain seven times more available phosphate, eleven times more potash and five times more nitrogen than is contained in the upper six inches of soil"(AH, *War in the Soil,* 53). I am reminded of Molière, amazed to discover that he had been speaking prose all his life, when Albert writes: "It is by no means impossible

any farmer can possibly do is observe what Nature is already doing perfectly, and provide conditions to enable her to do her thing efficiently. "No more is attempted than Nature has already done." Scientists and farmers "have to go back to Nature and to copy the methods to be seen in her forest and prairie." Perhaps Albert has given Nature a nudge to speed things up: "The process, like most of our human cultivation processes as compared with wild processes, is a good deal more rapid than what Nature undertakes, who can afford at all times to operate with a sort of magnificent leisure." But, really, his discovery was there in the wild all along: "Imagine the carpet of the forest rolled up, all its great processes of fertility concentrated in smallish areas—you have the Indore heap."[210]

Albert's scientific studies of soil fertility get a value-added credibility when they are aligned with natural laws, not discovered or constructed, but just unavoidably *there.* Composting agricultural residues and returning them to farmers' fields respects as it "imitates" Nature's "great law of return": everything living eventually dies, decays, and becomes the raw stuff from which life begins anew.[211] Nature is also "one and indivisible," and abhors not just a vacuum but "monoculture" (planting only one crop, or divorcing animals from farms exclusively dedicated to a single crop).[212] The Indore heap requires a mix of many different kinds of vegetable and animal wastes: "A just balance between animals and crops is Nature's law and it cannot be evaded."[213] Where once Albert pursued the pure strain of a stable hybrid isolated from mere field varieties, now it is the promiscuous mix he seeks for a successful compost pile—because that is how Nature makes a fertile soil by herself. To be sure, Albert is not urging return to an absolute state of Nature (although strains of

that the process has been in unconscious operation on the grasslands of Great Britain for many years" (AH, "Waste Products of Agriculture," 111).

210. AH, *Agricultural Testament,* 144; AH, "Soil Fertility," 50, 41. The wisdom of the Orient is nothing more than a respect for Nature: "From time immemorial" the Chinese peasant "has modelled his agriculture on no less a pattern than Nature" (AH, "Humus: The Key to Prosperity," 8). "Agricultural systems which have passed the test of performance will be found to conform close[ly] with the examples provided by Nature" (AH, "Soil Fertility," 32–33).

211. AH, "Soil Fertility," 40; AH, introduction to *Pay Dirt: Farming and Gardening with Composts,* by J. I. Rodale (New York: Devin-Adair, 1945), v.

212. AH, *War in the Soil,* 64. "Nature has laid down the law of interdependence of all physical existences. Nature never separates her animal and her vegetable world: in their lives, as in their decay and deaths, beasts and plants are absolutely interlocked. She does not even recognize monoculture. Her sowings and harvests are mixed and intermingled to the last degree" (AH, "Soil Fertility, 43).

213. AH, "Humus: The Key to Prosperity," 10.

Rousseau will be heard later in his attacks on modern society), but rather, interventions in Nature's way must conform to her laws. "The character of the intervention undertaken is comprehended and . . . measures are initiated to restore the natural cycle in a proper way. . . . We must give back where we take out; we must restore what we have seized; if we have stopped the Wheel of Life for a moment, we must set it spinning again."[214] Returning composted organic waste to farmers' fields does that in a way that spreading a bag of N-P-K cannot.

Albert's displacement of science by Nature is complete when he describes pests, plants, cows, and other grazing quadrupeds as our real "professors of agriculture." Nature is configured not just as the supreme farmer, but as the "arch-economist" and the "arch-chemist."[215] If you ask the plant or animal, and listen, it will tell you everything you need to know about the fertility of its soil.[216] Disease is not a scourge to be eradicated with all of the pesticides and fungicides that science and chemical companies can muster, but a signal that something is wrong with the soil—that the soil needs fixing.[217] If one does not follow Albert's recipe for compost but piles up straight shit instead: "Nature . . . warns us by strong odors and clouds of noxious flies." The crafty anthropomorphism of Nature knows no limits: "Nature at once registers her disapproval [to poor soil conditions] through her Censors' Department. . . . In the conventional language of today the crop is attacked by disease."[218] If you really want to know whether natural or organic fertilizers are best, ask a cow, or a mole, a bird, a pig: "These experts . . . will invariably eat the humus plots down to the roots and only lightly pick over the chemically treated area."[219] Indeed, Nature—not the scientists—will decide the war

214. AH, *Soil and Health,* 195.

215. AH, *Agricultural Testament,* 161; AH, "Humus: The Key to Prosperity," 16.

216. "The plant or the animal will answer most queries about its needs if the questions are properly posed. The wise farmer, planter or gardener always deals with such responses with sympathy and respect" (AH, *Soil and Health,* 112).

217. "Nature has never found it necessary to design the equivalent of the spraying machine and the poison spray for the control of insect and fungous pests" (AH, *Agricultural Testament,* 4). "The appearance of disease is Nature's protest that all is not well with the soil, because when Mother Earth manages the land the birthright of every plant and every animal is health, and not disease" (AH, *War in the Soil,* 42). Speaking of potato eelworms, Albert argues: "It will obviously not advance matters if we try to slay the bearers of evil tidings. On the contrary we must remove the reason which underlies the despatch of those messengers of Nature by restoring soil fertility as rapidly as possible" (46).

218. AH, "Soil Fertility, 43; AH, *Agricultural Testament,* 156.

219. AH, *War in the Soil,* 17–18. Cows are more scientific than I ever thought they were: "Cows and bullocks are, perhaps, our best professors. . . . When these animals void their dung on a pasture, humus formation at once begins under and around the cowpat. But the proportion

in the soil: "The judge is Mother Earth, who is incorruptible and unpersuadable"; "Mother Nature, rather than the advocates of these various views, will in due course deliver her verdict." Downside risks are severe: "If we fail in this duty to the soil from which we all spring, Mother Earth will hit back and after a series of warnings will finally wipe us out of existence."[220]

Scapegoating; or, Why Scientists Scorn Albert and Abandon Nature

If only humans had listened to Mother Nature, then there would be no huge problem to keep Albert busy during his retirement. But others have chosen to abandon Nature, and the society Albert looks out on during the 1930s and 1940s appears unbalanced, off-track. If Albert is to change the world by getting everybody to compost, he must first identify the evil forces that have intervened in Nature's patterns so disastrously, creating the mess his organicism will clean up. Put another way, Albert must provide an account for why agriculture no longer emulates the forest floor, and why conventional science has not recognized his claims about soil fertility—grounded as they are in Nature—as the credible truth. Advancing the cause of science (here, Albert's holistic kind) also involves the identification and castigation of forces said to impede its progress.

The list of villains is long and varied, but might be summed up as "modernity." Few of the sweeping social changes since, say, the seventeenth century, are exempt from responsibility for moving agriculture away from the laws of Nature. The abandonment of Nature's lessons probably started in earnest with the Industrial Revolution, which required that the "processes of growth" be "speeded up to produce the food and raw materials needed by the population and the factory." Urbanization has also contributed to the decline: towns and cities have become "parasites" that violate the law of return by extracting a good deal from the soil (in the form of food for their inhabitants) while returning nothing of value.[221] Machines, which have replaced animals on the farms,

---

of animal to vegetable waste is here in excess. . . . The oxygen supply is totally inadequate. The consequence is that high-quality humus is impossible. . . . The grazing animal confirms this loss of quality by leaving the grass round the cowpat uneaten" (AH, "Grassland Management," *Soil and Health* 1 [1946], 145).

220. AH, "Editorial [Origin and Purpose]," 4; AH, *Agricultural Testament,* 59; AH, *War in the Soil,* 60.

221. AH, *Agricultural Testament,* ix, 104. The kind of "modern" agriculture that Albert had earlier described derisively as "extensive" is the result of "the hunger of the growing urban areas, the population of which is unproductive from the point of view of soil fertility" (17).

are tarred with the same brush: "Machines do not void urine and dung and so contribute nothing to the maintenance of soil fertility." Maybe machines and cities are corruptions of Nature's way only because they are embedded in capitalism: in theorizing the decline and fall of the Roman Empire, Albert detects "a capitalist system" whose "apparent interests were fundamentally opposed to a sound agriculture."[222] Cutting very near to the bone, Albert lists imperialism as another villain: "The accumulated fertility of those distant regions of the earth which have produced the materials for the oil-cake [for example] is being robbed in order to bolster up a worn-out European soil."[223] Possibly not all industry or capitalists are guilty, but certainly blame must be assigned to those who benefit immediately from the replacement of organic manures by artificials: "The war in the soil is the result of a conflict between the birthright of humanity—fresh food from fertile soil—and the profits of a section of Big Business in the shape of the manufacturers of artificial fertilizers and their satellite companies who produce poison sprays."[224] Government shows itself to be complicit in this process: during World War II, the Ministry of Agriculture joins forces with the manufacturers of chemical fertilizers to deal with food shortages, hastening the dependence of farmers on chemical fertilizers, pesticides, and herbicides.[225]

Albert saves his nastiest words for the very idea of profit, especially as it has insidiously worked its way into the minds of farmers and the social organization of agriculture. Profit is the enemy of Nature and truth: "It has been said that even the principle of gravitation would have had a hard row to hoe, had it stood in the way of the pursuit of profit and the operations of big business." Sadly, "the idea of profit, that lusty child of the industrial era, has inevitably begun to dominate the agricultural outlook." On the farm, an exclusive focus on the bottom line is shortsighted: "The soil's capital—soil fertility—[is] left out of the reckoning." Worse, it turns the farmer into an "expert bandit" who exploits the land for maximum yield while returning nothing natural—a "profiteer at the expense of posterity."[226]

222. AH, *Agricultural Testament,* 18, 8.

223. AH, *Soil and Health,* 64.

224. AH, *War in the Soil,* 3. Albert suggests that factories that developed a means to fix atmospheric nitrogen for making explosives during World War I needed a profitable market after the armistice, and so turned to nitrogenous fertilizers (AH, *Agricultural Testament,* 18).

225. "The financial resources of a great nation were available to help the farmers to purchase these chemical stimulants and thus indirectly to subsidize the artificial manure industry itself; the staffs of these vested interests were at the disposal of the Ministry of Agriculture" (AH, *Soil and Health,* 76).

226. AH, *War in the Soil,* 37; AH, "Soil Fertility, 34, 35; AH, *Agricultural Testament,* 199.

None of this might have happened had "authority," by which Albert means science, not "abandoned the task of illuminating the laws of Nature."[227] Conventional science has allowed agriculture and the rest of society to escape from the laws of Nature, with disastrous consequences. Albert can almost excuse industrialists for doing what they must to remain profitable, and he sympathizes with farmers for being squeezed by the demands of hungry cities to exploit the soil by resorting to chemical quick fixes. But he cannot forgive scientists of the Rothamsted kind, who—because of vested interests of their own, and because of the social organization of their research—have failed to warn society of the imminent perils of ignoring soil fertility, composting, and humus in its farms and gardens. Albert will not let his readers forget that the financial wherewithal for the Rothamsted station originally came from a patent on an artificial superphosphate—in other words, its scientists might still be in cahoots with the chemical fertilizer interests.[228] He comes close to suggesting that a conspiracy has slowed the march of his compost crusade: "The disciples of Rothamsted, which include the Ministry of Agriculture, the experiment stations, and the agricultural colleges, have combined forces with the vested interests concerned with the production and sale of chemical fertilizer and protective poisons for the crop to deflect the onward march of organic farming and gardening." He also thinks it possible that conventional scientists resist his unorthodox science because their professional livelihood depends on its being false: "In reality all [these studies] show is how employment can be found for a number of specialists for quite a long time, and indeed what a lot of scientific work can be done by competent workers with purely negative results as far as yield and the quality of the crop are concerned."[229]

The objectivity of conventional science may be compromised by economic and professional interests, but it is thoroughly polluted by a mode of social organization that defines it as, well, "conventional." Albert continues his invective against its specialization and fragmentation (always contrasted to the holistic science established at Indore): "The natural universe, which is one, has been halved, quartered, fractionized, and woe betide the investigator who looks at any segment other than his own!"[230] Albert now adds two new attacks on conventional science—in part to

227. AH, *Soil and Health,* 91.

228. "The manufacturers of artificial manures including one of their body—the founder of Rothamsted—were naturally desirous of exploiting a profitable business" (AH, *War in the Soil,* 10).

229. AH, *Soil and Health,* 260; AH, *Agricultural Testament,* 194.

230. AH, *Soil and Health,* 77.

blame it for the mess the earth is in, and in part to account for the scorn the Rothamsted gang has heaped on his Indore heaps.

First, conventional scientists are said to be obsessed with numbers, with statistics—quantifying everything, ignoring quality (or anything that resists easy enumeration). In their work, "the yield was everything; quality was sacrificed for quantity."[231] (When Albert justifies organic farming as a panacea for our human health problems, he will fasten onto this distinction between quality and quantity: it is not the amount of bread and fruit we eat, but its taste and nutritional value that makes people resistant to disease.) So many important things about agriculture lie beyond the precision of statistical rendering, a method that by definition should be suspect since "by a judicious selection . . . it is possible to prove or disprove anything or everything."[232] Moreover, statistical analyses are redundant in the face of Nature herself: "Elaborate statistics will be superfluous as the improved health of these communities will speak for itself and will need no support from numbers, tables, curves and higher mathematics. Mother Earth in the appearance of her children will provide all that is necessary."[233] In fashioning a response to his Rothamsted critics, who demand statistical evidence showing the superiority of organic manures over artificial ones, Albert draws on Napoleon for help: "Never do what your enemy wants you to do, if for no other reason than he wishes you to do it." He rejects the presumption that the war in the soil must be fought with statistical weapons, and instead proposes that Rothamsted's insistence for quantitative data is a tactic for stalling the inevitable collapse of the N-P-K "mentality" into an esoteric quagmire of figures and graphs.[234] Did Albert ever look back on the careful statistical analyses he and Gabrielle had published from Pusa?

231. Ibid., 63. Related to this is Albert's argument that conventional scientists have looked only at growth (easily measurable and quantifiable) and have ignored decay—although they are equal components of the "wheel of life." In developing and promoting artificial fertilizers to maximize yield without returning anything natural to the soil, they have thus speeded up one half of the wheel with no regard for the other half: "The steering is bound to be erratic, the sense of direction is certain to be lost" (AH, "The Manufacture of Humus by the Indore Process," *Journal of the Royal Society of Arts* 84 (1935): 51.

232. AH, *Agricultural Testament,* 185. "Many of the things that matter on the land, such as soil fertility, tilth, soil management, the quality of produce, the bloom and health of animals, the general management of livestock, the working relations between master and man, the esprit de corps of the farm as a whole, cannot be weighed or measured" (196).

233. Ibid., 180.

234. "This amounts to a claim that the war in the soil must be fought on conventional lines as laid down by one of the contestants. It is forgotten that wars are not won by outworn methods but by the discovery of new and original devices. . . . What the artificial manure interests desire at all costs is delay. The accumulation of statistics would obviously consume much time. . . . The

Second, conventional scientists have convinced themselves that they should dominate Nature rather than listen to and obey her laws. Albert in effect chides these "laboratory hermits" for not being farmers—for never having to put their ideas into actual practice on a real farm. Instead, they are consumed by "the temptation to grow a few isolated plants in pots filled with sand—watered by a solution containing the requisite amount of NPK." How could such an artificial arrangement possibly hope to emulate Nature: no worms could get inside those pots![235] Rothamsted experimental procedures are so removed from natural conditions that Albert calls them "superb arrogance," and their new inquiries into hydroponics (growing plants in fertilized water, without soil) becomes "science gone mad."[236] He connects the Broadbalk experimental plots to what had become, by the 1940s, the ultimate evil: The plot "only represents itself—a small pocket handkerchief of land in charge of a jailor intent on keeping it under strict lock and key for a century; in other words, it has fallen into the clutches of a Gestapo agent."[237] The only remedy is for conventional scientists to leave their labs and stats behind and become unconventional. In short, they must return to the land, to Nature, for "to investigate plant diseases without any first-hand experience of growing the plant is to play Hamlet without the Prince of Denmark."[238] In Albert's new science, it seems numbers have been traded for metaphor.

The costs of conventional scientists' failure to warn us about the dan-

---

whole effort to supply convincing figures might very easily peter out in a technical wrangle—one of those infructuous discussions between experts described as a controversy, in which nothing is ever settled" (AH, *War in the Soil,* 13).

235. AH, *Soil and Health,* 72. The quantophrenia of conventional science is closely coupled with the fallacy (in Albert's view) that Nature can be understood only when it is brought under experimental control in the lab or test plot: "In an evil moment were invented the replicated and randomized plots, by means of which the statisticians can be furnished all the data needed for their esoteric and fastidious ministrations. . . . The new authoritarian doctrine demands that [the investigator] shut himself up in a study with a treatise on mathematics and correct his first results statistically" (78).

236. "There was a kind of superb arrogance in the idea that we had only to put the ashes of a few plants in a test tube, analyze them, and scatter back into the soil equivalent quantities of dead minerals." But science still has much to learn, for "the apparent submission of Nature has turned out to be only a great refusal to have so childish manipulations imposed on her" (AH, *Soil and Health,* 71, 194).

237. AH, *War in the Soil,* 8.

238. AH, *Soil and Health,* 104. "They should at once leave their comfortable laboratories and libraries, take up a piece of land, learn all about compost and its manufacture, and then watch how Nature, the old nurse, looks after the health of crops when we faithfully comply with her great law of return" (AH, *War in the Soil,* 43).

gers of detaching agriculture from Nature are enumerated with gravity and urgency. High-yield varieties of crops are found to "run out" after a sustained use of artificial chemicals, and the use of chemical N-P-K is said to kill the good humus-making worms while preventing the potato plant from resisting the bad eelworm.[239] The American "dust bowl" is symptomatic of how "Mother Earth has been deprived of her manurial rights; the land is going on strike."[240] "Bastard nitrogen" leads to "sterility in plants and animals," and (in a remarkable chain of translations, even for Albert) "artificial manures lead inevitably to artificial nutrition, artificial food, artificial animals and finally to artificial men and women."[241] How to get back from artificial to natural? "The good earth lies bruised and broken," while conventional science has "sunk to the inferior and petty work of photographing the corpse."[242]

Reducing, Displacing, Enlisting

Albert knows well what needs to be done to agriculture: imitate Nature, employ the Indore process, apply humus to get fertile soil. But the forces arrayed against organic agricultural practices are more powerful than farmers, and he cannot rely on them alone to change the rest of the world. Agriculture has become trapped in a nest of evils—industrialization, urbanization, capitalism, and a science blind to it all. Even if farmers suddenly decided to give up their bags of artificial fertilizers for the compost heap, they would be resisted by hungry urbanites, ruthless promoters of chemical interests, and Rothamsted scientists who would accuse them of mysticism. To restore farming to organic principles, Albert must convince the rest of society that such a return to Nature matters *for them.* The advancement of science also depends, at times, on framing the problems of everybody as uniquely solvable by whatever fact or technique the scientist might happen to have on hand. Albert effects this reduction in textbook fashion: If you want a healthy and sane world, don't come through my laboratory but through my "humus—the key to prosperity."[243] Absolutely nothing is irrelevant to soil fertility: "Everything on this planet is dependent on the way mankind makes use of this green carpet."

239. AH, *Agricultural Testament,* x; AH, *War in the Soil,* 15; AH, "Manufacture of Humus," 50.

240. AH, *Agricultural Testament,* 19–20.

241. AH, "The Nitrogen Problem," *Soil and Health* 1 (1946): 196; AH, *Agricultural Testament,* 37.

242. AH, "Humus: The Key to Prosperity," 5; AH, *Soil and Health,* 81.

243. AH, "Humus: The Key to Prosperity," 8; also Latour, "Give Me a Laboratory," and *Pasteurization of France.*

No one can fail to be saved by the Indore process of composting: "In consideration of soil fertility many things besides agriculture proper are involved—finance, industry, public health, the efficiency of the population and the future of civilization."[244]

At a time when fascism had (for the moment) been quelled, at huge cost to human life and property, it must have given some Britishers pause to hear Albert say that the "real Arsenal of Democracy is a fertile soil." What with his earlier characterization of the Broadbalk plots as Hitlerian enslavements of Nature, Albert seems to force his listeners into a choice between composting and Nazism. Such scare tactics will not help Albert much to change the world unless he can also tie humus to the more immediate and practical ambitions of the populace. For that, he returns to his oldest scientific fascination—disease—only now he talks about human health rather than sugarcane pests. From the mid-1930s on, Albert works closely with what would become the "whole foods movements" in England, based on the premise that more nutritious and fresher food makes for healthier bodies—and greater resistance to all sorts of maladies. Albert's theory of disease in plants—resistance depends on the quality of the "food" farmers put in their soil (organic matter)—allows him to see connections between soil fertility and the health of humans: soil with humus-enriched good heart yields high-quality and nutritious food for healthier human bodies. The dirty old compost heap becomes the mother of healthy humans: "The birthright of all living things is health. This law is true for soil, plant, animal and man; the health of these four is one connected chain [and the] general failure in the last three links is to be attributed to failure in the first link, the soil: the undernourishment of the soil is at the root of all."[245]

Albert spends much time elaborating upon these equivalences. We hear, for example, that contagious abortions and foot-and-mouth disease are reduced when cows feed in pastures enriched with composted organic waste, and that the incidence of colds and measles in children declines when foods are "raised from fertile soil and eaten fresh."[246] It is not too much for him to say that "on the efficiency of this mycorrhizal association the health and well-being of mankind must depend."[247] Future studies of human health must, he says, start from a new baseline: soil fertility. (By this, I guess, Albert means that epidemiological stan-

244. AH, *Soil and Health,* 21; AH, *Agricultural Testament,* 219.

245. AH, *Soil and Health,* 14, 12.

246. AH, "Waste Products of Agriculture," 112; AH, *Soil and Health,* 166; AH, *Agricultural Testament,* 83.

247. AH, *Agricultural Testament,* 25.

dards and goals should be calibrated initially from health patterns expected in a society fed only from fields treated with compost and not with chemical N-P-K.) Moreover, not only physical but psychological well-being depends upon soil fertility.[248]

Everybody presumably has a stake in better health, but Albert also pitches his humus as a solution for diverse problems peculiar to one or another segment of society. British farmers (if organic) could stave off competition from imported prepared food by convincing consumers that their wares are fresher and more nutritious because they were grown on fertile soil. Those planning new towns could save money and help the earth by installing what we now would call bioremediation facilities for the disposal of town waste and municipal sludge.[249] Following up on his Indore foray into labor-management relations, Albert believes that these may be improved if workers were to eat only foods from fertile soil: noting the "growing industrial unrest and frequent strikes, . . . some of the reformers are convinced that at the root of these manifestations lies the malnutrition resulting from murdered food and from poisoned and worn-out soil." Even local governments could save the money they now waste in removing high weeds from roadsides if instead nearby residents would see the weeds as green gold—just waiting for the compost pile, and later, the backyard garden. Indeed, even the villainous Imperial Chemical Industries could find an earth-friendly profit center in the conversion of town waste to dried and processed organic fertilizer.[250] Is there anyone left *not* ready to benefit from a compost-enriched fertile soil?

Several folks do stand to lose business if Albert's reduction is adopted as a standard operating principle. If soil fertility becomes the key to human health, for example, whither conventional physicians? Albert would move them out of the way: "The general outlook of our medical men is . . . pathological. . . . There is now little or no training for positive health."[251] Like the errant entomologist who believes that the remedy for

248. AH, "Humus: The Key to Prosperity," 7. "The psychological effect of [growing food organically] on laborers has been remarkable. . . . Further, their physical health and their efficiency as laborers rapidly improve" (AH, *Agricultural Testament*, 77). "Upon a timely recognition of the urgency of arriving at the truth in this matter depends the health and perhaps the sanity of the human race" (AH, "Humus: The Key to Prosperity," 7).

249. AH, "Waste Products of Agriculture," 112–13; AH, "Manufacture of Humus," 48.

250. AH, *War in the Soil*, 5; AH, *Soil and Health*, 214; AH, "Manufacture of Humus," 48.

251. AH, *Soil and Health*, 174. "Medical investigations should be deflected from the sterile desert of disease to the study of health—to mankind in relation to his environment. Agricultural research . . . should start afresh from a new baseline—soil fertility—and so provide the raw

plant disease is fumigation of the pest with poison, doctors should also turn to the conditions of health: fertile soil for the plant, which provides more nutritious food for people.[252] "The foundation of the public health system of tomorrow is a fertile soil." Look who else would be displaced by the Indore process: "I had learnt how to grow healthy crops, practically free from disease, without the slightest help from mycologists, entomologists, bacteriologists, agricultural chemists, statisticians, clearing-houses of information, artificial manures, spraying machines, insecticides, fungicides, germicides, and all the other expensive paraphernalia of the modern Experimental Station."[253] Albert tells a story from Pusa about how he convinced a vet to refrain from inoculating his farm animals, asking him instead how he might improve the quality of their food and the pastures on which they grazed.[254]

Finally, Albert seeks one last displacement—special for me, because his conceptual language of reduction, enrollment, and displacement evokes Bruno Latour's study of how Pasteur built his science by seizing anthrax (and other diseases) from veterinarians and public health officials. Bacteriology too must move over, says Albert, "in the vain hope that laboratory science will find a remedy for what common-sense should prevent. . . . The microscope and methods of Pasteur and of his successors can never hope to achieve a permanent and effective cure of disease."[255] To keep the credibility of his claims alive and—oh yes—to save the world, Albert must argue that only his scientific space—promiscuously hybridized, holistic—can find truth, even if this requires him to deny cultural authority to the paragons of science who preceded him.

---

material for the nutritional studies of the future—fresh produce from fertile soil" (AH, *Agricultural Testament,* 180).

252. "Large sums of money are being spent every year on entomological schemes, from which the men who contribute the cotton cess are not likely to deliver any great benefit" (AH, "Manufacture of Humus," 40).

253. AH, *War in the Soil,* 42; AH, *Agricultural Testament,* 161. Social service agencies would also go by the board: "When the finance of crop production is considered together with that of the various social services which are needed to repair the consequences of an unsound agriculture, and when it is borne in mind that our greatest possession is a healthy, virile population, the cheapness of artificial manures disappears altogether" (AH, *Agricultural Testament,* 38).

254. "The next step was to discourage the official veterinary surgeons who often visited Pusa from inoculating these animals with various vaccines and sera to ward off the common diseases. I achieved this by firmly refusing to have anything to do with such measures, at the same time asking these specialists to inspect my animals and to suggest measures to improve their feeding, management and housing. . . . This carried the day" (AH, *Soil and Health,* 159).

255. AH, "The Manufacture of Humus," 51; Latour, *Pasteurization of France.*

Testifying

But is Albert *right?* Is the mycorrhizal association essential for effective plant growth? Do organic manures provide the soil and plant with vital ingredients not offered by artificials? Is the Indore process the most efficient recipe for making humus from organic residues? Is human civilization really dependent upon soil fertility? Should we trust him—and if so, why? In attacking the statistics and controlled laboratory experiments of conventional science, Albert cedes a powerful discourse for truth making. What else could he use? He takes a page from George Combe's book (chapter 3), and turns to the people. From his days at Wye listening to the Kentish hop-farmers to his years in India learning from (some) cultivators, Albert has always believed that science has no monopoly on the production of useful and truthful knowledge about Nature. Now he is ready to turn that populist epistemology into a powerful tool for enhancing his own credibility: he risks his facts by turning over their adjudication to everybody else. See for yourself, it works! Farmers, public health officials, veterinarians, scientists, experiment station workers, mothers of sick children need not wait for Nature's final verdict: supportive evidence rolls in, not from labs and small plots, but from farms and schools and sanitation engineering offices.

Albert frequently boasts about the widespread adoption of the Indore process for making compost: farmers on four continents have reported success with it, along with municipal waste officials in South Africa, and representatives of tea plantations in Asia and coffee plantations in Costa Rica. Their continued practice of organic methods is not based on his recommendation as a scientist, but on their own evaluations of its utility. This is how it should be, says Albert: the question "depends at least as much on the plain efforts of the plain man in his own farm, garden or allotment as on all the expensive paraphernalia, apparatus and elaboration of the modern scientist." Moreover, "the day for handing over the right of decision to rather effete and very conventional institutions and very narrow-minded authorities has passed."[256] Albert even suggests that an unwillingness of conventional scientists to respect the experiences of the untutored might signal a lack of conviction in their theories or recom-

256. AH, *Soil and Health,* 13; AH, "Editorial [Origin and Purpose]," 4. "No farmer of a Western country should permit himself to adopt the Process as a piece of blind routine: it is his privilege . . . to be able to understand some of the scientific truths which constitute the laws to which he is working" (AH, "Soil Fertility," 42).

mended practices.[257] Albert himself has nothing at all to hide: "No patents, no special materials to be sent for, and there is nothing secret about [the Indore process]." It happens on the forest floor, and in Japan and China for time immemorial: "Any idea of patent would be a fraud on the cultivator."[258]

Part of the explanation for the widespread practical adoption of the Indore process is the ease with which it can be instituted. Indeed, Albert reports, "my wife [Louise] turned a heap of about four tons in the course of two days without undue exertion." The process is not costly, especially when balanced against expenses that no longer need to be reckoned with (artificial manures, pesticides, for example). Nor is the recipe particularly complicated: "The original Indore technique has been simplified so that the ordinary illiterate African can keep it in his head."[259] Composting and organic agriculture is snowballing *because of* the many testimonials to its success from people who have tried it. I append the barest few:[260]

> Dr. G. B. Chapman, Mt. Albert Grammar School, Auckland, New Zealand: "The first thing to be noted during the twelve months following the change-over to garden produce from our humus-treated soil was the declining catarrhal condition among the boys. There was also a very marked decline in colds and influenza."

> Rev. G. M. Kerr, Leper Home and Hospital, Dichpali, India: "From the beginning, the effectiveness of the compost produced was evident. Growth was twice to four times better in plots to which properly prepared humus was applied than in those equally well worked but left uncomposted. . . . Our 500 patients do all the labor required, one large gang working exclusively at composting. . . . Universal use of this method of humus preparation might be nothing less than revolutionary in agricultural practice in India. . . . This is

257. Albert asks "whether agricultural research in adopting the esoteric attitude, in putting itself above the public and above the farmer whom it professes to serve, in taking refuge in the abstruse heaven of the high mathematics, has not subconsciously been trying to cover up what must be regarded as a period of ineptitude and of the most colossal failure" (AH, *Soil and Health,* 81).

258. Ibid., 212; AH, "Manufacture of Humus," 28.

259. AH, *Soil and Health,* 220; AH, "Manufacture of Humus," 44.

260. The sources for these sample testimonials are cited here in sequence, all but the final two from Albert's own works: *Soil and Health,* 184; *Agricultural Testament,* 80; *Soil and Health,* 168; "Waste Products of Agriculture," 106; *Soil and Health,* 182; "Manufacture of Humus," 54; *Soil and Health,* 132; *War in the Soil,* 61; F. Newman Turner, "New Life to Land and Live Stock," *Soil and Health,* Memorial Number (1948): 34; J. W. Scharff, "Improvement in Health of a Labor Corps," *Soil and Health,* Memorial Number (1948): 77.

> not a scientific conclusion according to your usual methods of reckoning, but it is the practical issue as it appeals to us."
>
> Mr. James Insch, Scotland: "Rats broke into his granary and devoured the produce of the mucked field and left the other [artificially fertilized produce] severely alone."
>
> Lieut.-Colonel Tyrell, Inspector-General of Hospitals and Director of Public Health, Holkar State, India: "The comparison between the new system and the old method with its enormous dumps of refuse, requiring many months to disintegrate and forming a breeding place for flies and rats, is striking."
>
> The Rev. W. S. Airy, St. Martin's School: "Our soil, which has always been dressed annually with some ten or twelve tons of farmyard manure, the contents of poultry houses on the grounds, and two compost heaps, enjoys immunity from insect pests and disease. . . . We had many lads who came to us as weaklings and left hearty and robust; they never looked back in point of health."
>
> Mr. F. A. Secrett: "I notice that in Covent Garden and in the large provincial markets those stands are favored where the produce has come from farms which have received organic animal manure. Although higher prices are charged for this produce, it is sold out first."
>
> Mr. J. G. D. Hamilton, Jordans, Buckinghamshire: "In Wiltshire, I was told by two old thatchers in different parts of the country that the straw they had to work with now was not nearly so good as that which they had had in years gone by. Both gave as the reason the modern use of artificials in place of farmyard manure."
>
> Mr. N. W. Ayson, South Africa: "At the beginning my garden was just pure yellow stiff clay . . . , now it is an almost black heavy loam and does not require any more manure for some years. Thinking to get even better crops I have added to beans, mealies, beet and so on artificial fertilizers containing phosphates, but this had no effect whatever as against those parts not so treated."
>
> F. Newman Turner, a British farmer: "Kept under this regime, the sterile animals I was advised to have slaughtered have come back to breeding again. . . . Had I taken the veterinary surgeon's advice I should have been ruined, but instead Nature and Sir Albert have made me into that rare specimen, a happy farmer."
>
> J. W. Scharff, M. D., Chief Health Officer, Singapore: "Compost making had been started on a large scale. [The children's] movements had become livelier; their eyes were bright with happiness; their whole expression and outlook showed the new-found thrills and pleasure they were getting out of life."

Albert admits that these testimonials to the success of Indore-made humus are a different sort of evidence than the evidence of conventional scientists: "Examples such as these quoted do not, of course, conform with the standards deemed essential by the laboratory worker and by the statistician. Nevertheless, they are of the greatest value as indicators of results which are being obtained all over the world when organic gardening is practiced on a large scale." Such experiences are a confirmation of the truth and practical utility of Albert's ideas no less scientific, no less authoritative, than his own careful experiments. Moreover, his theories and recommendations are made even more credible precisely because others—not he—have performed the tests. "One small example outweighs a ton of theory."[261]

ONE SCIENTIST, MANY SCIENCES

With this last new science, does Albert succeed in changing the world? Have we become a world of composters? (I wish . . .) That Albert's legacy was an uneven one is apparent even in bits and pieces of evidence from his final years. In his contribution to the memorial edition of the magazine *Soil and Health* (published at Albert's death), Moubray writes rosily from Rhodesia: "We are on the eve of the compost era." J. I. Rodale gathers 60,000 subscribers to his American magazine *Organic Farming,* for which Albert served as an editorial advisor.[262] Lady Eve Balfour, a contributor to *Soil and Health* during its two-year run, forms the Soil Association in order to promote the cause of organic gardening and agriculture in Britain. And later, beginning in the 1960s, green parties and environmental movements spread Albert's message, and the U.S. Congress (never reluctant to take on more boundary-work) sponsors hearings on organic agriculture. Even Prince Charles builds a completely organic farm at Highgrove, on Nature's principles, with a refined mix of animals and plants, and compost heaps galore.

But there is plenty of evidence pointing the opposite direction: "The consumption of superphosphate rose from an average of 42.8 thousand tons per month in 1935 to 109.3 thousand tons in March 1945. The nitrogen content of artificials such as sulphate of ammonia, nitro-chalk, and so forth was 9.39 thousand tons per month in 1940. By March 1945, a monthly average of over 17 thousand tons had been reached." Albert, his own death not far off, reports grimly that "the invasion of the ocean by artificial fertilizers has begun," with early efforts at fish farming. Even

261. AH, *Soil and Health,* 168, 13.

262. AH, *Agricultural Testament,* 234; AH, "The Process of Organic Agriculture in the U.S A.," *Soil and Health* 2 (1947): 219.

worse, "the artificial manure industry has set foot upon Indian soil and a whole new town is being created to serve the interests of those who would bring the cultivators of India onto the conventional chemical highroad to the destruction of soil fertility." What Albert opposes so vigorously is a "proposed £7,000,000 project for establishing a fertilizer plant at Sindhri in Bihar."[263] This factory did not, like the Union-Carbide plant at Bhopal, explode and spread its poison over the surrounding town, but it might easily have done so. The nearly three thousand deaths at Bhopal in 1984 would not have happened if India had followed Albert's natural path: pests should be heeded as a signal of infertile soil. Less than a decade after Albert's death, India's "green revolution" coupled together again the forces of modernity condemned by Albert as promoting an agriculture dependent upon artificial manures: the chemical industry, a state bent on industrialization and urbanization, leading agricultural scientists supported by the Rockefeller Foundation. Y. D. Wad, his Indian collaborator, wrote of Albert on his death: "May his soul rest in eternal peace, confident that he lighted a torch for a world drifting and groping in darkness to discover the right path."[264] None of Albert's several sciences have made it possible for us to end the drifting and groping.

The war in the soil goes on without Albert and Gabrielle Howard, but much can be learned from their lives about how difficult it is to achieve sane policies and practices for a fragile earth. I think of them now and then as I turn over that inexact pile of leaves and grass and kitchen waste in my backyard: they got it right (and that is one reason why I chose to retrieve their lives and works from lost history). Much can also be learned from Albert and Gabrielle about the sociology of scientific authority. Surely some of the credibility of scientists results from their skill in (re)-framing the edges of their cultural space flexibly, as they respond to and also realign human interests, desperate problems, and possible constituencies.

The vastly different sciences that the Howards practiced throughout their separate and joint careers measure the diverse means that scientists have used to win our trust. Albert in Barbados grounded credibility in a conventional scientific method—the language of experiment, the retreat from worldly prejudices. At Pusa and Quetta, Albert and Gabrielle offered science as uniquely capable of looking at problems from all of the myopic standpoints of others—relativizing reality without taking a stand-

263. AH, "How the Artificial Manure Industry Has Grown," *Soil and Health* 1 (1946): 70; AH, "Facilis Descensus Averno," *Soil and Health* 2 (1947): 35.

264. Y. D. Wad, "The Work at Indore," *Soil and Health,* Memorial Number (1948): 31.

point of its own. Indore was a scientific experiment on science—in effect, an argument for continuing trust in scientific practices and institutions that need not remain fixed in patterns that have defined them since the seventeenth century. Albert and Gabrielle's new science melded the standpoints of others into its own holistic gaze: credibility came from culturally hybridizing indigenous wisdom and experiment, India and Europe, humans and dirt. With retirement back in England, Albert redrew science yet again as he buttressed the credibility of his beliefs about the saving grace of humus—not by taking them outside or beyond the authoritative space for science (though that is where they were put by conventional scientists)—but by bringing the whole world into science, by giving Everyman the chance to judge his truths, by demonstrating that myriad human problems had one solution, by allowing Nature the last word.

HOME TO ROOST: SCIENCE WARS AS BOUNDARY-WORK

As I was hunting down the last few references for this book, I received a letter from a prominent philosopher of science asking if I would serve on the advisory editorial board for a new book series, "Science and Technology Studies." Such are the routine burdens of the scholar, but this particular invitation seemed to me a sign of the times: "We hope that the series will help raise the level of the current debate on science and technology, by sticking to the standards of rationality and the concern for empirical tests that characterize modernity." Translation: "We believe that you stand with us in the science wars, on the side of reason, empiricism, and truth of science, and against postmodernism, relativism, and radical social constructivism." How should I respond? There was evidently more at stake here than the time and energy I might have to read an occasional manuscript. Where do *I* stand with respect to reason and empiricism, relativism and constructivism? Where was *I* on this philosopher's cultural map of science? The cartographic quandary was now no longer George Combe's or Albert Howard's or John Tyndall's, but mine: Boundary-work comes home to roost.

This epilogue ends with my answer to this loaded request. My decision may surprise, and so I offer first a brief contextualization—one that will also serve to review the central message of this book. Why bother to bring up "current events"—the science wars rage as I write—at the tail end of a painstaking and dispassionate sociological study of old credibility contests in which rival parties manipulate the boundaries of science in order to legitimate their beliefs about reality and secure for their knowledge making a provisional epistemic authority that carries with it influence, prestige, and material resources?[1] I do so because these science wars are credibility contests in which rival parties manipulate the boundaries of science in order to legitimate their beliefs about reality and secure for their knowledge making a provisional epistemic authority that carries with it influence, prestige, and material resources. I seek to make the science wars historically mundane by showing that they are of a piece with the five episodes of cultural cartography I have just explored. By removing the science wars from the heat of battle and by cooling them off with a douse in sociological theory and precedent, I hope to show the commonplace character of our most recent, but surely not our last, episode of boundary-work. What good is sociology if it cannot pick out otherwise overlooked but consequential patterns from the ground of history and use them to make sense of where we are now?

It is difficult, especially from the inside, for anyone to put a finger on the start of the science wars or to locate their many and diverse incidences or to say with confidence just what they are all about. *In medias res* is hardly a propitious time for summing up. Blood has been shed, but nobody has won anything yet, and it is not even clear what would count as victory for either side—let alone how to pursue it. What we have so far is a scattered set of maybe disconnected happenings afterward lumped together as constituting today's "science wars."

- Biologist Paul Gross and mathematician Norman Levitt defend science in their 1994 book *Higher Superstition,* an admittedly polemical assault on anti-science attitudes proffered by a so-called academic left of social constructivists, cultural theorists, feminists, multiculturalists, and some extreme environmentalists.[2]
- Also in 1994, a permanent exhibition "Science in American Life" opens at the National Museum of American History (part of the Smithsonian chain) in Washington, D.C. Less than one year later, the

1. Barbara Herrnstein Smith *(Belief and Resistance)* offers a measured assessment of intellectual controversy in general, with its tendencies toward polarization, reflexive loops, and asymmetries (I'm right because that is the way reality is; you are wrong because you are demented or demonic).

2. Paul Gross and Norman Levitt, *Higher Superstition: The Academic Left and its Quarrels with Science* (Baltimore: Johns Hopkins University Press, 1994). Starting the war here may be Amerocentrist: publication of Lewis Wolpert, *The Unnatural Nature of Science* (London: Faber and Faber, 1992), and especially the exchange between Wolpert and Harry Collins in the *Times Higher Education Supplement* (September 16, 1994), anticipated many of the arguments that would become common after *Higher Superstition.* Because they are "data," these traces of the science wars appear only in footnotes and are not listed in the bibliography of secondary works.

American Chemical Society, which provided financial backing for the show, in effect disowns the exhibition's representation of science—a representation they characterize as skewed toward the evils of science (atomic destruction, chemical pollution) rather than toward its palpable benefits. The anti-science tone is blamed on postmodernist sentiments among some curators and members of the advisory board.[3]

- The conference "Flight from Science and Reason," held in the spring of 1995 at the New York Academy of Sciences, draws participants from all around the academy—natural scientists, social scientists, humanists. Most have come to decry the assault on rationalist ideals of the Enlightenment fomented by postmodernists, (de)constructivists, and others who would take up "cudgels against science."[4]
- In a special issue entitled "Science Wars," the cultural studies journal *Social Text* publishes the paper "Transgressing the Boundaries: Toward a Transformative Hermeneutics of Quantum Gravity," by physicist Alan Sokal. Full of references to leading figures in science studies (Bloor, Haraway, Latour, Pickering, Woolgar) and replete with hard-to-chew passages from Derrida, Lacan, Lyotard, and Irigary, Sokal's paper is revealed to be a hoax. He admits almost immediately in the gossip rag *Lingua Franca* that he had written it to demonstrate the cronyism, decline of academic standards, pseudoradicalism, and unintelligibility rampant in cultural studies of science.[5]
- On May 16, 1997, the *Chronicle of Higher Education* runs "The Science Wars Flare at the Institute for Advanced Study." The article describes how the appointment of Princeton University historian of science Norton Wise (coauthor of a prize-winning biography of Lord Kelvin, with doctorates in both history and physics) to the permanent faculty of the School of Social Science was blocked by critics inside and outside the Institute "hostile to science studies." The article also

3. I was a member of that advisory board, and twice wrote up my experiences: Gieryn, "Policing STS [Science and Technology Studies]" and "Balancing Acts."

4. The phrase is taken from an advertising blurb announcing arrival of the conference proceedings in book form in late 1996 (Paul Gross et al., eds., *Flight from Science and Reason* [New York Academy of Sciences, 1996]; reissued under the same title by the Johns Hopkins University Press, 1997). The conference was reviewed in David H. Guston, "The Flight from Reasonableness" *Technoscience* 8 (1995): 11–14.

5. Alan Sokal, "Transgressing the Boundaries: Toward a Transformative Hermeneutics of Quantum Gravity," *Social Text* 14 (spring/summer 1996): 217–52; Sokal, "A Physicist Experiments with Cultural Studies," *Lingua Franca* 6 (May/June 1996): 62–64.

points out that six years earlier, Bruno Latour's candidacy for the same position met the same fate for (evidently) the same reason.[6]

- Two edited collections of papers help to harden the divisions between "sides" in the "war": Andrew Ross, coeditor of *Social Text,* augments the special issue on the science wars for stand-alone publication; philosopher of science Noretta Koertge parries with *A House Built on Sand: Flaws in Postmodernist Accounts of Science.*[7]
- Whispers of reconciliation or truce are nevertheless heard: conferences at Durham in December 1994 and at Kansas in early 1997 bring scientists together with those who study their history and sociology; a "Science Peace Workshop" is held at Southampton in July 1997; chemist Jay Labinger publishes a "view from the petri dish" in *Social Studies of Science;* exchanges between physicist David Mermin and science studies warriors Trevor Pinch and Harry Collins appear in *Physics Today;* calls for pax on both houses are made in an editorial in *Nature;* even *Newsweek* discusses a group of Princeton lab scientists and historians of science (called "Reality Check") who meet "once a month to talk about how science can be skewed by ideology and how to make it more objective."[8]

It is not plain who makes up the sides of the science wars and what beliefs about science separate them. It is far too simple to see natural scientists lined up against those who do history or sociology or cultural

6. Liz McMillen, "The Science Wars Flare at the Institute for Advanced Study," *Chronicle of Higher Education,* May 16, 1997, A13. Cf. Colin Macilwain, "'Science Wars' Blamed for Loss of Post," *Nature* 387 (May 22, 1997): 325.

7. Andrew Ross, ed., *Science Wars* (Durham: Duke University Press, 1996); Noretta Koertge, *A House Built on Sand: Flaws in Postmodernist Accounts of Science* (New York: Oxford University Press, forthcoming).

8. Jay Labinger, "Science as Culture," *Social Studies of Science* 25 (1995): 285–306; N. David Mermin, "The Golemization of Relativity," *Physics Today,* April 1996, 11–12 and "What's Wrong with This Sustaining Myth?" *Physics Today,* March 1996, 11–12; Harry Collins and Trevor Pinch, N. David Mermin, letters, in "Sociologists, Scientist Continue Debate about Scientific Process," *Physics Today,* July 1996, 11, 13, 15; "Science Wars and the Need for Respect and Rigour," editorial in *Nature* 385 (January 30, 1997): 373; Sharon Begley, "The Science Wars," *Newsweek,* April 21, 1997, 54–56 (Begley, along with Adam Rogers, had already written a piece on Alan Sokal's hoax: "'Morphogenic Field' Day," *Newsweek* June 3, 1996, 37). For an overview of the issues, see Colin Macilwain, "Campuses Ring to a Stormy Clash over Truth and Reason," *Nature* 387 (May 22, 1997): 331–33. In trying to ring a conciliatory note, the editorial in *Nature* argued: "Where public perceptions of science are undermined by slipshod scholarship and misrepresentation, let the battle continue. But scientists who reflect at all about the wider significance of their work stand to benefit from a sharpened awareness of the genuine insights that science studies can offer" (373).

studies of science, just as it is far too simple to see in "science" truth, reason, and empiricism lined up against relativism, constructivism, and ideology. Pro-science versus anti-science is about as misleading in this context as pro-life versus anti-life is in the American controversy over abortion; nor do participants fall out cleanly along liberal-conservative political lines. There are too many Simmelian crosscutting alliances and fifth columns for the science wars ever to be a neat battle to the finish: an *n*-sided, ever shifting mess is more likely. Still, the apparent antagonists must be labeled something, and so—at the risk of reifying or homogenizing a mixed bag and polarizing the sides even further—I propose "science defenders" and "science studies," to refer respectively to (a) those who see science as under attack by (b) those who examine science as a historical, sociological, and cultural phenomenon.

Though the wars are a muddle just now, it is nevertheless possible to look at them sociologically and ask: What are they a specimen of? The answer is no surprise. Look back at the introduction to this book for the five telltale markings of cultural cartographies involving science.[9]

1. CONTESTED CREDIBILITY

Boundary-work is brought on by disputes over credibility: Who has the legitimate power to represent a sector of the universe—on what grounds? by what methods or virtues? in which circumstances? Cultural maps showing a space for science get drawn when it is not clear how epistemic authority is to be allocated among a variety of claims makers. The maps are interpretative justifications of the restriction of legitimate knowledge to this space and its denial to other spaces, served up as

9. In making these current events into a specimen of the cultural cartography of science, I restrict my attention to materials produced explicitly for the science wars—mostly published commentaries (web postings number in the thousands). I have not followed Paul Gross back into his biological work, nor have I followed Harry Collins, Bruno Latour, or Donna Haraway back into their scholarly studies of technoscience. What interests me here—as in the five episodes that precede this discussion—is how science gets rhetorically constructed as a cultural space, in public contests for credibility and wars over representational legitimacy. Emphatically, I do not respect the full range of conceptualizations of science floating about in science studies—constructivists, feminists, Mertonians, Edinburghians, manglers, Latourians—but pay attention only to what is said about science (and about science studies) by those of my peers actively engaged in the science wars. In the same way that representations of phrenology in testimonials for Combe are not faithful mirrors of what actually happened when fingers met skull, so too are these wartime representations of science by those from science studies an unfaithful portrait of what it is in the relatively pacific but divergent scholarly writings of Collins, Latour, and Haraway.

guides for those who wish to know what to believe and how to locate the credible knowledge on which they can base practical decisions. In this contest, each side draws a different map to create "science" as a distinctive ontological preserve over which they have legitimate claim to authoritative representation. The science wars are "a rumbling debate about the nature of scientific practice and knowledge, about who is qualified to pronounce on either."[10] Each side makes science into something best understood through the methods and tools it uniquely brings to the job: "science is nature" for its defenders; "science is culture" for historians and sociologists.

Science studies brings hermeneutic skills to the study of science: historians and sociologists are adept at deciphering the meaning of texts or material objects, discerning unanticipated consequences of ordinary practices, locating events in a historical flow or at a particular place, interpreting the institutional structures that bind and steer chosen lives and decisions. Many of those in science studies would accept Richard Levins's proposition that "the pattern of knowledge and ignorance in science is not dictated by nature but is structured by interest and belief." Nature typically (not always) becomes beliefs-about-nature or accounts-of-nature when the focus is on facts as collective suppositions—suppositions that are consensually stabilized over time and space via through contingent processes of laboratory tinkering, or inscription making, or ally enlisting. Literary critic Stanley Fish distinguishes social constructivism from nihilism: "What sociologists of science say is that of course the world is real and independent of our observations but that accounts of the world are produced by observers and are therefore relative to their capacities, education, training, etc. It is not the world or its properties but the vocabularies in whose terms we know them that are socially constructed—fashioned by human beings." On this, Marxist and feminist agree: Stanley Aronowitz writes "that all processes of knowledge, including science, are mediated . . . by the social and cultural context within which it has developed, [and] its truths are inevitably relational to the means at hand for knowing"; and Ruth Hubbard acknowledges that "though nature is material and real, our descriptions and understandings of it are necessarily mediated by the culture of science."[11]

10. Macilwain, "Campuses Ring to a Stormy Clash," 331.

11. Richard Levins, "Ten Propositions on Science and Antiscience," in Ross, *Science Wars,* 183; Stanley Fish, "Professor Sokal's Bad Joke," *New York Times,* May 21, 1996, A23; Stanley Aronowitz, "Alan Sokal's 'Transgression,'" *Dissent* 44 (winter 1997): 107, 110; Ruth Hubbard, "Gender and Genitals: Constructs of Sex and Gender," in Ross, *Science Wars,* 169.

With this constructed and artifactual character, science is just like all the other things explored by social scientists and historians—politics, crime, stratification, religion, gender, power. There is nothing distinctive about science as institution, organized practices, networks, collective beliefs, or symbols that would preclude its description and understanding with time-tested empirical methods used effectively by these disciplines for other social and cultural objects—ethnography, interviews, interpretation of documents, discourse analysis, surveys, and censuses. "In this respect," as sociologist Trevor Pinch argues, "science is no different from any craft activities like pottery, carpentry, and cooking. . . . Knowledge always goes with people." Neither gene nor quark has much interest for sociologists and historians of science except as something endowed with meaning—put into language or other symbol, manipulated by hand or machine, embedded in networks of resources and power, the object of thought and human action. Donna Haraway puts it this way: "Questions about what counts as knowledge need to be examined in terms of practice, institutions, people, funding and language." But in rejecting the idea that science is "a version of some universal truth that is the same in all times and places," as Andrew Ross does, science studies does not necessarily deny the value of scientific knowledge—as Katherine Hayles suggests: "It is a fallacy . . . to think that culturally contingent knowledge is not reliable. . . . Knowledge is useful to us because, not in spite of, the fact that it is limited, partial, and perspectival."[12]

Defenders of science argue that all of this misses what makes science science: nature. Indeed, if science were nothing but a discourse—nothing but practices and texts, material instruments and artifacts, networks of power and resources, institutions and organizations, beliefs and actions—then it would in many ways be indistinguishable from politics, religion, or poetry. Only by considering how nature intrudes in science can its distinctiveness be seen—and, in turn, the grounds for its distinctive epistemic authority appreciated. Michael Holmquest streamlines the defenders' position: "Science is nature, and therefore the very opposite of culture." Philosopher of science Michael Ruse concurs: "With its aim of yielding real insight into the nature and workings of the world, science uniquely transcends culture." Physicist Steven Weinberg suggests that the truth or falsity of scientific claims is not a matter of negotiation or

12. Trevor Pinch, "Science as Golem," *Academe* 82 (January/February 1996): 16; Donna Haraway quoted in Liz McMillen, "Science Wars Flare," A13; Andrew Ross quoted in Janny Scott, "Postmodern Gravity Deconstructed, Slyly," *New York Times,* May 18, 1996, A1; N. Katherine Hayles, "Consolidating the Canon," in Ross, *Science Wars,* 233.

hope but of the evidential facts of the case: "If scientists are talking about something real, then what they say is either true or false. If it is true, then how can it depend on the social environment of the scientist? . . . The correct answer . . . is what it is because that is the way the world is." Science is the progressively better (more accurate, more complete, more parsimonious, more efficient) representation, explanation, and prediction of natural reality, despite the fact that it is never finished and always corrigible, as Alan Sokal points out: "Every scientist knows perfectly well that our knowledge is always partial and subject to revision—which does not make it any less objective." New facts replace earlier illusions-passing-as-facts when nature does not do what scientists suppose it should do; concepts like evolution and instruments like particle accelerators are created to tighten the grip of science on nature, but they last only until nature makes them obsolete; facts and evidence are decided by nature, not constructed by people. For chemist Dudley R. Herschbach, "science . . . exalts Nature: she is the boss; we try to discover her rules; she lets us know the extent to which we have done so."[13]

It is trivially true that science has culture and a past: its symbols, instruments, ambitions, resources, collaborations, and accomplishments are indelibly human. Sociologist of science Stephen Cole, speaking for the defense: "Clearly social factors play an important role in the evaluation of new knowledge; but so does evidence obtained from the natural world. . . . Yes science is socially constructed, but yes how it is constructed is to various degrees and extent constrained by nature." But the social goings-on are prologue to what matters about science: the gradual nonlinear convergence of knowledge and nature, the winnowing away of beliefs incongruent with reality. There is room for sociologists to ply their trade in the prologue, but what they can say about the justification of scientific truth is tiny, according to Gross and Levitt: "Perspectivism has interesting things to say about the history of science, the shape of modern

13. Michael Holmquest, "Sokal's Hoax: An Exchange," *New York Review of Books,* October 3, 1996, 54; Michael Ruse, "Struggle for the Soul of Science," *Sciences* 34 (November/December 1994): 41; Steven Weinberg, "Sokal's Hoax," *New York Review of Books,* August 8, 1996, 14; Alan Sokal, "Pourquoi j'ai écrit ma parodie," *Le Monde,* January 31, 1997 (author's translation); Dudley Herschbach, "Imaginary Gardens with Real Toads," in Gross et al., *Flight from Science and Reason,* 18. Philosopher Susan Haack distinguishes belief from justification: "Warrant is social in the sense that talk of how warranted a scientific claim is, is elliptical for talk of how justified a scientific community is in accepting it; but how justified they are in accepting it does not depend on how justified they *think* they are, but on how good their evidence is" ("Towards a Sober Sociology of Science," in Gross et al., *Flight from Science and Reason,* 262; cf. Gross and Levitt, *Higher Superstition,* 17).

science as a social institution, the rhetoric of scientific debates. When it comes to the core of scientific substance, however, and the deep methodological and epistemological questions—above all, the incredibly difficult ontological questions—that arise in scientific contexts, perspectivism can make at best a trivial contribution." Practices at the bench (or the drafting of technical papers) are of little consequence except as they yield claims or interpretations that stand up to nature better or worse than rivals. Constructivists especially have a tendency to slice into the process of fact making too early—when everything does indeed look messy and "cultural"—but lose sight of the "big picture" of stable background truths and new evidential findings that eventually settle the matter into truth. Physicist N. David Mermin suggests that "certain sociologists . . . confuse the scaffolding with the self-supporting edifice." Defenders also easily admit that science again becomes a cultural entity when facts leave labs and journals to enter the public.[14]

The breach is opened, the gauntlet thrown down: in these tales from the front, nature is excluded from science by those in science studies (or just made into the upshot of meaningful practices); culture is excluded from science by its defenders (or just made into an upstream prelude for the later meetings with nature that truly matter). Each side in the science wars defends its epistemic authority over the representation of science by making it into something for which its tools are the right ones for the job. If science is culture, the best tools for its understanding are interpretative, hermeneutic, ethnographic; if science is nature, the best tools

14. Stephen Cole, "Voodoo Sociology," in Gross et al., *Flight from Science and Reason,* 284; Gross and Levitt, *Higher Superstition,* 40; Mermin letter, in "Sociologists, Scientist Continue Debate," 15.

Physicists Gottfried and Wilson imply that Edinburghian constructivists unjustifiably extend their (useful) studies of scientific *practice* to a social explanation of the content of scientific *facts* visible only in the end. So Andrew Pickering gets into trouble when "the strong programmes's methodology is applied by them to scientific knowledge, and not just to practice, because now the sociologist is now really acting as the judge of scientific knowledge, assumes the power to stop the clock at an arbitrary point, thereby ignoring subsequent evidence as to whether some bandwagon fell over the cliff or stayed on track" (Kurt Gottfried and Kenneth G. Wilson, "Science as a Cultural Construct," *Nature* 386 [April 10, 1997]: 545–47). Dudley R. Herschbach uses the same image of science to separate the cultural stuff upstream from the nature-driven evidential facts that come out at the end of the day: "At its frontiers, science-in-the-making is inevitably a messy and uncertain business, easily misunderstood by policy makers, funding agencies, reporters, students, and sometimes even the researchers. Those who decry science often confuse its rude frontiers with its civilized domains or foundations" ("Imaginary Gardens," 15–16). Gottfried and Wilson suggest that constructivism may be more applicable to "science in the public arena: courtrooms, Chernobyl, and so on" than to the facts inside real science ("Science as a Cultural Construct," 547). My own willingness to wade into the science wars rests exactly on the sociological poverty of essentializing a boundary between science in the lab (real) and science in public arenas (merely cultural).

are quark, accelerator, and PCR machines. Characterizations of the other side become caricature: science studies is said to deny altogether that natural reality exists, just as defenders are said to deny the existence of anything cultural or historical. Sokal wonders: "Is it now dogma in cultural studies that there exists no external world?" Popperian Noretta Koertge says that "there are also extreme social constructivists who engage in a form of 'biodenial' whereby they deliberately downplay or even totally ignore the role of nonsocial elements." And Herschbach tosses out the archetypal there-are-no-relativists-at-30,000-feet: "When asked to refute Bishop Berkeley, Ben Johnson simply kicked a rock. To a cultural constructionist, we might likewise point out an airplane, say." If science studies is said to live with denatured reality, science defenders must live in a decultured world. Norton Wise says that "Weinberg presents us with an ideology of science, an ideology which radically separates science from culture," while feminist philosopher Sandra Harding suggests that defenders' views "have been obscured by preoccupations with representations of ideal science as an undistorting mirror of a fixed and perfectly coherent nature."[15]

Both sides see the science wars as a battle for the legitimate right to represent just what science is—culture, nature, a little of both? Haack lays bare the contested credibility: "And as for those who argue that since scientific knowledge is nothing but a social construction, the physical sciences must be subordinate to the social sciences, the only reply needed is that it isn't, so—thank goodness!—they aren't." From the other side, anthropologist George Levine suggests that science defenders are engaged "in a self-defeating crusade to keep people who are affected by science but don't 'do' science from having anything to say about it," and Hilary Rose agrees: "Science is one of few cultural activities where the practitioners have always sought (indeed, rather successfully) to stay in charge of the story about science. . . . There is more than a whiff of an ideology of the authority of the ultimate expert who is alone qualified to say what is and what is not science." This is all boundary-work of a sort: culture and nature are demarcated, and science is put on one side or the other. So whose science is it?[16]

15. Sokal, "A Physicist Experiments," 62; Noretta Koertge, "Wrestling with the Social Constructor," in Gross et al., *Flight from Science and Reason,* 267; Herschbach, "Imaginary Gardens,"18; Norton Wise, "Sokal's Hoax: An Exchange," *New York Review of Books,* October 3, 1996, 55; Sandra Harding, "Science Is 'Good to Think With,'" in Ross, *Science Wars,* 23.

16. Haack, "Towards a Sober Sociology of Science," 264; George Levine, in "The Sokal Hoax: A Forum," *Lingua Franca* 6 (July/August 1996): 64; Hilary Rose, "My Enemy's Enemy Is—Only Perhaps—My Friend," in Ross, *Science Wars,* 86.

2. MAPPING OUT

This boundary-work is then used to delegitimate the other side's pretensions to official renderings of science: cartographic exclusions become the order of the day, buttressed by a strong insiderism (it takes one to know one).[17] It is said that historians and sociologists of science do not practice science and therefore cannot know it; that natural scientists do not know history or social science and therefore cannot do it.

Defenders of science effect a volatile double exclusion: science studies is located outside the cultural space for "science," which raises doubts both about its ability to comprehend or interpret that which it is not a part of and about its general ability to produce credible (accurate, reliable, honest) accounts of anything—on the assumption that such credibility marks the genuinely scientific. A penumbra of invidious places is found for science studies: theology, philosophy, mysticism, ideology, charlatanism and nonsense, smug "smartasserie," Nazism.[18] Because science studies is outside science, it cannot hope to get it right. Gross and Levitt suggest that the field concocts "a model of 'science' [that] is a lot like the wicker-and-mud mock-up of a C-47 [aircraft] built by the cargo cultists." They elaborate: "A serious investigation of the interplay of cultural and social factors with the workings of scientific research . . . above all . . . requires an intimate appreciation of the science in question, of its inner logic and of the store of data on which it relies, of its intellectual and experimental tools. . . . We are saying, in effect, that a scholar devoted to a project of this kind must be, inter alia, a scientist of profes-

17. Merton, "The Perspectives of Insiders and Outsiders," in *Sociology of Science,* chap. 5.

18. Collins and Pinch are said to impose "theological standards" (Belver C. Griffith, letter, in "Discussion of Nature of Science Provokes Hit-or-Myth Debate," *Physics Today,* January 1997, 13). "We should evaluate Collins and Pinch's arguments by treating them as philosophy, rather than trying to subject them to the methods used in science to discover truth" (Joseph F. Dolan, letter, in "Discussion of Nature of Science Provokes Hit-or-Myth Debate," *Physics Today,* January 1997, 11). "Mysticism or freewheeling, intellectual deceit or antiintellectualism . . . " (Mario Bunge, "In Praise of Intolerance to Charlatanism in Academia," in Gross et al., *Flight from Science and Reason,* 96). "We must talk about the intellectual virtues that are constantly guiding science and not cede the moral high ground to ideologues" (Koertge, "Wrestling with the Social Constructor," 272). For science studies as charlatanism and nonsense, see Alan Sokal, in "Mystery Science Theatre [Symposium]," *Lingua Franca* 6 (July/August 1996): 57; its characterization as "smartasserie" is from Norman Levitt, "More Higher Superstitions," *Skeptic* 4 (1996): 82. Arguing that existentialism is among the roots of today's science studies, Bunge maintains: "Heidegger was a Nazi ideologist and militant. . . . Existentialism is no ordinary garbage: it is unrecyclable rubbish. Its study in academic courses is justified only as an illustration of, and warning against, irrationalism, academic imposture, gobbledygook, and subservience to reactionary ideology" ("In Praise of Intolerance," 97).

sional competence, or nearly so." Other defenders agree: "It is unlikely that any account from the outside can ever capture what is really happening on the inside." This ignorance of how science works makes those in science studies equally unequipped to map its borders and territories: "Because the constructivist-relativists ignore science, they are incapable of distinguishing it from pseudoscience;" or maybe such inaccurate cartography is intended: "Postmodernism blur[s] the distinction between [science] and 'other ways of knowing'—myth and superstition, for example." Although science studies cannot know science as only scientists can, they are nonetheless dead set against it: "Our antiscience colleagues are characterized by their appalling ignorances of the very object of their attack, namely science." Historian of science Gerald Holton also puts radical constructivists in the territory of anti-science and describes them as a "challenge to the very legitimacy of science," while Cole suggests that science studies takes glee in giving natural scientists their comeuppance: "Many correctly perceived the constructivist approach as an attack on the natural sciences and were pleased to see these sciences, which have long lorded it over the social sciences, knocked off their pedestal." Herschbach takes Sandra Harding's equivalence of Newton's *Principia* and "a rape manual" as evidence that science studies is out to "deprecate not only science but all objective scholarship and public discourse."[19]

Science defenders are simultaneously lumpers and splitters in their representations of the other side. Lumping together relativists and constructivists, feminists and Marxists, history of science and cultural studies, sociologists and afrocentrists, postmodernists and environmentalists tars them all with the sins of the most strident. Gross and Levitt create an altogether new nation labeled "the academic left," defined by its prominent landmarks: antipathy toward real science and resentment of its authority, success, or resources; ignorance about what science really is; old leftist political projects retreaded and now passing as scholarly advance. Boundaries that would separate academic historical studies of science from tree-hugging or faith healing are drawn ever so faintly by Gross and Levitt—overshadowed by the thick black line between science and its non- or anti-. It matters little, in this war of maps, that animal

19. Gross and Levitt, *Higher Superstition,* 41, 235; Anthony G. Basile, letter, in "Discussion of Nature of Science Provokes Hit-or-Myth Debate," *Physics Today,* January 1997, 13; Paul Boghossian, "What the Sokal Hoax Ought to Teach Us," *Times Literary Supplement,* December 13, 1996, 15; Bunge, "In Praise of Intolerance," 105 and 101; Gerald Holton, "Science Education and the Sense of Self," in Gross et al., *Flight from Science and Reason,* 552; Cole, "Voodoo Sociology," 276.

rights advocates and Bruno Latour may have different intellectual or political ambitions, or that they choose different strategies for pursuing them: they are one in their threat to science, alike in the illegitimacy of their representations of science, misguided in their designs on our future.[20]

Casual modifiers—"extreme" relativists, "radical" constructivists—expose a different (but just as effective) battle plan, whereby defenders of science divide science studies in order to conquer it. Hoping to turn the other side against itself, they insert boundaries between domesticated historical or sociological studies (no threat to science and maybe even exemplifying its normative standards of truth making) and a lunatic fringe. For example, Haack locates the boundary this way: "Bad sociology of science is thus *purely* sociological, whereas good sociology of science, acknowledging the relevance of evidential considerations, is not. Good sociology of science, in consequence, requires some grasp of scientific theory and evidence, while bad sociology of science does not."[21] The potential slippery slope is high and steep in such boundary-work, obvious in the incidents at Princeton's Institute for Advanced Study. Some defenders took Latour's inventive ontologies and playful prose as warrant for his placement on the fringe outside the Institute's School of Social Science. But could the same as easily be said of Norton Wise, with a graduate degree in physics, whose sober book on Kelvin was thought by some historians to be too full of formulae, who heads a doctoral program at Princeton that requires its historians to have training in the science they hope to study? The lunatic fringe was stretched so far in order to gobble up Wise that many were left to wonder if that was all there was to science studies.[22] In the Smithsonian controversy, the territory of the

20. "Differences [among the academic left] are soft-pedaled in the interest of an overriding common purpose, which is to demystify science, to undermine its epistemic authority and to valorize 'ways of knowing' incompatible with it" (Gross and Levitt, *Higher Superstition,* 11).

21. Haack, "Towards a Sober Sociology of Science," 260. Gottfried and Wilson, both physicists, distinguish the cultural studies ("a fringe group, ill-informed about science") from "the more sophisticated and troubling view of scientific knowledge as put forward by . . . the 'Edinburgh' school of sociology" ("Science as a Cultural Construct," 545).

Such boundary-work does not prevent at least one defender from pointing out that natural scientists actually *need* those who live on the fringe of biology, chemistry, or physics: "Frequently, what might appear as the most promising path up the mountain does not pan out; there are unanticipated roadblocks. Then it is vital to have some scientists willing to explore unorthodox paths, perhaps straying far from the route favored by consensus. By going off in what is deemed the wrong way, such a maverick may discern the right path" (Herschbach, "Imaginary Gardens," 16).

22. Perhaps relevant is Wise's active involvement in the science wars during the months before his fate was decided: his review of Gross and Levitt in *Isis* was caustic, and he went after

unacceptable and threatening expanded to include not just radical constructivist science studies and their more moderate cousins, but all of the cognate disciplines that feed into studies of science: the American Chemical Society requested that all displays of *social science* be removed from the story "Science in American Life."[23]

Savage cultural cartography like this is not the monopoly of defenders of science. Soldiers from science studies create spaces for the other side just as useful for delegitimating *their* accounts of science: guard dogs, fundamentalists, priests, academic police, Inquisitors, apocalyptics, and apologists.[24] I do this myself, in a review of Michael Friedlander's *At the Fringes of Science,* a book I characterize as "missionary work" (although I have to point out that Friedlander used the phrase first). Friedlander, I write, "plays science cop, patrolling and protecting its frontiers not just from pretenders to its authority over nature but from those outside science who mistakenly believe that they too have warrant to decide for themselves which science is good or bad, real or pseudo." What a grenade! I locate Friedlander adjacent to Velikovsky, because both give "off the appearance of science . . . but lacking endorsement from specialists in the many fields" they traverse. I deny this physicist the epistemic authority to speak the sociology of science, just as he denies Velikovsky the legitimacy to speak cosmology, astronomy, and geophysics: on this map of mine, both belong to pseudo-land. Science defenders are banished

---

Nobel physicist Steven Weinberg in the *New York Review of Books.* For public confessions of why Wise was not enough for the institute, see McMillen, "Science Wars Flare."

23. Marvin Lang, a chemist and colleague of mine on the exhibition advisory board, complained about input from "social scientists and pseudo-scientists who had no idea of how science worked" (quoted in Colin Macilwain, "Smithsonian Heeds Physicists' Complaints," *Nature* 374 [March 16, 1995]: 307).

24. Speaking about encounters with Gross and Levitt, Harry Collins recalls: "In the early days, it was as if we were entering this native village, and all the guard dogs came out and started biting our ankles. . . . It was a mistake to try to have a discussion with the guard dogs" (in Macilwain, "Campuses Ring," 332). The descriptions "fundamentalist" and "priest" come from Collins and Pinch, who add: "We are falling into a style of discourse more appropriate to a McCarthy-style hearing or a religious inquisition"(letter, in "Sociologists, Scientist Continue Debate," 11). On the Sokal hoax, Latour writes: "L'affaire me paraît beaucoup plus intéressante qu'une simple question de police académique. . . . Ce n'est plus la guerre contre les Soviétiques, mais celle contre les intellectuels 'postmodernes' venus de l'étranger" ("Y a-t-il une science après la guerre froide?" *Le Monde,* January 18, 1997). Sharon Traweek wonders "if we're back to the Inquisition?" (in McMillen, "Science Wars Flare," A9). "Apocalyptics" is from Ross's introduction to *Science Wars,* 11; and "apologists" is from George Levine, "What Is Science Studies for and Who Cares?" in Ross, *Science Wars,* 133. For good measure, Sarah Franklin writes that "Gross and Levitt espouse a paternalistic Right-to-Life discourse concerning the vital essence of the scientific ethos" ("Making Transparencies: Seeing through the Science Wars," in Ross, *Science Wars,* 165).

from science studies—no training, no license, no endorsement from insiders—to a territory reserved for those ignorant of its aims and accomplishments, peopled by aging natural scientists looking back wistfully (but not analytically or theoretically), by journalists who parody and caricature, and by anyone else who shoots from the hip (without reading Latour and Wise, or getting them right). Sokal is thus seen as having tried "to ventriloquize work he [doesn't] fully understand"; Wolpert may be "a knowledgeable and competent scientist [but] this should not lead us to think that this qualifies him as an authority in the philosophy of science"; and Gross and Levitt's "criticisms [are] questions not of fact but of interpretation in history, sociology and especially philosophy in which [they have] no special competence."[25] Evidence for the defenders' incompetence at science studies is their "lumping" inability to recognize and appreciate huge differences separating the field's various battalions.[26]

25. Thomas F. Gieryn, review of *At the Fringes of Science,* by Michael Friedlander, *Isis* 87 (1996): 767–68; George Levine, "Sokal's Hoax: An Exchange" *New York Review of Books,* October 3, 1996, 54; Michael Lynch, "Detoxifying the 'Poison Pen Effect'" in Ross, *Science Wars,* 241; Roger Hart, "The Flight from Reason: *Higher Superstition* and the Refutation of Science Studies," in Ross, *Science Wars,* 284. Stanley Aronowitz notices the obvious asymmetry in the defenders' position: "While everybody, including physicists and molecular biologists, is qualified to comment on politics and culture, nobody except qualified experts should comment on the natural science[s]" ("Politics of the Science Wars," in Ross, *Science Wars,* 203).

Norton Wise's review of *Higher Superstition* suggests that its authors understand neither the history of science nor its sociological analysis: "Here colorless white is 'Enlightenment'—not the historical Enlightenment, made up of all colors of the spectrum, but a stereotyped essence in disembodied realm of ideas." And then, after recounting Gross and Levitt's injunction that those who do history and sociology of science must know the science first, Wise hoists them by their own petard: "As their book so painfully demonstrates, this same need for professional competence applies to historical research and writing" ("The Enemy Without and the Enemy Within," *Isis* 87 [1996]: 323, 327). After Wise was denied the position at the Institute for Advanced Study, Frederick Gregory (as president of the History of Science Society) wrote to the *Chronicle of Higher Education:* "I can only conclude that the persons responsible for this regrettable mistake in judgment either deliberately ignored or are not familiar with the balance achieved by Professor Wise in his scholarship" (unpublished letter).

And, in a nice reflection of Gross's C-47 science constructed by the cargo cult also known as science studies, we have Lynch's comment on how "Gross and Levitt construe the constructivist stance (or, rather, [how] they *mock it up*)" (Lynch, "Detoxifying," 249; emphasis mine).

26. Wise continues his assault on *Higher Superstition:* "The rhetorical technique is always the same. Pick out a stereotyped set of figures, make them stand in for a wide range of different and often conflicting positions in a complex movement, and ignore all aspects of that movement that do not fit the stereotype" ("The Enemy Without," 324). In the midst of a battle with Norman Levitt, Richard Olson writes about his and Gross's tactics: "Nearly all persons who accept the notion that cultures have any bearing on the content of science in any degree are caricatured by identifying their views with the most radical cultural constructionists, post-modernists, academic feminists, and ecologists. ("Where Is Knowingness to Be Found?" *Skeptic* 4 [1996]: 83). Anthropologist Emily Martin says: "None of us recognize ourselves in these diatribes" (in

3. WHAT SCIENCE BECOMES HERE AND NOW

In struggles for credibility, the cultural space for science takes diverse pragmatic shapes, and distinctive landmarks are located within or without, as contestants squeeze and stretch its borders in order to best justify their own reality claims as legitimate and persuasive. As it happens, all contestants in these science wars—defenders and even supposed attackers—want to locate their own inquiries inside science for the same reason that Howard, Pons, Fleischmann, Combe, Tyndall, and sociologists lacking professional self-esteem do, to benefit from the long historical association of "science" with credibility and (though not necessarily their explicit intent) to secure that epistemic authority until at least the next round of boundary-work. Moreover, there is remarkable agreement on both sides of the battle lines about how—methodologically—proper science should be practiced. They disagree, of course, on just who is being proper.

Many of those from science studies are at pains to show that they are not only pro-science, but that they *do* the science of science properly. Barry Barnes told *Nature* that the goal of Edinburgh's "strong programme" was "to understand the nature of knowledge in a scientific manner," and his former colleague Steven Shapin adds: "We have great respect for what scientists actually do. . . . If our fundamental philosophical position was anti- anything, it was anti-rationalist philosophy, not anti-science." Norton Wise puts science studies within the same cultural space as Steven Weinberg's physics: "All historians, philosophers and sociologists that I know share Weinberg's hope for rational understanding and . . . none of them deny objective reality as Weinberg presents it." Levine says that "nobody [in science studies] is trying to shut science down," and Nelkin suggests that this is because the field—as much as physics or chemistry—relies on its "scientificity" to sustain its credibility: "The role of science as a model of rationality in human affairs is not really in question. In particular, the historians and sociologists who study science have always had to justify their work to scientists and to validate its credibility in terms of scientific standards." Sometimes the self-location of science studies within science is achieved through the deportation of a lunatic fringe. For the Marxist Aronowitz, "the point is not to debunk

---

McMillen, "Science Wars Flare," A9). Biologist Richard Lewontin joins the parade: "The hopeless muddle they make of the category renders the term *academic left* useless for any analytic purpose, yet it appears over and over" ("À la recherche du temps perdu: A Review Essay," in Ross, *Science Wars,* 296).

science or to 'deconstruct' it in order to show it is merely a fiction. This may be the postmodern project, but it is not the project of science studies." For feminist Evelyn Fox Keller, "it is not science that is threatened by the hapless publication of gibberish; it is science studies itself."[27]

These maps—showing the history and sociology of science snuggled up inside science with physics and chemistry—may be as hard a sell these days as they were just after World War II (chapter 2). Prominent defenders parade their hard scientific credentials (Gross is a biologist, Levitt a mathematician, Sokal a physicist) to deny authority to soft impostors. Still, those from science studies may have learned a thing or two from their studies of rhetoric, and they devise a sly cartographic response that at once preserves the appearance of science as rational and objective while at the same time excluding Gross, Levitt, Sokal & Co. In this counterargument, the boundary-work of science defenders is severed from whatever good and hard science they do in their day jobs. "Science" now becomes a space for fair debate, the proper tribunal for respectfully sorting out contrasting claims, where those who disagree can engage each others' work honestly and collectively, avoiding misrepresentation or stereotype. Those high-profile science cops do not fit science so drawn. David Edge complains that Paul Gross treats his adversaries in the science wars with "contempt and derision": "abandoning all pretense of trust and respect," he does not engage in "fair, honest and well-informed disputation," and because of this "demean[s] (and will eventually destroy) the very science and reason that we are all so anxious to conserve and extend." After accusing science defenders of conducting a witch-hunt against Norton Wise at the Institute for Advanced Study, Bruno Latour says that it is the "Sokalites" "who are anti-science . . . against the objectivity of science, against intellectual freedom. . . . They are not to be found in the ranks of science studies." Collins and Pinch add that defenders "hide behind creation myths, witch-hunts, censorship and suppression [and] in doing so . . . will destroy the democratic foundation of the science they say they love." Wise writes that *Higher Superstition* "will be deeply saddening to anyone committed to the free and open investigation of the workings of science." Lewontin believes that the

27. Barnes and Shapin in David Dickson, "Champions or Challengers of the Cause of Science?" *Nature* 387 (May 22, 1997), 333; Wise, "Sokal's Hoax: An Exchange," 55; Levine, "What Is Science Studies For?" 128; Dorothy Nelkin, "The Science Wars: Responses to a Marriage Failed," in Ross, *Science Wars,*" 121; Aronowitz, "Alan Sokal's 'Transgression,'" 110; Keller, in "The Sokal Hoax: A Forum," *Lingua Franca* 6 (July/August 1996): 58.

book is so far outside science that its authors do not even get the science right: "The 'science' of Gross and Levitt is something out of a high school textbook."[28] Sokal's hoax becomes a deceptive "breach of ethics" and "caricatur[e] of complex scholarship" rather than an open and informed confrontation of ideas. Even worse, as the "experiment" he hoped for, it is poorly designed and thus proves nothing.[29] Meanwhile, sociologists of science give their field features that should make any bench scientist feel as if they are working shoulder-to-shoulder: their work is realist, "maxi-

28. David Edge, letter, in "Evolution Teaching," *Science* 274 (November 8, 1996): 904; Bruno Latour, "Science as Culture Newsgroup," electronic posting, May 21, 1997; Collins and Pinch letters, in "Sociologists, Scientist Continue Debate," 13; Wise, "The Enemy Without," 323; Lewontin, "À la recherche du temps perdu," 297.

Basil O'Neill writes that Gross and Levitt suffer from a "failure to read properly," and they "tear passages out of context, and corral thinkers into positions that they then attack. . . . What is more surprising is [their] incapacity to be logical" ("Here Be Dragons," *Times Higher Education Supplement*, July 1, 1994, 23). Hayles accuses them of "a systematic pattern of misleading and unfair quotation, a failure to read accurately, a failure to grasp an argument's main thrust, opportunistic and biased use of sources, and the use of character assassination and verbal abuse rather than reason to discredit an opponent's work" ("Consolidating the Canon," 227). For Lynch, *Higher Superstition* displays "a severely uncharitable reading of the texts, focusing on factual errors, tendentious remarks, and outrageous claims" ("Detoxifying," 248); while for Hart "omission, misrepresentation, caricature, ad hominem remarks and altered quotations combine to form [their] attacks" ("Flight from Reason," 273). I have not seen these sins turn up *inside* anybody's map of science.

29. Bruce Robbins and Andrew Ross, in "Mystery Science Theatre [Symposium]," 54; Ross in Scott, "Postmodern Gravity," A1; Stanley Fish suggests that "Professor Sokal's legacy" is "fraud" in that he goes "beyond error to erode the foundation of trust on which science is built" ("Professor Sokal's Bad Joke," A23).

Here are the weaknesses of Sokal's experimental design: he identified himself to the editors at *Social Text* as a physicist, which allowed them later to say that they had accepted the piece—warts and all—because it came from a scientist who was at least trying to consider cultural studies on its own terms, although "his article would have been regarded as somewhat outdated if it had come from a humanist or a social scientist" (Robbins and Ross, in "Mystery Science Theatre," 55). Boghossian suggests that Ross and the editorial board of the journal may even have bowed to the physicist's authority: "The prospect of being able to display in their pages a natural scientist—a physicist, no less—throwing the full weight of his authority behind their cause was compelling enough for them to overlook the fact that they didn't have much of a clue exactly what sort of support they were being offered" ("What the Sokal Hoax Ought to Teach Us," 14). Robbins admits that "we thought he was a progressive scientist, a physicist who was willing to be publicly critical of scientific orthodoxies" ("Reality and *Social Text*," *In These Times*, July 8, 1996, 28).

Moreover, if Sokal's objective was to discredit the entire field of the cultural studies of science (and not just *Social Text* or its editors), he should have remained silent about the hoax—to see if readers and other practitioners would also have read his paper as sincere. Sokal later said that he went public with the ruse because word of it had leaked out, and the beans would have been spilled almost immediately either way (see Babette E. Babich, "Physics vs. *Social Text:* Anatomy of a Hoax," *Telos,* no. 107 [spring 1996]: 45).

mally objective, accurate and comprehensive"; they are "simply subjecting science to the same ruthless criticism that corresponds to the scientific ideal of self-critical inquiry."[30]

Now if we look at the defenders' maps of science, the landmarks of good scientific practice are no different, but it is of course science studies that does not measure up. Boundary-work is epitomized in Gross and Levitt's contrast of real science with the academic left:[31]

Academic left	Science
muddleheadedness (1)	reliable and rests on a sound methodology (2)
insularity and ignorance (7)	reliable factual knowledge (12)
mythmaking, symbolic wish-fulfillment (8)	open-endedness (17)
sloganeering (8)	powerful, systematic, and ever-expanding (17)
weakness of fact and logic (8)	deep epistemological skepticism (21)
polemical (11)	internal logical consistency, empirical verification (24)
radical epistemological skepticism (11)	science works (49)
unprovable and bootless speculation (12)	objective truth about the world (52)
impressionistic description (12)	cogent, self-consistent, logically coherent, requiring patience, diligence, humility, and intellectual energy (53)
subjective hermeneutics (12)	nature might provide a template (58)
trendy doctrine, windy generalization (37)	deep and surprising predictions about the real world (62)
perverse theories (43)	exacting logical analysis of abstract models (62)
rush hell-for-leather toward unalloyed twaddle (43)	rigorous (81)
tooth-fairy hypothesis (47)	driven by the unyielding contours of reality (81)
sloppy, full of holes (48)	verification and falsification (81)
covert appeals to emotion and prejudices (48)	hard-won truth (85)
turgid and opaque (50)	severe tests of meaningfulness (86)
incoherent (51)	epistemic dignity (90)
hallucinatory (52)	
vulgarizations; amateurs (52)	
hermeneutic hootchy-koo (53)	
paradox and contrarian whimsy (60)	

30. Harding, "Science Is 'Good to Think With,'" 20; Aronowitz, "Politics of the Science Wars," 205. Science studies may even be as committed to "realism" as any physicist: "Voilà une discipline [l'histoire sociale des sciences] à peu près inconnue, qui propose de l'activité scientifique une vision enfin réaliste, dans tous les sens du mot" (Latour, "Y a-t-il une science après la guerre froide?").

31. These terms all appear in Gross and Levitt's *Higher Superstition,* to which the page numbers in parentheses refer.

fallacious (65)
ideological (68)
priesthood (73)
pretentiousness (79)
untrammeled relativism (84)
narratives of superstition (85)
conceptual freak show (88)
pseudoscientific (90)
philosophical styrofoam (98)
intellectual tinsel (100)
febrile delusion (103)
dogmatisms (112)
metaphors (112)
old-time camp meeting (113)
overdue for deep psychoanalysis (121)
multicultural gravy train (131)
epistemological Merry Pranksters (136)
academic inner city (136)
millenarian longings (169)
moonbeams and fairy-dust (176)
unsupported by any evidence (211)
pathetic gullibility (212)
psychotalk (239)
sheer puffery (246)
incipient Lysenkoism (252)

vast and serious (122)
self-correcting (139)
rationality itself (165)
controlled experiments of rational design (186)
knowledge (195)
reality-driven (234)
reliable, profound, and productive (256)

Science as such retains its worthy virtues, its putative grounds for credibility. But now the proponents of science studies are "the enemies of learning, rigor, and empirical evidence . . . who proclaim that there is no objective truth, whence 'anything goes,' [and] who pass off political opinion as science and engage in bogus scholarship." Sokal writes that "Aronowitz distorts [his critic's] positions," that "much of his essay is based on setting up and demolishing straw opponents," that "such investigations need to be [but are not] conducted with due intellectual rigor." Gross responds to Edge's criticism of *Higher Superstition:* Edge "cannot have read the book" and serves up "ad hominem arguments." Cole takes their illogic and duplicity as a sign that those in science studies are simply wrong: "They know that what they say cannot hold water and, when pushed to its real foundations, is logically absurd. Therefore, the only way to defend themselves is to say that they never said what I said they said (or if they said it they did not mean it)."[32] So although rivals in the science wars may be divided by their ontological commitments to science-as-nature or science-as-culture, they come together in agreeing that science-the-method is objective, fair, reliable, and honest. But only the good guys do science the way it is supposed to be done.

4. CONTINGENT STAKES

Cultural cartography happens when there is something valued on the line: material resources, prestige, the truth of a cherished claim, power. If the present science wars are anything like the five episodes of boundary-work that came before, this is not a merely academic or scholastic exercise. What is it about science just now—politically, economically, symbolically—that gives these wars their vitriol? As in my other case studies, the stakes for all sides become part of the cartography: interests are attached (to others) or denied (on our side) in order to legitimate our map as an accurate rendition, rather than some self-interested distortion, of a "real" culturescape.

According to science studies, these may not be the best of times for scientists—politically or economically. That assertion then becomes a premise for delegitimating the defenders' cultural maps as poorly disguised efforts to secure more reliable funding for science or a more salutary climate of public opinion (by scapegoating science studies). Sociologist Dorothy Nelkin suggests that the science wars are "really" about "strains on federal support for science," and Richard Olson elaborates: "The passing of the 'Golden Age' of research and development associated with the Cold War [has] created a real short term threat to the economic health of the science and engineering communities, with physics being hit particularly hard." Anthropologist George Levine links this retrenchment to the science wars: "Counter-aggression of scientists hostile to 'postmodernism' is surely the consequence of an economic pinch hurting them." Levine continues elsewhere: "The squabbles that followed have now been inflated to holy wars because there is so much at stake: intellectual authority, educational direction, disciplinary turf, the allocation of

32. Bunge, "In Praise of Intolerance," 96; "Alan Sokal Replies," *Dissent* 44 (winter 1997): 111; Paul Gross, letter, in "Characterizing Scientific Knowledge," *Science* 275 (January 10, 1997): 142; Cole, "Voodoo Sociology," 281.

Bunge singles out feminists for special exclusion from science: "They denounce precision—in particular, quantitation, rational argument, the search for empirical data, and the empirical testing of hypotheses as so many tools of male domination" ("In Praise of Intolerance," 100).

big money." Thinking about sciences like molecular biology, Ross suggests that "a sequestered craftlike pursuit has been undermined by the wholesale proletarianization of scientific labor in commercial production." Nelkin adds: "The growing importance of industry-university collaborations has left a public impression that science is for hire, that some scientists are simply indentured scholars to a corporate entity, and that scientific information is less a public resource . . . than a private commodity." Philosopher Silvan S. Schweber puts it all together: "Doubtless part of the reason for their participation [in the science wars] lies in the post-cold-war marginalization of physicists both within the academy and industry, as evidenced by the cancellation of the Superconducting Super Collider and the 'downsizing' physics departments and industrial laboratories." Well-publicized instances of research misconduct and fraud are said to have tarnished the public image of science as well.[33]

Such a bleak picture is vigorously denied by science defenders as a cause of the science wars. It stands to reason: their boundary-work is compromised if maps of a wonderful science—surrounded by fierce wolves in skins of postmodernism—were shown to be merely self-serving cartographic instrumentalities designed to restore science budgets or to justify questionable research on human fetuses. No, structural contractions of science (however severe or imagined to be) are pushed to the background. Paul Gross asks (and answers): "Is it reasonable to think that people like [physicist Steven] Weinberg started to make a fuss about [science studies] because he was getting his funding cut? It's ridiculous." Weinberg himself is only a little facetious when he admits: "In years of lobbying for federal support of scientific programs, I never heard anything remotely postmodern or constructivist from a member of Congress." Alan Sokal "dismisses the criticism that his concern about the growing influence of . . . 'constructivist' ideas about science reflects worries about a decline in both funding for physics and [its] social status with the end of the Cold War." Although these political and economic stakes are far from the minds of science defenders, they are seen as an essential element of the other side: "Social constructivism is not simply an intellectual movement, a way of looking at science, but it is an interest group that tries to monopolize rewards for its members or fellow travellers and

33. Nelkin, "What Are the Science Wars Really About?" *Chronicle of Higher Education,* July 26, 1996, A52; Richard Olson, "Whose Is the Higher Superstitution?" *Skeptic* 4 (1996): 33; Levine, "Sokal's Hoax," 54, and "What Is Science Studies For?" 125; Ross, introduction to *Science Wars,* 9; Nelkin, "Science Wars," 119; Schweber speaks reconciliation and peace: "Reflections on the Sokal Affair: What Is at Stake?" *Physics Today,* March 1997, 73.

exclude from recognition those who question any of its dogma."[34] This argument thus completes the mutual imputation of polluting self-interest.

Science defenders offer their own cultural cartography as an altruistic mission designed to rescue the epistemically gullible from seductions of relativism, constructivism, or radical environmentalism. The "real" stakes of the science wars are reason, logic, evidence, and objectivity—virtues secure for now within the walls of science, but imperiled outside among impressionable students and fad-hungry journalists (to whom postmodernists can feed tales of the Enlightenment's complicity with power and implication in environmental degradation, racial and gender inequities, colonial domination). Gross and Levitt make this plain in *Higher Superstition:* "The academic left's rebellion against science is unlikely to affect scientific practice and content. . . . The danger, for the moment at least, is not to science itself. What is threatened is the capability of the larger culture, which embraces the mass media as well as the more serious process of education, to interact fruitfully with the sciences, to draw insight from scientific advances, and, above all, to evaluate science intelligently." Sokal is more pointed: "My goal isn't to defend science from the barbarian hordes of lit crit (we'll survive just fine, thank you)." Bunge believes that the world as we know it is down the tubes if science studies succeeds: "Spoil the charlatan and put modern culture at risk. Jeopardize modern culture and undermine modern civilization. Debilitate modern civilization and prepare for a new Dark Age."[35]

The stakes for science studies may be more desperate: survival. In retrenched universities, an old union rule is sometimes enforced: last aboard, first to go. Interdisciplinary programs and departments in science and technology studies are relatively vulnerable newcomers. The judicious Schweber drops the other shoe in suggesting that cultural cartography by science studies is no more easily detached than that of its detractors from struggles for funding and a secure base on campus: "The current science wars and culture wars will make support for the humanities and the social sciences more difficult." Sheila Jasanoff implies that the threats to science and technology studies from the science wars are far greater than the threats of constructivism or feminism to physics and

34. Gross quoted in Macilwain, "Campuses Ring," 333; Weinberg, "Sokal's Hoax: An Exchange," 56; Sokal quoted in David Dickson, "The 'Sokal Affair' Takes Transatlantic Turn," *Nature* 385 (January 30, 1997): 381; Cole, "Voodoo Sociology," 274.

35. Gross and Levitt, *Higher Superstition,* 4; Sokal, in "Mystery Science Theatre [Symposium]," 57; Bunge, "In Praise of Intolerance," 110.

chemistry. "When money is tight, you tend to cut out the things that are considered non-essential. The curriculum in academic life is becoming more market driven than 'intellect' driven. The sociology of scientific knowledge is an easy thing to target, and appears to stand for a lot of things that some people say are going wrong in society." The implication is that efforts, for example, by right-wing academic watchdogs (such as the National Association of Scholars) to root out any challenge to Enlightenment virtues of truth and reason—like feminism, postmodernism, or constructivism—is itself an abrogation of those virtues (which, of course, science studies demonstrably embodies). Thus both sides in effect must argue that *everybody* will suffer if the other side of the science wars is victorious.[36]

5. OUTCOMES: WINNERS AND NON-LOSERS?

Given obvious power differences between natural scientists and those who study them, it is a bit odd that this war has gone on for as long as it has (and that it seems unlikely to cease any time soon).[37] Really, how many physicists does it take to quash a sociologist or two? Could it possibly be—in contrast to the five other episodes of cultural cartography, where epistemic authority typically was a zero-sum game—that the science wars are a win-win deal? Maybe science defenders and science studies both have reasons to perpetuate a contrived pseudo-war that belies the interests they share. Indeed, such talk of common interests has lately been heard. Levine says that both sides will sink or swim together amid government budget cutting (and, one might add, attacks from anti-intellectualist extremists): "We" should take "this awkward moment as an occasion for recognizing common interests." Enemies can be allies

36. Schweber, "Reflections," 74; Jasanoff in David Dickson, "Science Studies Braces for the Fall-Out" *Nature* 387 (May 22, 1997): 332. For example, remembering that Sokal fashioned himself as the "real" Left pitted against postmodern impostors, Schrecker writes: "the Right is fighting a broad-based campaign to demonize those sectors of the academic community that encourage critical thinking and offer an alternative perspective on the status quo. . . . Sokal's merry prank may well backfire" (in "The Sokal Hoax: A Forum," 61).

37. Stop press! The *Chronicle of Higher Education* reports on its "Daily News" Web site (November 13, 1997) that the science wars were a factor in the "retirement" of Katherine Livingston, long-time book review editor at *Science.* Ongoing battles continue to receive high-profile media coverage: Madhusree Mukerjee, "Undressing the Emperor," *Scientific American* 278 (March 1998): 30–31; "You Can't Follow the Science Wars without a Battle Map," *Economist* 145 (December 13, 1997): 77–79 (this title confirms my choice to present the phenomenon as an instance of cultural cartography). For more vitriol from a science defender, see Jean Bricmont, "Science Studies—What's Wrong?" *Physics World* 10 (December 1997): 15–16.

against other enemies. Pinch sees a silver lining: "A lot of the shouting that's gone on has been most unfortunate. . . . But I see this as a great opportunity to address issues that need to be addressed and haven't been." His collaborator Harry Collins agrees: "The whole timbre of the discussion has become useful and productive."[38] Useful and productive for what, for whom?

For a fledgling academic pursuit like science studies, the limelight is preferable to obscurity: we sociologists get giddy just to be noticed in *Nature* or *Science.* Because of the science wars, science studies makes it to the front page of the *New York Times* (courtesy of the Sokal hoax), becomes a noticed blip on radar screens of university administrators and department heads (thanks to the *Chronicle of Higher Education*), and assumes a reality for many natural scientists who have learned only lately what was being said about them (as they read *Physics Today*). The attention alone may be salutary for science studies, even if it comes at a risk that university administrators (and the publics in line behind them) will believe every nasty word that Gross and Levitt have written.

Science defenders may also benefit from protracted science wars—even if they should somehow "lose." That is, the image that science studies constructs for science may—in a not so unlikely scenario—add still more legitimacy to scientists' cultural hold on epistemic authority. To make science "culture"—a human endeavor, and therefore hesitant, flawed, fallible, incomplete, uncertain, corrigible—might get scientists off the Superman hook as they enter techno-debates over public policy. No physicist or chemist or biologist can possibly live up to the heroic myth in which science speaks with absolute finality and assuredness, lifts every veil of complexity and confusion, guarantees inevitable outcomes. Science studies sustains scientists' epistemic authority by replacing these unrealistic images with useful facts about the provisional qualities of scientific knowledge and the bricolage of scientific practice. Philosopher Steve Fuller believes that science studies could "lead the public to have saner expectations of science. Words like *truth, rationality* and *objectivity* have [in the past] inspired unrealistic hopes for what science can accomplish." Hayles says it better: "Properly framed, the challenge that the cultural and social studies of science pose to objectivism should make science stronger, not weaker, by clarifying its connections to the complex-

38. Levine, "Sokal's Hoax: An Exchange," 54; Pinch and Collins quoted in Macilwain, "Campuses Ring," 331. "I urge that we find out who our allies are in various disciplines, including the sciences, and work to forge alliances with them" (Hayles, "Consolidating the Canon," 233).

ities of instantiated and situated human life." With that, even physicist Mermin could agree: "Scientists who set themselves up as sorcerers are a menace to the public and to science."[39]

In short, the skill and expertise of scientists will still be worth seeking out even after "the public . . . understand[s] that expert disagreement is part and parcel of scientific life. Scientists then are skilled experts like any other skilled experts, such as realtors, plumbers and chefs." A peacemaking *Nature* editorial suggests that science studies can help natural scientists as they enter public controversies: "Those who have been developing our knowledge of science from this perspective are playing an increasingly important role in mediating the relationship between science and society [and] only a deep understanding of science as a social (as well as intellectual) process will enable us to strengthen the bridge between the worlds of science and politics." Levine sums up: "The best thing that could happen to science, if it wants to convince society as a whole that it deserves support, would be to reject the arrogant language of holy war and humanize itself." Science studies can help by showing "how science is involved in culture, how science *is* culture."[40]

Enough symmetry. To study boundary-work does not preclude the need (at times) to do boundary-work. And so in *defense* of science studies, and

39. Steven Fuller, "Does Science Put an End to History, or History to Science? Or, Why Being Pro-science Is Harder than You Think," in Ross, *Science Wars,* 49; Hayles, "Consolidating the Canon," 235.; Mermin, "What's Wrong with This Sustaining Myth?" 11.

40. Pinch, "Science as Golem," 18; "Science Wars and the Need for Respect and Rigour," 373; Levine, "What Is Science Studies For?" 124. There is, of course, another side to this coin, as Langdon Winner cynically reminds us: "An upper-level NSF administrator . . . noted that EVIST [the funding program Ethics and Values in Science and Technology] had gained support within the foundation because it helped the scientific community respond to political pressures from Congress and the general public. As problems in ethics or values of science arose in matters of, say, nuclear power or molecular biology, scientists could respond that qualified experts in the humanities and social sciences were looking into the matter" ("The Gloves Come Off: Shattered Alliances in Science and Technology Studies," in Ross, *Science Wars,* 107).

Not everybody in science studies is content with this cooptation of their field—or is "sucking up" the term?—in the service of saving science from the mythic images that scientists themselves have elsewhere so often trotted out to legitimate their claims or to protect their autonomy from scrutiny. Those who fault constructivism for its supposed critical impotence—like Winner—are inclined to see science wars as deflecting attention away from what is really troubling about science: not the epistemology of its fact making, but the political and human consequences of its facts. But even such a political critique of science demands science: "Concepts such as objectivity, rationality, good method, and science need to be reappropriated, reoccupied and reinvigorated for democracy-advancing projects" (Harding, "Science Is 'Good to Think With,'" 19–20).

as a self-exemplifying distillation of everything I have said in this book about where epistemic authority comes from, here is my reply to the philosopher who asked me to join his editorial team.

June 30, 1997

Dear Professor ——,

I am pleased to accept your invitation to join the Advisory Editorial Board for the new book series on Science and Technology Studies.

I share your desire to improve the quality of debate in our field, and—being a scientist who happens to study science as a cultural and historical phenomenon—I heartily endorse your efforts to maintain standards of rationality, empirical test and modernity.

However, precisely because of my scientific studies of science, it is clear to me that the meaning of such terms as "rational," "empirical," "modern"—and even "science"—are highly variable, negotiated and contingent (rather than universal or transcendent).

Indeed, I have come to see such concepts not as a set of rules for proper fact-construction, but as rhetorical tools deployed in the pursuit or defense of epistemic authority, or in efforts to deny legitimacy to rival claims.

I look forward to working with you.

Sincerely,

Thomas F. Gieryn
Professor of Sociology

Bibliography of Secondary Works

Abbott, Andrew. 1988. *The System of Professions.* Chicago: University of Chicago Press.

———. 1995. "Things of Boundaries." *Social Research* 62:857–82.

Adas, Michael. 1989. *Machines as the Measure of Men.* Ithaca: Cornell University Press.

Alter, Peter and Angela Davies. 1987. *The Reluctant Patron: Science and the State in Britain, 1850–1920.* London: Berg.

Aronowitz, Stanley. 1988. *Science as Power: Discourse and Ideology in Modern Society.* Minneapolis: University of Minnesota Press.

Babbage, Charles. 1970. *Reflections on the Decline of Science.* New York: A. M. Kelley. Original edition, 1830.

Babbage, Charles, et al. 1975. *Debates on the Decline of Science.* New York: Arno Press. Originally published 1830–.

Baber, Zaheer. 1996. *The Science of Empire: Scientific Knowledge, Civilization, and Colonial Rule in India.* Albany: State University of New York Press.

Bakan, David. 1966. "The Influence of Phrenology on American Psychology." *Journal of the History of the Behavioral Sciences* 2:200–220.

Balmer, Brian. 1996. "The Political Cartography of the Human Genome Project." *Perspectives on Science* 4:249–82.

Barnes, Barry. 1974. *Scientific Knowledge and Sociological Theory.* London: Routledge.

Barnes, Barry, David Bloor, and John Henry. 1996. *Scientific Knowledge: A Sociological Analysis.* Chicago: University of Chicago Press.

Barnes, Barry, and David Edge. 1982. "The Interaction of Science and Technology." In *Science in Context,* 147–54. Cambridge: MIT Press.

———, eds. 1982. *Science in Context.* Cambridge: MIT Press.

Bartlett, William V. 1995. "Preaching Science: John Tyndall and the Rhetoric of Victorian Scientific Naturalism." Ph.D. dissertation, Rutgers University.

Barton, Ruth. 1987. "John Tyndall, Pantheist: A Rereading of the Belfast Address." *Osiris,* 2d series, 3:111–34.

———. 1998. "Just before *Nature:* The Purposes of Science and the Purposes of Popularization in Some English Popular Science Journals of the 1860s." *Annals of Science* 55:1–33.

Basalla, George. 1967. "The Spread of Western Science." *Science* 156:611–22.

Basalla, George, William Coleman, Robert H. Kargon, eds. 1970. *Victorian Science.* New York: Doubleday.

Bauer, Henry H. 1984. *Beyond Velikovsky.* Urbana: University of Illinois Press.

Beck, Ulrich. 1992. *Risk Society: Towards a New Modernity.* London: Sage. Original edition, 1986.

Becker, Howard S. 1982. *Art Worlds.* Berkeley: University of California Press.

———. 1986. *Doing Things Together.* Evanston: Northwestern University Press.

Beisel, Nicola. 1992. "Constructing a Shifting Moral Boundary: Literature and Obscenity in Nineteenth-Century America." In *Cultivating Differences,* ed. Michèle Lamont and Marcel Fournier, 104–28. Chicago: University of Chicago Press.

Ben-David, Joseph. 1991. *Scientific Growth: Essays on the Social Organization and Ethos of Science.* Berkeley: University of California Press.

Ben-Yehuda, Nachman. 1985. *Deviance and Moral Boundaries.* Chicago: University of Chicago Press.

Berman, Morris. 1978. *Social Change and Scientific Organization: the Royal Institution, 1799–1844.* Ithaca: Cornell University Press.

Bernal, J. D. 1970. *Science and Industry in the Nineteenth Century.* Bloomington: Indiana University Press. Original edition, 1953.

Boguslaw, Robert. 1965. *The New Utopians: A Study of System Design and Social Change.* Englewood Cliffs, N.J.: Prentice-Hall.

Boring, Edwin G. 1957. *A History of Experimental Psychology.* New York: Appleton-Century-Crofts.

Bourdieu, Pierre. 1975. "The Specificity of the Scientific Field and the Social Conditions of the Progress of Reason." *Social Science Information* 14:19–47.

———. 1977. *Outline of a Theory of Practice.* Cambridge: Cambridge University Press.

Bowker, Geoff. 1992. "What's in a Patent?" In *Shaping Technology/Building Society,* ed. Wiebe E. Bijker and John Law, 53–74. Cambridge: MIT Press.

Boyer, Paul. 1985. "Social Scientists and the Bomb." *Bulletin of the Atomic Scientists* 41 (October): 31–36.

Brain, David. 1989. "Discipline and Style: The Ecole des Beaux-Arts and the Social Production of an American Architecture." *Theory and Society* 18:807–68.

Brassley, Paul. 1995. "Agricultural Research in Britain, 1850–1914: Failure, Success, and Development." *Annals of Science* 52:465–80.

Brezina, Dennis W. 1968. *The Congressional Debate on the Social Sciences in 1968.* Staff Discussion Paper, no. 400. Washington, D.C.: George Washington University Program of Policy Studies.

Brock, William, and Roy MacLeod. 1976. "The Scientists' Declaration: Reflexions on Science and Belief in the Wake of *Essays and Reviews,* 1864–65." *British Journal for the History of Science* 9:39–66.

Brock, W. H., and N. D. C. Mollan, eds. 1981. *John Tyndall: Essays on a Natural Philosopher.* Dublin: Royal Dublin Society.

Bronk, Detlev W. 1975. "The National Science Foundation: Origins, Hopes, and Aspirations" *Science* 188 (May 2): 409–14.

Brooke, John H. 1991. *Science and Religion: Some Historical Perspectives.* Cambridge: Cambridge University Press.

Brown, Alan Willard. 1947. *The Metaphysical Society: Victorian Minds in Crisis, 1869–1880.* New York: Columbia University Press.

Bruce, Robert V. 1987. *The Launching of Modern American Science, 1846–1876.* New York: Knopf.

Buck, Peter. 1985. "Adjusting to Military Life: The Social Sciences Go to War." In *Military Enterprise and Technological Change,* ed. Merritt Roe Smith, 203–52. Cambridge: MIT Press.

Burchfield, Joe. 1981. "John Tyndall—a Biographical Sketch." In *John Tyndall: Essays on a Natural Philosopher,* ed. W. H. Brock and N. D. C. Mollan, 1–13. Dublin: Royal Dublin Society.

Burns, Gene. 1990. "The Politics of Ideology: The Papal Struggle with Liberalism." *American Journal of Sociology* 95:1123–52.

Cambrosio, Alberto, and Peter Keating. 1983. "The Disciplinary Stake: The Case of Chronobiology." *Social Studies of Science* 13:323–53.

Cambrosio, Alberto, Peter Keating, and Michael Mackenzie. 1990. "Scientific Practice in the Courtroom: The Construction of Sociotechnical Identities in a Biotechnological Patent Dispute." *Social Problems* 37:275–93.

Cameron, Iain, and David Edge. 1979. *Scientific Images and Their Social Uses: An Introduction to the Concept of Scientism.* London: Butterworths.

Campbell, John Angus. 1986. "Scientific Revolution and the Grammar of Culture: The Case of Darwin's Origin." *Quarterly Journal of Speech* 72:351–76.

Cannon, Susan Faye. 1978. *Science in Culture: The Early Victorian Period.* New York: Science History Publications.

Cantor, G. N. 1975. "The Edinburgh Phrenology Debate, 1803–1828." *Annals of Science* 32:195–218.

Capshew, James H. 1998. *Psychologists on the March.* Cambridge: Cambridge University Press.

Capshew, James H., and Karen A. Rader. 1992. "Big Science: Price to the Present." *Osiris,* 2d series, 7:3–25.

Cardwell, D. S. L. 1972. *The Organization of Science in England.* London: Heinemann.

Casper, Monica J. 1994. "At the Margins of Humanity: Fetal Positions in Science and Medicine." *Science, Technology, and Human Values* 19:307–23.

Chalkey, Lyman. 1951. "Prologue to the U.S. National Science Foundation." National Science Foundation, Washington, D.C. Mimeograph.

Chartier, Roger. 1988. *Cultural History.* Ithaca: Cornell University Press.

Clarke, Adele, and Theresa Montini. 1993. "The Many Faces of RU486: Tales of Situated Knowledges and Technological Considerations." *Science, Technology, and Human Values* 18:42–78.

Clemens, Elisabeth S. 1986. "Of Asteroids and Dinosaurs: The Role of the Press in the Shaping of Scientific Debate." *Social Studies of Science* 16:421–56.

Close, Frank. 1991. *Too Hot to Handle: The Race for Cold Fusion.* Princeton: Princeton University Press.

Collins, H. M. 1975. "The Seven Sexes: A Study in the Sociology of a Phenomenon, or the Replication of Experiments in Physics." *Sociology* 9:205–24.

Collins, Harry, and Trevor Pinch. 1993. *The Golem: What Everyone Should Know about Science.* Cambridge: Cambridge University Press.

Comrie, John C. 1927. *History of Scottish Medicine to 1860.* London: Wellcome Museum.

Cooter, R. J. 1976. "Phrenology: The Provocation of Progress." *History of Science* 14:211–34.

———. 1980. "Deploying 'Pseudoscience': Then and Now." In *Science, Pseudo-Science, and Society,* ed. Marsha P. Hanen, Margaret J. Osler, and Robert G.Weyant, 237–72. Waterloo, Ont.: Wilfrid Laurier University Press.

———. 1984. *The Cultural Meaning of Popular Science.* Cambridge: Cambridge University Press.

———. 1989. *Phrenology in the British Isles: An Annotated, Historical Bibliography and Index.* Metuchen, N.J.: Scarecrow.

Cooter, Roger, and Stephen Pumfrey. 1994. "Separate Spheres and Public Places: Reflections on the History of Science Popularization and Science in Popular Culture." *History of Science* 22:237–67.

Crease, Robert P., and N. P. Samios. 1989. "Cold Fusion Confusion." *New York Times Magazine,* September 24, 34–38.

Crosland, Maurice. 1986. "Assessment by Peers in Nineteenth-Century France: The Manuscript Reports on Candidates for Election to the Académie des Sciences." *Minerva* 24:413–32.

Crowther, J. G. 1968. *Scientific Types.* London: Barrie and Rockliff.

Curfs, Garrit Thomas. 1990. "Experiment as Rhetoric in the Cold Fusion Controversy." Master's thesis, Virginia Polytechnic Institute and State University.

Darnton, Robert. 1984. *The Great Cat Massacre and Other Episodes in French Cultural History.* New York: Basic Books.

Davie, George Elder. 1961. *The Democratic Intellect: Scotland and Her Universities in the Nineteenth Century.* Edinburgh: Edinburgh University Press.

Davies, John C. 1955. *Phrenology Fad and Science: A Nineteenth Century American Crusade.* New Haven: Yale University Press.

DeGiustino, David. 1975. *Conquest of Mind: Phrenology and Victorian Social Thought.* Totowa, N.J.: Rowman and Littlefield.

Derbyshire, Ian. 1995. "The Building of India's Railways: The Application of Western Technology in the Colonial Periphery 1850–1920." In *Technology and the Raj: Western Technology and Technical Transfers to India, 1700–1947,* ed. Roy MacLeod and Deepak Kumar, 177–215. Thousand Oaks, Calif.: Sage.

Derksen, Maarten. 1997. "Are We Not Experimenting Then? The Rhetorical Demarcation of Psychology and Common Sense." *Theory and Psychology* 7:435–56.

Desmond, Adrian. 1997. *Huxley: From Darwin's Disciple to Evolution's High Priest.* Reading, Mass.: Addison-Wesley.

Dickson, David. 1988. *The New Politics of Science.* Chicago: University of Chicago Press. Original edition, 1984.

DiMaggio, Paul. 1987. "Classification in Art." *American Sociological Review* 52:440–55.

Douglas, Mary. 1966. *Purity and Danger: An Analysis of Concepts of Pollution and Taboo.* London: Routledge and Kegan Paul.

———. 1986. *How Institutions Think.* Syracuse: Syracuse University Press.

Drees, Willem B. 1996. *Religion, Science, and Naturalism.* New York: Cambridge University Press.

Dupree, A. Hunter. 1986. *Science in the Federal Government.* Baltimore: Johns Hopkins University Press. Original edition, 1957.

Durkheim, Emile. 1961. *Moral Education.* New York: Free Press. Original edition, 1925.

England, J. Merton. 1976. "Dr. Bush Writes a Report: *Science—the Endless Frontier.*" *Science* 191 (9 January): 41–47.

———. 1982. *A Patron for Pure Science: The National Science Foundation's Formative Years, 1945–57.* Washington, D.C.: National Science Foundation.

Epstein, Steven. 1996. *Impure Science: AIDS, Activism, and the Politics of Knowledge.* Berkeley: University of California Press.

Evans, Robert. 1997. "Soothsaying or Science? Falsification, Uncertainty, and Social Change in Macroeconomic Modelling." *Social Studies of Science* 27:395–438.

Eve, A. S., and C. H. Creasey. 1945. *Life and Work of John Tyndall.* London: Macmillan.

Ezrahi, Yaron. 1990. *The Descent of Icarus: Science and the Transformation of Contemporary Democracy.* Cambridge: Harvard University Press.

Figert, Anne E. 1996. *Women and the Ownership of PMS: The Structuring of a Psychiatric Disorder.* New York: Aldine de Gruyter.

Flugel, J. C. 1933. *A Hundred Years of Psychology, 1833–1933.* New York: Macmillan.

Fortun, Michael. 1998. "The Human Genome Project and the Acceleration of Biotechnology." In *Private Science: Biotechnology and the Rise of the Molecular Sciences,* ed. Arnold Thackray, 182–201. Philadelphia: University of Pennsylvania Press.

Foucault, Michel. 1980. *Power/Knowledge: Selected Interviews and Other Writings, 1972–1977.* New York: Pantheon.

Friedlander, Michael W. 1995. *At the Fringes of Science.* Boulder: Westview.

Galloway, J. H. 1996. "Botany in the Service of Empire: The Barbados Cane-Breeding Program and the Revival of the Caribbean Sugar Industry, 1880s–1930s." *Annals of the Association of American Geographers* 86:682–706.

Gardner, Martin. 1952. *Fads and Fallacies in the Name of Science.* New York: Putnam's.

———. 1981. *Science—Good, Bad, and Bogus.* Buffalo: Prometheus.

Garfinkel, Harold. 1956. "Conditions of Successful Degradation Ceremonies." *American Journal of Sociology* 61:420–24.

Garfinkel, Harold, Michael Lynch, and Eric Livingston. 1981. "The Work of Discovering Science Construed with Materials from the Optically Discovered Pulsar." *Philosophy of the Social Sciences* 11:131–58.

Garroutte, Eva Marie. 1993. "Language and Cultural Authority: Nineteenth-Century Science and the Colonization of Religious Discourse." Ph.D. dissertation, Princeton University.

Gaziano, Emanuel. 1996. "Ecological Metaphors as Scientific Boundary Work: Innovation and Authority in Interwar Sociology and Biology." *American Journal of Sociology* 101:874–907.

Geertz, Clifford. 1973. *The Interpretation of Cultures.* New York: Basic Books.

———. 1983. *Local Knowledge.* New York: Basic Books.

Gerstein, Dean. 1986. "Social Science as a National Resource." In *The Nationalization of the Social Sciences,* ed. Samuel Z. Klausner and Victor M. Lidz, 247–64. Philadelphia: University of Pennsylvania Press.

Gibbon, Charles. 1878. *The Life of George Combe.* London: Macmillan.

Giddens, Anthony. 1976. *New Rules of Sociological Method.* New York: Basic Books.

———. 1984. *The Constitution of Society.* Cambridge: Polity Press.

Gieryn, Thomas F. 1983. "Boundary-Work and the Demarcation of Science from Non-science: Strains and Interests in Professional Interests of Scientists." *American Sociological Review* 48:781–95.

———. 1987. "Science Pure and Applied, or How the Autonomy and Patronage of Scientists Became a Matter of National Security." *Quarterly Journal of Ideology* 11:1–16.

———. 1988. "Distancing Science from Religion in Seventeenth-Century England." *Isis* 79:582–93.

———. 1992. "The Ballad of Pons and Fleischmann: Experiment and Narrative in the (Un)Making of Cold Fusion." In *The Social Dimensions of Science,* ed. Ernan McMullin, 217–43. Notre Dame, Ind.: University of Notre Dame Press.

———. 1994. "Objectivity for These Times." *Perspectives on Science* 2:324–49.

———. 1995. "Boundaries of Science." In *Handbook of Science and Technology Studies,* ed. Sheila Jasanoff et al., 393–443. Thousand Oaks, Calif.: Sage.

———. 1996. "Policing STS [Science and Technology Studies]: A Boundary-Work Souvenir from the Smithsonian Exhibition on 'Science in American Life.'" *Science, Technology, and Human Values* 21:100–115.

———. 1998. "Balancing Acts: Science, Enola Gay, and History Wars at the Smithsonian." In *The Politics of Display,* ed. Sharon MacDonald, 197–228. London: Routledge.

Gieryn, Thomas F., George M. Bevins, and Stephen C. Zehr. 1985. "Professionalization of American Scientists: Public Science in the Creation/Evolution Trials." *American Sociological Review* 50:392–409.

Gieryn, Thomas F., and Anne Figert. 1986. "Scientists Protect Their Cognitive Authority: The Status Degradation Ceremony of Sir Cyril Burt." In *The Knowledge Society,* ed. Gernot Böhme and Nico Stehr, 67–86. Dordrecht: D. Reidel.

———. 1990. "Ingredients for a Theory of Science in Society: O-rings, Ice Water, C-clamp, Richard Feynman, and the Press." In *Theories of Science in Society,* ed. Thomas F. Gieryn and Susan E. Cozzens, 67–97. Bloomington: Indiana University Press.

Gilbert, G. Nigel, and Michael Mulkay. 1984. *Opening Pandora's Box: A Sociological Analysis of Scientists' Discourse.* Cambridge: Cambridge University Press.

Gilmartin, David. 1994. "Scientific Empire and Imperial Science: Colonialism and Irrigation Technology in the Indus Basin." *Journal of Asian Studies* 53:1127–49.

Goodstein, David. 1994. "Pariah Science: Whatever Happened to Cold Fusion?" *American Scholar* 63 (autumn): 527–41.

Gottschalk, Bobbie Jo. 1991. "Cold Fusion: A Descriptive Case Study of By-Passing the Due Process of Scientific Peer Review." Master's thesis, Purdue University.

Graham, Milton. 1954. *Federal Utilization of Social Science Research.* Washington, D.C.: Brookings Institute.

Grant, Sir Alexander. 1884. *The Story of the University of Edinburgh,* 2 vols. London: Longmans, Green.

Grant, A. C. 1966. "George Combe and the 1836 Election for the Edinburgh University Chair of Logic." *Book of the Old Edinburgh Club* 32:174–84.

———. 1968. "New Light on an Old View." *Journal of the History of Ideas* 26:293–301.

Greenberg, Daniel. 1967. *The Politics of Pure Science.* New York: New American Library.

Gross, Alan. 1995. "Renewing Aristotelean Theory: The Cold Fusion Controversy as a Test Case." *Quarterly Journal of Speech* 81:48–62.

Gusfield, Joseph R. 1981. *The Culture of Public Problems: Drinking-Driving and the Symbolic Order.* Chicago: University of Chicago Press.

Guston, David H., and Kenneth Keniston. 1994. "Updating the Social Contract for Science." *Technology Review* 97 (November/December): 60–68.

Hakfoort, Casper. 1995. "The Historiography of Scientism: A Critical Review." *History of Science* 33:375–95.

Hamlin, Christopher. 1986. "Scientific Method and Expert Witnessing: Victorian Perspectives on a Modern Problem." *Social Studies of Science* 16:485–513.

Haraway, Donna J. 1991. *Simians, Cyborgs, and Women: The Reinvention of Nature.* New York: Routledge.

Harris, Fred R., ed. 1970. *Social Science and National Policy.* Chicago: Aldine.

Haskell, Thomas L., ed. 1984. *The Authority of Experts: Studies in History and Theory.* Bloomington: Indiana University Press.

Heims, Steve Joshua. 1991. *Constructing a Social Science for Postwar America: The Cybernetics Group, 1946–53.* Cambridge: MIT Press.

Henry, R. J. 1995. "Technology Transfer and Its Constraints: Early Warnings from Agricultural Development in Colonial India." In *Technology and the Raj: Western Technology and Technical Transfers to India, 1700–1947,* ed. Roy MacLeod and Deepak Kumar, 51–77. Thousand Oaks, Calif.: Sage.

Herman, Ellen. 1995. *The Romance of American Psychology.* Berkeley: University of California Press.

Hess, David J. 1993. *Science in the New Age: The Paranormal, Its Defenders and Debunkers, and American Culture.* Madison: University of Wisconsin Press.

Hilgartner, Stephen. 1990. "The Dominant View of Popularization: Conceptual Problems, Political Uses." *Social Studies of Science* 20:519–39.

Hill, Gillian. 1978. *Cartographical Curiosities.* London: British Museum Publications.

Holmquest, Anne. 1990. "The Rhetorical Strategy of Boundary-Work." *Argumentation* 4:235–58.

Horowitz, Irving Louis, ed. 1967. *The Rise and Fall of Project Camelot.* Cambridge: MIT Press.

———. 1970. "The Life and Death of Project Camelot." In *The Values of Social Science,* ed. Norman K. Denzin, 159–84. Chicago: Aldine.

Horstman, Klasien. 1997. "Chemical Analysis of Urine for Life Insurance: The Construction of Reliability." *Science, Technology, and Human Values* 22:57–78.

Houghton, Walter E. 1957. *The Victorian Frame of Mind.* New Haven: Yale University Press.

Howard, Louise E. 1953. *Sir Albert Howard in India.* London: Faber and Faber.

Huizinga, John R. 1993. *Cold Fusion: The Scientific Fiasco of the Century.* New York: Oxford University Press.

Hunt, Lynn, ed. 1989. *The New Cultural History.* Berkeley: University of California Press.

Indyk, Debbie, and David A. Rier. 1993. "Grassroots AIDS Knowledge: Implications for the Boundaries of Science and Collective Action." *Knowledge: Creation, Diffusion, Utilization* 15:3–43.

Irwin, Alan, and Brian Wynne, eds. 1996. *Misunderstanding Science? The Public Reconstruction of Science and Technology.* Cambridge: Cambridge University Press.

Jasanoff, Sheila. 1987. "Contested Boundaries in Policy-relevant Science." *Social Studies of Science* 17:195–230.

———. 1990. *The Fifth Branch: Science Advisers as Policymakers.* Cambridge: Harvard University Press.

———. 1995. *Science at the Bar: Law, Science, and Technology in America.* Cambridge: Harvard University Press.

Jasanoff, Sheila, Gerald Markle, James C. Petersen, and Trevor Pinch, eds. 1995. *Handbook of Science and Technology Studies.* Thousand Oaks, Calif.: Sage.

Jensen, C. Vernon. 1970. "The X Club: Fraternity of Victorian Scientists." *British Journal for the History of Science* 5:63–72.

Johnson-McGrath, Julie. 1995. "Speaking for the Dead: Forensic Pathologists and Criminal Justice in the United States." *Science, Technology, and Human Values* 20:438–59.

Karl, Barry. 1969. "Presidential Planning and Social Research." *Perspectives in American History* 3:347–409.

Keller, Evelyn Fox. 1983. *A Feeling for the Organism: The Life and Work of Barbara McClintock.* New York: Freeman.

Kerr, Anne, Sarah Cunningham-Burley, and Amanda Amos. 1997. "The New Genetics: Professionals' Discursive Boundaries." *Sociological Review* 45:279–303.

Kevles, Daniel J. 1977. "The National Science Foundation and the Debate over Postwar Research Policy, 1942–1945." *Isis* 68:5–26.

———. 1978. *The Physicists.* New York: Knopf.

Kim, Stephen S. 1996. *John Tyndall's Transcendental Materialism and the Conflict between Religion and Science in Victorian England.* Lewiston: Mellen University Press.

King, Geoff. 1996. *Mapping Reality: An Exploration of Cultural Cartographies.* New York: St. Martin's.

King, M. D. 1971. "Reason, Tradition, and the Progressiveness of Science." *History and Theory* 10:3–32.

Klausner, Samuel Z. 1986. "The Bid to Nationalize American Social Science." In *The Nationalization of the Social Sciences,* ed. Samuel Z. Klausner and Victor M. Lidz, 3–39. Philadelphia: University of Pennsylvania Press.

Klausner, Samuel Z., and Victor M. Lidz, eds. 1986. *The Nationalization of the Social Sciences.* Philadelphia: University of Pennsylvania Press.

Klein, Julie Thompson. 1996. *Crossing Boundaries: Knowledge, Disciplinarities, Interdisciplinarities.* Charlottesville: University Press of Virginia.

Kleinman, Daniel Lee. 1995. *Politics on the Endless Frontier: Postwar Research Policy in the United States.* Durham: Duke University Press.

Kleinman, Daniel Lee, and Mark Solovey. 1995. "Hot Science/Cold War: The National Science Foundation after World War II." *Radical History Review* 63:110–39.

Kline, Ronald. 1995. "Construing 'Technology' as 'Applied Science': Public Rhetoric of Scientists and Engineers in the United States, 1880–1945" *Isis* 86:194–221.

Knight, David. 1996. "Getting Science Across." *British Journal for the History of Science* 29:129–38.

Knorr[-Cetina], Karin. 1981. "Social and Scientific Method, or 'What Do We Make of the Distinction between the Natural and the Social Sciences?'" *Philosophy of the Social Sciences* 11:27–51.

Kohler, Robert E. 1982. *From Medical Chemistry to Biochemistry.* New York: Cambridge University Press.

Koontz, Louis Knott. 1946. "The Social Sciences in the National Science Foundation" *Pacific Historical Review* 15:1–30.

Kuhn, Thomas S. 1970. *The Structure of Scientific Revolutions.* 2d enlarged ed. Chicago: University of Chicago Press. Original edition, 1962.

Kuklick, Henrika. 1980. "Boundary Maintenance in American Sociology: Limitations to Academic 'Professionalization.'" *Journal of the History of the Behavioral Sciences* 16:201–19.

Kumar, Deepak, ed. 1991. *Science and Empire: Essays in Indian Context, 1700–1947.* Delhi: Anamika Prakashan.

———. 1996. "The 'Culture' of Science and Colonial Culture: India, 1820–1920." *British Journal for the History of Science* 29:195–209.

LaFollette, Marcel C. 1990. *Making Science Our Own: Public Images of Science, 1910–1955.* Chicago: University of Chicago Press.

Lamont, Michèle. 1992. *Money, Morals, and Manners.* Chicago: University of Chicago Press.

Lamont, Michèle, and Marcel Fournier, eds. 1992. *Cultivating Differences: Symbolic Boundaries and the Marking of Inequality.* Chicago: University of Chicago Press.

Langmuir, Irving. 1953. "Pathological Science." Colloquium given at the Knolls Research Laboratory, December 18. Transcribed and edited by R. Hall. General Electric Research and Development Center Report 68-C-035, April 1968.

Larsen, Otto N. 1992. *Milestones and Millstones: Social Science at the National Science Foundation, 1945–1991.* New Brunswick, N.J.: Transaction.

Latour, Bruno. 1983. "Give Me a Laboratory and I Will Raise the World." In *Science Observed,* ed. Karin Knorr-Cetina and Michael Mulkay, 141–70. London: Sage.

———. 1987. *Science in Action.* Cambridge: Harvard University Press.

———. 1988. *The Pasteurization of France.* Cambridge: Harvard University Press.

———. 1996. *Aramis, or The Love of Technology.* Cambridge: Harvard University Press.

Layton, Edward J. 1974. "Technology as Knowledge." *Technology and Culture* 15:31–41.

———. 1977. "Conditions of Technological Development" in *Science, Technology, and Society,* ed. Ina Spiegel-Rösing and Derek deSolla Price, 197–222. London: Sage.

Leahey, Thomas Hardy, and Grace Leahey. 1983. *Psychology's Occult Doubles: Psychology and the Problem of Pseudoscience.* Chicago: Nelson-Hall.

Lewenstein, Bruce V. 1992. "Cold Fusion and Hot History." *Osiris,* 2d series, 7:181–209.

———. 1995. "From Fax to Facts: Communication in the Cold Fusion Saga." *Social Studies of Science* 25:403–36.

———. 1995. "Do Public Electronic Bulletin Boards Help Create Scientific Knowledge? The Cold Fusion Case." *Science, Technology, and Human Values* 20:123–49.

Lievrouw, Leah A. 1990. "Communication and the Social Representation of Scientific Knowledge." *Critical Studies in Mass Communication* 7:1–10.

Lobeck, A. K. 1993. *Things Maps Don't Tell Us.* Chicago: University of Chicago Press. Original edition, 1956.

Lomask, Milton. 1975. *A Minor Miracle: An Informal History of the National Science Foundation.* Washington, D.C.: National Science Foundation.

Lynch, Michael. 1985. "'Here Is Adhesiveness': From Friendship to Homosexuality." *Victorian Studies* 29 (autumn): 67–96.

Lynch, Michael. 1993. *Scientific Practice and Ordinary Action.* Cambridge: Cambridge University Press.

Lyons, Gene M. 1969. *The Uneasy Partnership: Social Science and the Federal Government in the Twentieth Century.* New York: Russell Sage.

———. 1986. "The Many Faces of Social Science." In *The Nationalization of the Social Sciences,* ed. Samuel Z. Klausner and Victor M. Lidz, 197–208. Philadelphia: University of Pennsylvania Press.

MacKenzie, Donald. 1990. *Inventing Accuracy: A Historical Sociology of Nuclear Missile Guidance.* Cambridge: MIT Press.

MacLeod, Roy. 1969. "Science and Government in Victorian England: Lighthouse Illumination and the Board of Trade, 1866–1886." *Isis* 60:5–38.

———. 1970. "The X Club: A Scientific Network in Late-Victorian England." *Notes and Records of the Royal Society* (London) 24:305–22.

———. 1974. "The Ayrton Incident: A Commentary on the Relations of Science and Government in England, 1870–73." In *Science and Values: Patterns of Tradition and Change,* ed. Arnold Thackray and Everett Mendelsohn, 45–78. New York: Humanities Press.

———. 1976. "Science and the Treasury: Principles, Personalities, and Policies, 1870–85." In *The Patronage of Science in the Nineteenth Century,* ed. G. L'e. Turner, 115–72. Leyden: Noordhoff.

———. 1976. "John Tyndall." In *Dictionary of Scientific Biography,* vol. 13, ed. Charles C. Gillespie, 521–24. New York: Scribner's.

———. 1982. "'The Bankruptcy of Science' Debate: The Creed of Science and Its Critics, 1885–1900." *Science, Technology, and Human Values* 7:2–15.

MacLeod, Roy, and Deepak Kumar, eds. 1995. *Technology and the Raj: Western Technology and Technical Transfers to India, 1700–1947.* Thousand Oaks, Calif.: Sage.

Mallove, Eugene F. 1991. *Fire from Ice: Searching for the Truth behind the Cold Fusion Furor.* New York: Wiley.

Martin, Brian. 1991. *Scientific Knowledge in Controversy: The Social Dynamics of the Fluoridation Debate.* Albany: State University of New York Press.

———, ed. 1996. *Confronting the Experts.* Albany: State University of New York Press.

Marx, Karl. 1975. *Economic and Philosophical Manuscripts of 1844.* Vol. 20. London: Lawrence and Wishart.

———. 1978. "Theses on Feuerbach." In *The Marx-Engels Reader,* ed. Robert C. Tucker, 143–45. New York: Norton. Originally published 1845.

Mathias, Peter. 1972. "Who Unbound Prometheus? Science and Technical Change, 1600–1800." In *Science and Society, 1600–1900,* ed. Peter Mathias, 54–80. Cambridge: Cambridge University Press.

Mayr, Otto. 1982. "The Science-Technology Relationship." In *Science in Context,* ed. Barry Barnes and David Edge, 155–63. Cambridge: MIT Press. Original edition, 1976.

McAllister, James W. 1992. "Competition among Scientific Disciplines in Cold Nuclear Fusion Research." *Science in Context* 5:17–49.

McCloskey, D. 1985. *The Rhetoric of Economics.* Madison: University of Wisconsin Press.

McOmber, James B. 1996. "Silencing the Patient: Freud, Sexual Abuse, and 'The Etiology of Hysteria.'" *Quarterly Journal of Speech* 86:343–63.

Mendelsohn, Everett. 1964. "The Emergence of Science as a Profession in Nineteenth Century Europe." In *The Management of Scientists,* ed. Karl Hill, 3–48. Boston: Beacon.

———. 1977. "The Social Construction of Scientific Knowledge." *Sociology of the Sciences Yearbook* 1:3–26.

Merton, Robert K. 1973. *The Sociology of Science.* Chicago: University of Chicago Press.

Messer-Davidow, Ellen, David R. Shumway, and David J. Sylvan, eds. 1993. *Knowledges: Historical and Critical Studies in Disciplinarity.* Charlottesville: University Press of Virginia.

Milgram, Stanley. 1974. *Obedience to Authority.* New York: Harper and Row.

Miller, Roberta Balstad. 1982. "The Social Sciences and the Politics of Science: The 1940s." *American Sociologist* 17:205–9.

Monck, W. H. S. 1881. *Sir William Hamilton.* London: Sampson, Low, Marston, Searle, and Rivington.

Monmonier, Mark. 1996. *How to Lie with Maps.* 2d ed. Chicago: University of Chicago Press. Original edition, 1991.

Moore, Kelly. 1996. "Organizing Integrity: American Science and the Creation of Public Interest Organizations, 1955–1975." *American Journal of Sociology* 101:1592–1627.

Morrell, Jack, and Arnold Thackray. 1981. *Gentlemen of Science: Early Years of the British Association for the Advancement of Science.* Oxford: Oxford University Press.

Mukerji, Chandra. 1989. *A Fragile Power: Scientists and the State.* Princeton: Princeton University Press.

Mulkay, Michael. 1997. *The Embryo Research Debate: Science and the Politics of Reproduction.* New York: Cambridge University Press.

Myers, Greg. 1995. "From Discovery to Invention: The Writing and Rewriting of Two Patents." *Social Studies of Science* 25:57–105.

Nader, Laura, ed. 1996. *Naked Science: Anthropological Inquiry into Boundaries, Power, and Knowledge.* New York: Routledge.

National Science Foundation. 1969. *Knowledge into Action: Improving the Nation's Use of the Social Sciences.* Report of the Special Commission on the Social Sciences. Washington, D.C.: National Science Foundation.

———. 1969. *The Behavioral and Social Sciences: Outlook and Needs.* A report by the Behavioral and Social Sciences Survey Committee, under the auspices of the National Academy of Sciences and the Social Science Research Council. Englewood Cliffs, N.J.: Prentice-Hall.

Nippert-Eng, Christena E. 1996. *Home and Work.* Chicago: University of Chicago Press.

Numbers, Ronald L. 1992. *The Creationists: The Evolution of Scientific Creationism.* New York: Knopf.

Ofshe, Richard, and Ethan Watters. 1994. *Making Monsters: False Memories, Psychotherapy, and Sexual Hysteria.* New York: Scribner's.

Ogburn, William F. 1922. *Social Change with Respect to Culture and Original Nature.* New York: B. W. Huebsch.

Olby, Robert. 1991. "Social Imperialism and State Support for Agricultural Research in Edwardian Britain." *Annals of Science* 48:509–26.

Orlans, Harold. 1973. *Contracting for Knowledge.* San Francisco: Jossey-Bass.

Palladino, Paolo. 1990. "The Politics of Agricultural Research: Plant Breeding in Britain, 1910–1940." *Minerva* 28:446–68.

———. 1993. "Between Craft and Science: Plant Breeding, Mendelian Genetics, and British Universities, 1900–1920." *Technology and Culture* 34:300–323.

———. 1994. "Wizards and Devotees: On the Mendelian Theory of Inheritance and the Professionalization of Agricultural Science in Great Britain and the United States, 1880–1930." *History of Science* 32:409–44.

Palladino, Paolo, and Michael Warboys. 1993. "Science and Imperialism." *Isis* 84:91–102.

Parssinen, T. M. 1974. "Popular Science and Society: The Phrenology Movement in Early Victorian Britain." *Journal of Social History* 8:1–21.

Peat, F. David. 1989. *Cold Fusion: The Making of a Scientific Controversy.* Chicago: Contemporary Books.

Perenyi, Eleanor. 1981. *Green Thoughts: A Writer in the Garden.* New York: Random House.

Petitjean, Patrick, Catherine Jami, and Anne-Marie Moulin, eds. 1992. *Science and*

Empires: Historical Studies about Scientific Development and European Expansion. Boston: Kluwer.

Pickering, Andrew. 1995. *The Mangle of Practice.* Chicago: University of Chicago Press.

Pinch, Trevor J. 1979. "Normal Explanations of the Paranormal: The Demarcation Problem and Fraud in Parapsychology." *Social Studies of Science* 9:329–48.

———. 1992. "Opening Black Boxes: Science, Technology, and Society." *Social Studies of Science* 22:487–510.

Pinch, Trevor J., and Wiebe E. Bijker. 1987. "The Social Construction of Facts and Artifacts." In *The Social Construction of Technological Systems,* ed. Wiebe E. Bijker, Thomas P. Hughes, and Trevor J. Pinch. Cambridge: MIT Press.

Popper, Karl R. 1959. *The Logic of Scientific Discovery.* New York: Harper. Original edition, 1934.

Porter, Theodore M. 1995. *Trust in Numbers: The Pursuit of Objectivity in Science and Public Life.* Princeton: Princeton University Press.

Post, J. B. 1973. *An Atlas of Fantasy.* Baltimore: Mirage.

Pyenson, Lewis. 1993. "Cultural Imperialism and Exact Sciences Revisited." *Isis* 84:103–8.

Rasmussen, S. V. 1925. *The Philosophy of Sir William Hamilton.* London: Hachette.

Reichert, Dagmar. 1992. "On Boundaries." *Environment and Planning: Society and Space* 10:87–98.

Richards, Evelleen. 1988. "The Politics of Therapeutic Evaluation: The Vitamin C and Cancer Controversy." *Social Studies of Science* 18:653–701.

Riecken, Henry W. 1986. "Underdogging: The Early Career of the Social Sciences in the NSF." In *The Nationalization of the Social Sciences,* ed. Samuel Z. Klausner and Victor M. Lidz, 209–25. Philadelphia: University of Pennsylvania Press.

Rieff, Philip. 1969. *On Intellectuals.* Garden City, N.Y.: Doubleday.

Robinson, Arthur H., and Barbara Bartz Petchenik. 1976. *The Nature of Maps.* Chicago: University of Chicago Press.

Robinson, Arthur H., Randall D. Sale, Joel L. Morrison, and Phillip C. Muehrcke. 1984. *Elements of Cartography.* 5th ed. New York: John Wiley.

Robinson, Eric, and A. E. Musson. 1969. *James Watt and the Steam Revolution.* New York: Kelley.

Rose, Hilary. 1996. "My Enemy's Enemy Is—Only Perhaps—My Friend." In *Science Wars,* ed. Andrew Ross, 80–101. Durham: Duke University Press.

Ross, Dorothy. 1991. *The Origins of American Social Science.* Cambridge: Cambridge University Press.

Rothman, Milton A. 1990. "Cold Fusion: A Case History in 'Wishful Science'?" *Skeptical Inquirer* 14:161–70.

Rousseau, Denis L. 1992. "Case Studies in Pathological Science." *American Scientist* 80:54–63.

Russell, E. John. 1966. *A History of Agricultural Science in Great Britain, 1620–1954.* London: Allen and Unwin.

Saunders, Laurance James. 1950. *Scottish Democracy, 1815–1840: The Social and Intellectual Background.* Edinburgh: Oliver and Boyd.

Schutz, Alfred. 1971. *Collected Papers.* Vol. 1. The Hague: Martinus Nijhoff.

Schwartz, Barry. 1981. *Vertical Classification: A Study in Structuralism and the Sociology of Knowledge.* Chicago: University of Chicago Press.

Serres, Michel. 1982. *Hermes: Literature, Science, Philosophy.* Baltimore: Johns Hopkins University Press.

Sewell, Willam H., Jr. 1992. "A Theory of Structure: Duality, Agency, and Transformation." *American Journal of Sociology* 98:1–29.

Shackley, Simon, and Brian Wynne. 1996. "Representing Uncertainty in Global Climate Change Science and Policy: Boundary-Ordering Devices and Authority." *Science, Technology, and Human Values* 21:275–302.

Shapin, Steven. 1975. "Phrenological Knowledge and the Social Structure of Early Nineteenth Century Edinburgh." *Annals of Science* 32:219–43.

———. 1979. "Homo Phrenologicus: Anthropological Perspectives on an Historical Problem." In *Natural Order: Historical Studies of Scientific Culture,* ed. Barry Barnes and Steven Shapin, 41–71. Beverly Hills, Calif: Sage.

———. 1979. "The Politics of Observation: Cerebral Anatomy and Social Interests in the Edinburgh Phrenology Disputes." In *On the Margins of Science: The Social Construction of Rejected Knowledge,* ed. Roy Wallis, 139–78. Keele, U.K.: University of Keele Press.

———. 1982. "History of Science and Its Sociological Reconstruction." *History of Science* 20:157–211.

———. 1989. "The Invisible Technician." *American Scientist* 77:554–63.

———. 1994. "Cordelia's Love: Credibility and the Social Studies of Science." *Perspectives on Science* 3:255–75.

———. 1994. *A Social History of Truth.* Chicago: University of Chicago Press.

———. 1995. "Here and Everywhere: Sociology of Scientific Knowledge." *Annual Review of Sociology* 21:289–321.

Shapin, Steven, and Simon Schaffer. 1985. *Leviathan and the Air-Pump.* Princeton: Princeton University Press.

Sherwood, Morgan. 1968. "Federal Policy for Basic Research: Presidential Staff and the National Science Foundation, 1950–1956." *Journal of American History* 55:599–615.

Silber, Ilana Friedrich. 1995. "Spaces, Fields, Boundaries: The Rise of Spatial Metaphors in Contemporary Sociological Theory." *Social Research* 62:322–55.

Simon, Bart. 1990. "Voices of Cold (Con)Fusion: Pluralism, Belief, and the Rhetoric of Replication in the Cold Fusion Controversy." Master's thesis, University of Edinburgh.

———. 1998. "Undead Science: Making Sense of Cold Fusion after the (Arti)Fact." *Social Studies of Science.* In press.

Sjoberg, Gideon, ed. 1967. *Ethics, Politics, and Social Research.* Cambridge, Mass.: Schenkman.

Smith, Barbara Herrnstein. 1997. *Belief and Resistance: Dynamics of Contemporary Intellectual Controversy.* Cambridge: Harvard University Press.

Smith, Roger, and Brian Wynne, eds. 1989. *Expert Evidence: Interpreting Science in the Law.* London: Routledge, Chapman and Hall.

Solomon, Shana M., and Edward J. Hackett. 1996. "Setting Boundaries between Science and Law: Lessons from *Daubert v. Merrell Dow Pharmaceuticals, Inc.*" *Science, Technology, and Human Values* 21:131–56.

Star, Susan Leigh, and James R. Griesemer. 1989. "Institutional Ecology, 'Translations,' and Boundary Objects: Amateurs and Professionals in Berkeley's Museum of Vertebrate Zoology, 1907–39." *Social Studies of Science* 19:387–420.

Starr, Paul. 1982. *The Social Transformation of American Medicine.* New York: Basic Books.

Stine, Jeffrey K. 1986. *A History of Science Policy in the United States, 1940–1985.* Report prepared for the Task Force on Science Policy, Committee on Science and Technology, House of Representatives, 99th Congress, 2d Session. Washington, D.C.: Government Printing Office.

Stirling, James Hutchison. 1865. *Sir William Hamilton: Being the Philosophy of Perception, An Analysis.* London: Longmans, Green.

Sullivan, Dale L. 1994. "Exclusionary Epideictic: Nova's Narrative Excommunication of Fleischmann and Pons." *Science, Technology, and Human Values* 19:283–306.

Takacs, David. 1996. *The Idea of Biodiversity: Philosophies of Paradise.* Baltimore: Johns Hopkins University Press.

Taubes, Gary. 1993. *Bad Science: The Short Life and Weird Times of Cold Fusion.* New York: Random House.

Taylor, Charles Alan. 1991. "Defining the Scientific Community: A Rhetorical Perspective on Demarcation." *Communication Monographs* 58:402–20.

———. 1996. *Defining Science: A Rhetoric of Demarcation.* Madison: University of Wisconsin Press.

Temkin, Owsei. 1947. "Gall and the Phrenological Movement." *Bulletin of the History of Medicine* 21:275–321.

Thompson, Donald. 1957. "John Tyndall and the Royal Institution." *Annals of Science* 13:9–22.

Thrower, Norman J. W. 1996. *Maps and Civilization.* Chicago: University of Chicago Press. Original edition, 1972.

Timmermans, Stefan. 1996. "Saving Lives or Saving Multiple Identities? The Double Dynamic of Resuscitation Scripts." *Social Studies of Science* 26:767–97.

Toumey, Christopher P. 1996. *Conjuring Science: Scientific Symbols and Cultural Meanings in American Life.* New Brunswick, N.J.: Rutgers University Press.

Troyer, Ronald J., and Gerald E. Markle. 1983. *Cigarettes: The Battle over Smoking.* New Brunswick, N.J.: Rutgers University Press.

Turnbull, David. 1993. *Maps Are Territories: Science Is an Atlas.* Chicago: University of Chicago Press. Original edition, 1989.

Turner, Frank M. 1974. *Between Science and Religion: The Reaction to Scientific Naturalism in Late Victorian England.* New Haven: Yale University Press.

———. 1974. "Rainfall, Plagues, and the Prince of Wales: A Chapter in the Conflict of Religion and Science." *Journal of British Studies* 13:46–95.

———. 1978. "The Victorian Conflict between Science and Religion: A Professional Dimension." *Isis* 69:256–76.

———. 1980. "Public Science in Britain, 1880–1919." *Isis* 71:589–608.

———. 1981. "John Tyndall and Victorian Scientific Naturalism." In *John Tyndall: Essays on a Natural Philosopher,* ed. W. H. Brock and N. D. C. Mollan, 169–80. Dublin: Royal Dublin Society.

———. 1993. *Contesting Cultural Authority: Essays in Victorian Intellectual Life.* New York: Cambridge University Press.

Turner, Stephen P. 1990. "Forms of Patronage." In *Theories of Science in Society,* ed. Thomas F. Gieryn and Susan E. Cozzens, 185–211. Bloomington: Indiana University Press.

Turner, Stephen P., and Jonathan H. Turner, *The Impossible Science: A Institutional Analysis of American Sociology.* Newbury Park, Calif.: Sage.

Tyabji, Nasir. 1995. "The Genesis of Chemical-Based Industrialization: Oilseeds in Madras." In *Technology and the Raj: Western Technology and Technical Transfers to India, 1700–1947,* ed. Roy MacLeod and Deepak Kumar, 78–111. Thousand Oaks, Calif.: Sage.

Vaughan, Diane. 1996. *The Challenger Launch Decision.* Chicago: University of Chicago Press.

Veitch, John. 1869. *Memoir of Sir William Hamilton, Bart.* London: William Blackwood.

Wallis, Roy. 1985. "Science and Pseudo-science." *Social Science Information* 24:585–601.

———, ed. 1979. *On the Margins of Science: The Social Construction of Rejected Knowledge. Sociological Review* monograph. Keele, U.K.: University of Keele Press.

Walsh, Anthony A. 1970. "Is Phrenology Foolish? A Rejoinder." *Journal of the History of the Behavioral Sciences* 6:358–61.

Wang, Jessica. 1995. "Liberals, the Progressive Left, and the Political Economy of Postwar American Science: The National Science Foundation Debate Revisited." *Historical Studies in the Physical Sciences* 26:139–66.

Watson, Graham. 1984. "The Social Construction of Boundaries between Social and Cultural Anthropology in Britain and North America." *Journal of Anthropological Research* 40:351–66.

Weart, Spencer R. 1988. *Nuclear Fear: A History of Images.* Cambridge: Harvard University Press.

Weber, Max. 1946. "Science as a Vocation." In *From Max Weber,* ed. Hans Gerth and C. Wright Mills, 129–56. Oxford: Oxford University Press. Original edition, 1918.

———. 1949. *The Methodology of the Social Sciences.* New York: Free Press. Original edition, 1903–17.

———. 1978. *Economy and Society.* Ed. Guenther Roth and Claus Wittich. Berkeley: University of California Press.

White, Paul Stuart. 1996. "Making a 'Man of Science': The Boundaries of Scientific Identity and Learned Community in the Life of Thomas Huxley." Ph.D dissertation, University of Chicago.

Wilson, James Q. 1997. "Keep Social-Science 'Experts' Out of the Courtroom." *Chronicle of Higher Education,* June 6, A52.

Winch, Samuel P. 1997. *Mapping the Cultural Space of Journalism: How Journalists Distinguish News from Entertainment.* Westport, Conn.: Praeger.

Winkler, John K., and Walter Bromberg. 1939. *Mind Explorers.* New York: Reynal and Hitchcock.

Wolfe, Alan. 1997. "Public and Private in Theory and Practice: Some Implications of an Uncertain Boundary." In *Public and Private in Thought and Practice: Perspectives on a Grand Dichotomy,* ed. Jeff Weintraub and Krishan Kumar, 182–203. Chicago: University of Chicago Press.

Wolfle, Dael. 1986. "Making a Case for the Social Sciences." In *The Nationalization of the Social Sciences,* ed. Samuel Z. Klausner and Victor M. Lidz, 185–96. Philadelphia: University of Pennsylvania Press.

Wood, Denis. 1992. *The Power of Maps.* New York: Guilford.

Woolgar, Steve. 1981. "Playing with Relativism: Boundary Work as a Feature of Argumentative Discourse in the Sociology of Science." Paper presented at the meeting of the Society for Social Studies of Science, Atlanta, November 6.

———. 1988. *Science: The Very Idea.* London: Tavistock.

Woolgar, Steve, and Dorothy Pawluch. 1985. "Ontological Gerrymandering: The Anatomy of Social Problems Explanations." *Social Problems* 32:214–27.

Wrobel, Arthur. 1987. "Phrenology as Political Science." In *Pseudo-Science in Nineteenth Century America,* ed. Arthur Wrobel, 122–43. Lexington: University Press of Kentucky.

Yang, Anand. 1989. *The Limited Raj: Agrarian Relations in Colonial India, Saran District, 1793–1920.* Berkeley: University of California Press.

Yearley, Steven. 1988. "The Dictates of Method and Policy: Interpretational Structures in the Representation of Scientific Work." *Human Sciences* 11:341–59.

Yeo, Richard R. 1993. *Defining Science: William Whewell, Natural Knowledge, and Public Debate in Early Victorian Britain.* New York: Cambridge University Press.

Young, Donald. 1949. "Organization for Research in the Social Sciences in the United States." *International Social Sciences Bulletin* 1:99–107.

Young, Robert M. 1970. *Mind, Brain, and Adaptation in the Nineteenth Century: Cerebral Localization and Its Biological Context from Gall to Ferrier.* Oxford: Oxford University Press.

Youngson, A. J. 1966. *The Making of Classical Edinburgh.* Edinburgh: Edinburgh University Press.

Zachary, G. Pascal. 1997. *Endless Frontier: Vannevar Bush, Engineer of the American Century.* New York: Free Press.

Zehr, Stephen C. 1990. "Acid Rain as a Social, Political, and Scientific Controversy." Ph.D. dissertation, Indiana University.

———. 1994. "The Centrality of Scientists and the Translation of Interests in the U.S. Acid Rain Controversy." *Canadian Review of Sociology and Anthropology* 31:325–53.

———. 1994. "Accounting for the Ozone Hole: Scientific Representations of an Anomaly and Prior Incorrect Claims in Public Settings." *Sociological Quarterly* 35:603–19.

Zelizer, Viviana. 1994. *The Social Meaning of Money.* New York: Basic Books.

Zerubavel, Eviatar. 1991. *The Fine Line: Making Distinctions in Everyday Life.* New York: Free Press.

Zimmerman, Michael. 1995. *Science, Nonscience, and Nonsense: Approaching Environmental Literacy.* Baltimore: Johns Hopkins University Press.

Zuckerman, Harriet. 1988. "The Sociology of Science." In *Handbook of Sociology,* ed. Neil J. Smelser, 511–74. Newbury Park, Calif.: Sage.

Zuiches, James J. 1984. "The Organization and Funding of Social Science in the NSF." *Sociological Inquiry* 54:188–210.

Index